Robert J. Tomanek
Department of Anatomy and Cell Biology
University of Iowa
Iowa City, IA 52242-1109
USA

Raymond B. Runyan
Department of Cell Biology
and Anatomy
University of Arizona
Tucson, AZ 85724-5044
USA

Cover illustration: Heart tube formation during chick embryo development: Sagittal sections through 23- to 36-hour, 26- to 29-hour, and 33- to 38-hour embryos. See Chapter 2, "Myofibrillogenesis in the Heart," by Carol C. Gregorio and Parker B. Antin, for additional details. Artwork by Tania Yatskievych.

Library of Congress Cataloging-in-Publication Data

Formation of the heart and its regulation / edited by Robert J. Tomanek, Raymond B. Runyan.
p. cm.—(Cardiovascular molecular morphogenesis)
Includes bibliographical references and index.
ISBN 0-8176-4216-1 (h/c : alk. paper)
1. Heart. 2. Embryology, Human. I. Tomanek, Robert J. II. Runyan, Raymond B. III. Series.
[DNLM: 1. Heart—embryology. 2. Heart—growth & development. 3. Heart Defects, Congenital. 4. Morphogenesis—physiology. WG 201 F724 2001]
QP111.4 .F67 2001
612.1′7—dc21 00-046836

Printed on acid-free paper.

Birkhäuser

ISBN 0-8176-4216-1
ISBN 3-7643-4216-1 SPIN 10785212

Production managed by Louise Farkas; manufacturing supervised by Jeffrey Taub.
Typeset by Best-set Typesetter Ltd., Hong Kong.
Printed and bound by Maple-Vail Book Manufacturing Group, York, PA.
Printed in the United States of America.

9 8 7 6 5 4 3 2 1

Contents

Contributors

Parker B. Antin, Department of Cell Biology and Anatomy, University of Arizona, Tucson, AZ 85724, USA

David Bader, Stahlman Cardiovascular Research Laboratories, Vanderbilt University, Nashville, TN 37232-6300, USA

Thomas K. Borg, Department of Developmental Biology and Anatomy, University of South Carolina, School of Medicine, Columbia, SC 29208, USA

Angelique S. Boyer, Department of Cell Biology and Anatomy, University of Arizona, Tucson, AZ 85724-5044, USA

Stacey Carpenter, Stahlman Cardiovascular Research Laboratories, Vanderbilt University, Nashville, TN 37232-6300, USA

Wayne Carver, Department of Developmental Biology and Anatomy, University of South Carolina, School of Medicine, Columbia, SC 29208, USA

Edward B. Clark, Department of Pediatrics, University of Utah, Salt Lake City, UT 84113, USA

Jeffrey D. Croissant, Anterogen, Roxbury, MA 02119, USA

Adriana C. Gittenberger-de Groot, Department of Anatomy and Embryology, Leiden University Medical Center, 2300RC Leiden, The Netherlands

Edie C. Goldsmith, Department of Developmental Biology and Anatomy, University of South Carolina School of Medicine, Columbia, SC 29208, USA

Robert G. Gourdie, Department of Cell Biology and Anatomy, Medical University of South Carolina, Charleston, SC 29425, USA

Carol C. Gregorio, Department of Cell Biology and Anatomy, University of Arizona, Tucson, AZ 85724, USA

Jeanette Hyer, Department of Cell Biology, Cornell University Medical College, New York, NY 10021, USA

Bradley B. Keller, Department of Pediatrics, University of Kentucky, Lexington, KY 40536-0298, USA

Margaret L. Kirby, Developmental Biology Program, Institute of Molecular Medicine and Genetics, Medical College of Georgia, Augusta, GA 30912-2640, USA

Jenny L.-C. Lin, Department of Biological Sciences, University of Iowa, Iowa City, IA 52242-1324, USA

Jim J.-C. Lin, Department of Biological Sciences, University of Iowa, Iowa City, IA 52242-1324, USA

Roger R. Markwald, Cardiovascular Developmental Biology Center, Department of Cell Biology and Anatomy, Medical University of South Carolina, Charleston, SC 29425, USA

Takashi Mikawa, Department of Cell Biology, Cornell University Medical College, New York, NY 10021, USA

Daniel T. Monaghan, Department of Pharmacology, University of Nebraska Medical Center, Omaha, NE 68198-6260, USA

Robert E. Poelmann, Department of Anatomy and Embryology, Leiden University Medical Center, 2300 RC Leiden, The Netherlands

Jay D. Potts, Department of Developmental Biology and Anatomy, University of South Carolina School of Medicine, Columbia, SC 29208, USA

Robert L. Price, Department of Developmental Biology and Anatomy, University of South Carolina School of Medicine, Columbia, SC 29208, USA

Ann F. Ramsdell, Departments of Developmental Biology and Anatomy and Women's Studies, University of South Carolina, School of Medicine, Columbia, SC 29208, USA

Rebecca S. Reiter, Department of Biological Sciences, University of Iowa, Iowa City, IA 52242-1324, USA

Thomas H. Rosenquist, Department of Cell Biology and Anatomy, University of Nebraska Medical Center, Omaha, NE 68198-6395, USA

Raymond B. Runyan, Department of Cell Biology and Anatomy, University of Arizona, Tucson, AZ 85724-5044, USA

Kimiko Takebayashi-Suzuki, Department of Cell Biology, Cornell University Medical College, New York, NY 10021, USA

Louis Terracio, New York University College of Dentistry, New York, NY 10010-4086, USA

Thomas E. Thielen, Department of Developmental Biology and Anatomy, University of South Carolina School of Medicine, Columbia, SC 29208, USA

Robert J. Tomanek, Department of Anatomy and Cell Biology, University of Iowa, Iowa City, IA 52242-1109, USA

Mark Paul F.M. Vrancken Peeters, Department of Anatomy and Embryology, Leiden University Medical Center, 2300RC Leiden, The Netherlands

Da-Zhi Wang, Department of Molecular Biology and Oncology, University of Texas Southwestern Medical Center at Dallas, Dallas, TX 75235-9148, USA

Qin Wang, Department of Pharmacology, Vanderbilt University Medical Center, Nashville, TN 37232, USA

Andy Wessels, Cardiovascular Developmental Biology Center, Department of Cell Biology and Anatomy, Medical University of South Carolina, Charleston, SC 29425, USA

Halley S. Williams, Department of Biological Sciences, University of Iowa, Iowa City, IA 52242-1324, USA

H. Joseph Yost, Huntsman Cancer Institute, Center for Children, Departments of Oncological Sciences and Pediatrics, University of Utah, Salt Lake City, UT 84112-5330, USA

Series Preface

The overall scope of this new series will be to evolve an understanding of the genetic basis of (1) how early mesoderm commits to cells of a heart lineage that progressively and irreversibly assemble into a segmented, primary heart tube that can be remodeled into a four-chambered organ, and (2) how blood vessels are derived and assembled both in the heart and in the body. Our central aim is to establish a four-dimensional, spatiotemporal foundation for the heart and blood vessels that can be genetically dissected for function and mechanism.

Since Robert DeHaan's seminal chapter "Morphogenesis of the Vertebrate Heart" published in *Organogenesis* (Holt Rinehart & Winston, NY) in 1965, there have been surprisingly few books devoted to the subject of cardiovascular morphogenesis, despite the enormous growth of interest that occurred nationally and internationally. Most writings on the subject have been scholarly compilations of the proceedings of major national or international symposia or multiauthored volumes, often without a specific theme. What is missing are the unifying concepts that can make sense out of a burgeoning database of facts. The Editorial Board of this new series believes the time has come for a book series dedicated to cardiovascular morphogenesis that will serve not only as an important archival and didactic reference source for those who have recently come into the field but also as a guide to the evolution of a field that is clearly coming of age. The advances in heart and vessel morphogenesis are not only serving to reveal general basic mechanisms of embryogenesis but are also now influencing clinical thinking in pediatric and adult cardiology.

Undoubtedly, the Human Genome Project and other genetic approaches will continue to reveal new genes or groups of genes that may be involved in heart development. A central goal of this series will be to extend the identification of these and other genes into their functional role at the molecular, cellular, and organ levels. The major issues in morphogenesis highlighted in the series will be the local (heart or vessel) regulation of cell growth and death, cell adhesion and migration, and gene expression responsible for the cardiovascular cellular phenotypes.

Specific topics will include the following:

- The roles of extracardiac populations of cells in heart development.
- Coronary angiogenesis.

- Vasculogenesis.
- Breaking symmetry, laterality genes, and patterning.
- Formation and integration of the conduction cell phenotypes.
- Growth factors and inductive interactions in cardiogenesis and vasculogenesis.
- Morphogenetic role of the extracellular matrix.
- Genetic regulation of heart development.
- Application of developmental principles to cardiovascular tissue engineering.

Roger R. Markwald
Cardiovascular Developmental
Biology Center
Medical University of South Carolina
Charleston, South Carolina

Foreword

The ultimate goal of medicine extends beyond effective treatment to the ultimate prevention of disease. In this century, the disciplines of clinical epidemiology, pathology, and the basic sciences of cell biology and microbiology have yielded remarkable advances in children's health. Infant mortality has decreased from 1 in 10 to less than 5 in 1,000 over much of the United States.

As we enter the 21st century, the greatest single cause of infant mortality is birth defects, and the most prevalent birth defects involve the cardiovascular system (MMWR, 1998). Until recently the etiology of congenital cardiovascular malformations remained elusive. But now collaboration among clinicians and basic science investigators has made remarkable progress in identifying the causes of heart defects. This progress is detailed in this book, which emphasizes the importance of clinical and basic science collaborations. As this story unfolds, there are several truths to guide our thinking and analysis.

First, as humans we have two biologic histories: an ancient history of 3.8 billion years during which we evolved from single cell organisms through chordates to mammals to primates (Zimmer, 1998). Our evolutionary ancestors used common and now highly conserved gene pathways to form the four-chambered heart. This 3.8 billion-year history is compressed into the developmental processes that we study in fish, chick, and mouse as a window on the events of human cardiovascular development.

There is a genetic basis for many if not most types of congenital cardiovascular malformations. Some conotruncal defects, like truncus arteriosus commonus, closely resemble the normal arrangement of the heart and great vessels in lower vertebrates. Indeed, Dr. Helen B. Taussig (1988) proposed that human cardiovascular defects are atavisims, reoccurrences of an ancestral form of the heart.

Second, the environment is a key factor in cardiovascular development. Heart and vascular bed function begins and is sustained throughout the morphogenesis of the four-chambered heart. Beginning with the beating of the muscle-wrapped tubular heart, physical forces are integral to normal and abnormal development. Defining the form–function relationship is essential in understanding the pattern of gene expression and the translation of heart structure from the two-dimensional genetic code.

The micronutrient environment is also critical in normal development and is the basis of some forms of cardiovascular disease. Epidemiologic data suggest that during initial formation of the heart folate and vitamin A are critical for normal morphogenesis of the outflow tract (Hall and Solehdin, 1998). Adequate nutrition and fetal growth are also likely critical in determining the risk for late-onset cardiovascular disease (Barker, 1992). Defining the interplay among genetically determined pathways, cascades, and epigenetic factors is a key component of the developmental process.

So where are we going? The last 20 years have seen an explosion of information of the heart fueled by the technology of molecular genetics and cell biology. The challenge is now to integrate and transfer information from species to species and class to class. We must define common mechanisms and idiosyncratic factors in each model of cardiovascular development. The goal is to understand the complexity of human cardiovascular development.

Congenital cardiovascular defects are a major public health problem now attracting the attention of the scientific community. On the horizon is the prevention of heart defects. The contributions within these volumes show that goal is within our reach.

REFERENCES

Barker, D.J.P., ed. (1992). *Fetal and Infant Origins of Adult Disease. British Medical Journal*, 1992.

Hall, J.G., and Solehdin, F. (1998). Folate and its various ramifications. *Adv Pediatr* 45:1–35.

MMWR. (1998). Trends in infant mortality attributable to birth defects—United States, 1980–1995. *MMRW* 47(37):773–778.

Taussig, N.B. (1988). Evolutionary origin of cardiac malformations. *J Am Coll Cardiol* 12(4):1079–1086.

Zimmer, C. (1998). *At the Water's Edge*. The Free Press, New York.

Edward B. Clark
Department of Pediatrics
University of Utah
Salt Lake City, Utah

Preface

Cardiogenesis is a complex process involving temporal and spatial expression of genes and their products, and necessitating cell proliferation, migration, recognition, and signaling. Over the last 15 years there has been remarkable progress in the identification of mechanisms mediating cardiac development. Approaches that have contributed to new advancements in the field include genetic screening, gene targeting, cell, tissue, and organ cultures, confocal microscopy, and new animal models. The chapters in this book represent current perspectives into key topics in the field of heart development. Accordingly, this publication is intended to provide an updated and comprehensive review of the most recent advances in the field. There are other relevant topics that could have been included, but space limitations necessitate selectivity. Our intent in assembling the various chapters of this volume is to provide the reader with (1) a general sense of the field, (2) key references, and (3) insights into mechanisms underlying various events in heart development.

Chapter 1 is a comprehensive overview of the field of heart development, which raises fundamental questions that have stimulated research on heart development. It provides a temporal map of the key events contributing to formation of the heart. Chapters 2 to 7 address fundamental processes and events, including myofibrillogenesis, the epicardium, asymmetry and looping, atrial and ventricular differences, neural crest, and the conduction system. Chapters 8 to 13 are focused on the regulation of development by extracellular matrix, genes, growth factors, homocysteine, and mechanical factors. The editors are grateful to the contributing authors for their scholarly work and cooperation.

Iowa City, Iowa *Robert J. Tomanek*
Tucson, Arizona *Raymond R. Runyan*

CHAPTER 1

Overview of Heart Development

Roger R. Markwald and Andy Wessels

The fundamental questions that have driven the field of heart development over the past century continue to move the field today. These include but are not limited to the following: (1) Where in the postgastrulation embryo do the heart progenitor cells arise? (2) Do inductive interactions between germ layers influence their formation? (3) If so, what are the signals that regulate cell entry into a myocardial or endocardial cell lineage? (4) Is there a genetic cascade or heirarchy that regulates phenotypic expression in cells committed to a heart lineage? (5) How do cardiogenic cells integrate to form the two concentric epithelia—myocardium and endocardium—of the embryonic tubular heart? (6) Are all future regions of the mature heart represented in the primary embryonic heart (either before or after looping)? (7) How is pattern formation established in the primary heart tube such that the phenotype varies according to position along the rostrocaudal axis? (8) How do the specialized cells of the central and peripheral conduction systems arise and what regulates their formation and coordination? (9) How is the right-left axis of the heart established and by what developmental mechanisms do laterality genes modify pattern formation within the heart tube? (10) How does the single-circuited primary heart tube with a single lumen become transformed into a double-circuited, four-chambered heart in which each ventricle has a separate inlet and outlet? Answers to these questions are addressed in the chapters that follow. In this overview, we briefly summarize some of what has been learned about these ten broad-stroke questions and identify some of the potentially important future directions that might serve to continue to sustain progress in the field.

FORMATION OF CARDIOGENIC PRECURSOR CELLS

The mesodermal progenitors of the vertebrate tubular heart migrate from the rostral half of the primitive streak on either side of Hensen's node to form a pair of heart forming fields within the lateral plate mesoderm (Figure 1.1) (Rawles, 1943; Rosenquist and DeHaan, 1966; Garcia-Martinez and Schoenwolf, 1993). What distinguishes the heart-forming fields from other mesodermal cells is that they initiate expression of genes homologous to the *Drosophila tinman* gene

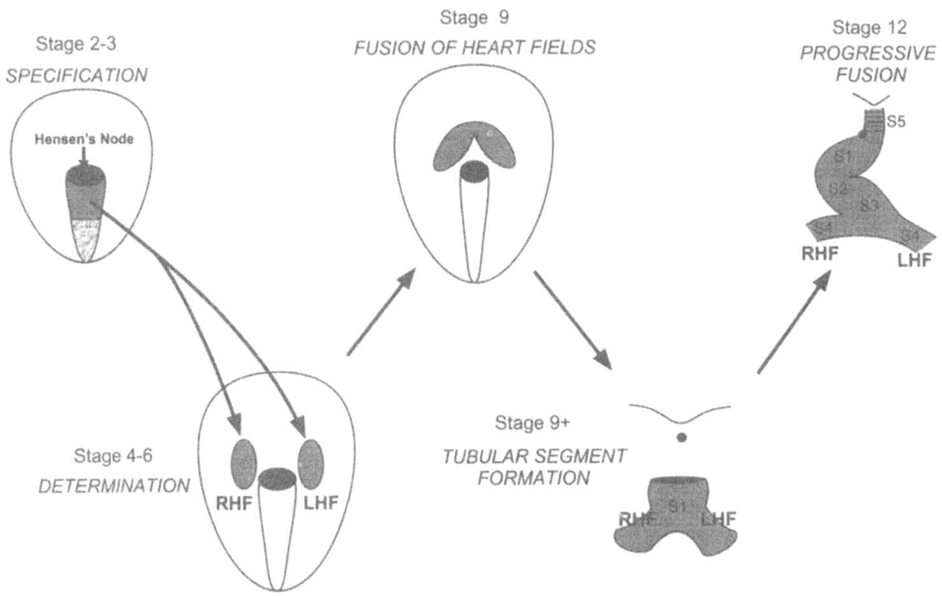

FIGURE 1.1. Diagram of early stages in the formation of the heart fields and heart tube in chick embryos. Cells specified or "reversibly committed" to a heart cell lineage migrate from the primitive streak to either side of Hensen's node to form two cohorts of precardiac mesodermal cells, termed the heart fields. Between stages 4 and 6, the right heart field (RHF) becomes dominant to the left (LHF) and each possesses cells now irreversibly committed or determined to a heart differentiation pathway. Fusion of the fields establishes a tubular heart, which is put together piece by piece over time. The straight heart tube is formed by segment one (S1), the future trabeculated region of the right ventricle. Other segments are added after looping, which occurs between S1 and S2 (future trabeculated region of the left ventricle). Segment 3 is the AV canal or future inlet for each ventricle. Note that segment 4 is still part of the unfused region of each heart field at stage 12. It will form the sinoatrium region of the venous pole. The last segment, S5, is the future outlet or conotruncal segment. Note that it begins its formation at stage 12 at the arterial pole. The solid circle represents an in vivo marker placed into the mesoderm located anterior to the straight heart tube at stage 9+ as part of an experiment described in de la Cruz (1998) and Markwald et al (1998). When incubation is continued until stage 12, the newly forming conus appears from tissue located anterior or rostral to the bead. Had the bead been located within or at the growing distal tip of the conus (S5), it would have suggested an origin from the existing heart tube rather than from mesodermal tissue located extrinsically to the rostral end of the heart tube.

(Bodmer, 1993), such as *Nkx2.5* (Lints et al, 1993) or *Nkx 2.8* (Brand et al, 1997). The latter, together with other transcriptions factors, e.g., *MEF-2c* (Lin et al, 1997) or *GATA-4/6* (Laverriere et al, 1994; Molkentin et al, 1994), regulate cardiomyocyte specific gene expression (for a review, see Olsen and Srivastava, 1996). In turn, morphogens secreted by the adjacent endoderm and ectoderm may act in concert to initiate expression of tinman homologues or other cardiogenic transcription factors. Two such categories of morphogens would include bone morphogenic proteins (Wall and Hogan, 1994) and wnt proteins (Nusse and Varmus, 1992). Bone morphogenetic proteins 2 and 4 have been shown to induce undifferentiated mesoderm to enter a cardiogenic lineage (Lough et al, 1996; Schultheiss et al, 1997; Ladd et al, 1998). Eisenberg et al (1997) have shown that wnt 11 is expressed in the posterior region of the heart fields in the area where it is proposed that undifferentiated mesoderm is being recruited into the heart fields (Markwald et al, 1998). When transfected into an undifferentiated mesodermal cell line, wnt 11 can induce a cardiomyocyte phenotype (Eisenberg et al, 1997). In vivo, sarcomeric myosin is first expressed at stage 7 in chick heart fields just prior to their fusion (Han et al, 1992; Montgomery et al, 1994). However, myosin at this time does not appear to be sarcomerically organized and appears to be iterated at sites where adherens junctions are established by *N*-cadherins and β-catenins (Linask, 1992; Linask and Lash, 1993; Soler and Knudsen, 1994). Cardiac specific actins are not expressed in the heart fields. The first muscle actin to be expressed is alpha smooth muscle actin, which is present in all segments of the tubular heart prior to any specific cardiac actin (Ruzicka and Schwartz, 1988). For a review see Linask and Lash (1998) or Chapter 2 in this volume.

While bone morphogenetic and wnt proteins may influence expression of a myocardial phenotype in cells determined ("irreversibly committed") to a heart lineage, they do not appear to be the signals that initially induce or regulate specification ("reversible commitment"). The identity of such a signal, be it autonomous or paracrine, remains unknown. What is clear is that specification occurs very early, perhaps within the epiblast or during movement of mesoderm through the rostral portion of the primitive streak (Antin et al, 1994; Cohen-Gould and Mikawa, 1996). From the very onset, cells fated (specified) and later determined to become heart cells appear to be associated with an extracellular antigen called JB3, recently identified as fibrillin-2 (Wunsch et al, 1994; Rongish et al, 1998). Whether this is a causal or casual relationship also remains to be determined.

A major morphogenetic event within the heart fields that precedes their fusion to form a definitive heart tube is the segregation of endocardial endothelial precursor cells from those committed to a myogenic lineage (Linask and Lash, 1993; Wunsch et al, 1994; Sugi and Markwald, 1996). While the heart fields (or precardiac splanchnic mesoderm) are considered to be epithelial, the fields are really an aggregated cluster of cells (Linask and Lash, 1998). Pre-endocardial cells appear to merely migrate away from the stratified layer of heart field epithelium and into the extracellular matrix associated with the basal surface of the endoderm. It remains to be determined if the heart field epithelium actually undergoes a transformation into free (i.e., mesenchymal) endocardial angioblasts. Most of the available data suggest that the heart fields are mostly unipotential cells committed to either a myocardial or endocardial phenotype, with few, if any, remaining bipotential (Linask and Lash, 1993; Cohen-Gould and Mikawa, 1996; Sugi and Markwald, 1996). Although few in number, the differentiation-inducible quail cell line, QCE-6, strongly argues for the existence of a multipotential mesodermal precursor cell within the stage 4 chick heart fields that can become endocardium, myocardium or even hematopoietic cell types (Eisenberg and Markwald, 1997). That there are signals emanating from the adjacent endoderm that initiate segregation of the two populations of unipotential cells is becoming increasingly apparent. In chick embryos, endodermally-derived secretory factors appear sufficient to initiate migration of endocardial precursor cells and their subsequent reintegration into an endocardial tube (Sugi and Markwald, 1996). Based on the phenotype of the knockout mouse, vascular endothelial growth factor (VEGF) is likely the endodermal factor required for the proliferation and integration of endothelial precursor cells (Dumont et al, 1999).

FUSION OF THE HEART FIELDS

The fate of the heart fields is to fuse ventrally beneath the foregut (DeRuiter et al, 1992). Fused heart fields create a crescent-shaped structure from which a linear or straight tube heart emerges (Figure 1.1) (Viragh et al, 1989). Extracellular matrices enriched in fibronectin are essential for the fusion, but whether the cells themselves actively direct the process is unknown (Linask and Lash, 1988). No gene transcription factor expressed within the heart fields, e.g., *Nkx2.5* (Lyons et al, 1995), *MEF-2c* (Lin et al, 1997), serum response factor, *GATA-4*, or *dHAND* (Srivastava et al, 1997), appears to affect fusion of the fields. However, the zebrafish mutation, *miles apart,* would suggest such a regulatory gene exists.

It is known that the heart fields acquire laterality prior to fusion, with the right heart field being dominant to the left with respect to influencing the direction of looping and variability of cell types (Hoyle et al, 1992; Sugi and Markwald, 1996). For a review, see Chapter 3 in this volume. Again, extracellular molecules sensitive to retinoic acid treatment appear important in maintaining sidedness (Smith et al, 1997). The *inverted (iv)* gene is a prominent candidate for determining laterality of the heart fields that later, after fusion, engender important defects if both copies of the gene are mutated (Brueckner and Horwich, 1989). However, the recent identification of the *iv* gene as left-right dynein (Supp et al, 1997) has not revealed why the failure to establish laterality within the heart fields subsequently translates into a future heart defect (Icardo and Sanchez, 1991). One clue may have been revealed decades ago by the studies of Stahlsberg (1969) which indicated that the two heart fields contribute differentially to the linear heart tube. This important but overlooked pioneering study has recently been confirmed and extended by using in vivo labeling techniques that indicate that the precursors for the two major prevalvular cushions of the atrioventricular (AV) canal are derived from different heart fields (Markwald et al, 1998; R. Thompson and R. Markwald, unpublished observations).

SEGMENTAL BASIS OF HEART TUBE FORMATION

A straight or linear heart tube emerges rostrally from the crescent formed by the fusion of the original paired heart fields (Figure 1.1). Two epithelia—endocardium and myocardium—separated by an acellular matrix (cardiac jelly) form the wall of the linear heart tube (Viragh et al, 1989). One of the most important questions concerning formation of the tubular heart is whether all future chambers of the mature heart have their primordium within the straight tube heart? In other words, are the future chambers of the heart arranged along a rostral-caudal sequence as frequently shown for the straight heart tube? Based on studies in living embryos, the answer is no. The only area of the four chambered heart represented in the linear heart tube is the future trabeculated region of the right ventricle (de la Cruz et al, 1989, 1998). Rather, when development is traced over time using in vivo markers, the heart tube appears to be put together "piece by piece" or segmentally (Markwald et al, 1998). Each segment arises at a different time point or birth date from the heart fields. In the chick, four segments arise sequentially from the fused heart fields between stages 9 and 14. Looping or bending of the heart tube begins at the seam or junction between the first two segments once the second is fully "clear" of the fused heart fields. The trabeculated regions of the right and left ventricles arise from these two segments respectively. The third segment—the AV canal—is initially the largest segment of the posterior limb of the looped heart (Gittenberger-deGroot et al, 1995). It forms the inlet region of each ventricle. The fourth segment—the sinoatrial segment—forms the future venous pole. The fifth and final segment—the conotruncus or outlet segment—develops between stages 12 and 24 at the distal end of the original first segment at the arterial pole (Figure 1.2; see color insert) and, as such, appears to the only segment that is *not* derived directly from primordial cells located within the original fused heart fields (de la Cruz et al, 1977). Thus, the segments have different birth dates and, perhaps for the outlet segment, birth places.

Significance of Segmentation

One implication of forming segments sequentially over time is that spatial and temporal "windows" may exist that can be developmentally regulated. If there are such "segmental" genes that regulate the formation and differentiation of segments, they could explain why some segments form cushion tissues and others myocardial invaginations (e.g., ventricular trabeculae or the primary atrial septum). Although the myocardial and endocardial epithelium blend imperceptibly between segments, there are several examples of genes that are expressed segmentally. For example, *dHAND* and *eHAND* are expressed in the future trabeculated regions of the right and left ventricles, respectively (Olsen and Srivastava, 1996). The *Irx* gene (Bao et al, 1999) is expressed in ventricular segments of the mouse, while murine tolloid-like-1 (mtl-1) (Clark et al, 1999) is expressed in the spina vestubuli of the atrium (Figure 1.2) and the interventricular primary septum. Conversely, in one or both cushion-forming segments (AV canal and outlet), the endocardium uniquely expresses the type 3 transforming growth factor (TGF) beta-receptor (Brown CB et al, 1999), whereas the adjacent myocardium expresses striated muscle LIM protein (SLIM 1) (Brown S et al, 1999), bone morphoqenetic protein (bmp)4 (Jones et al, 1991), and the heart defect (*hdf*) gene (Yamamura et al, 1997). Specific elements of a promoter have been shown to be able to direct (or restrict) expression of a gene to a particular segment of the heart (Ross et al, 1996; Franco et al, 1997). Whether such promoter elements normally do this in vivo is not known, but it demonstrates there is potential for even some globally expressed genes to become restricted to a specific segment. Conversely, in mammalian or chick embryos, exogenously administered retinoic acid can expand expression of genes like hdf or atrial vs. ventricular myosins across segmental boundaries (Yutzey et al, 1994; see also Chapter 5 in this volume). The effect of obscuring or neutralizing segmental identity is unclear; however, in those instances where retinoids modify laterality, the consequences can include abnormal looping and misalignment of internal septa (Bouman et al, 1995; Smith et al, 1997). In zebrafish, exogenously added retinoids selectively truncate the formation of the ventricle and bulbus arteriosus without affecting the sinus venosus, atrium, or AV canal (Stanier and Fishman, 1992). This further supports the hypothesis that segments are under different genetic regulation.

Another significance of segmental development as revealed by in vivo labeling is that no segment is the sole progenitor of an anatomic chamber of the mature heart (de la Cruz et al, 1991, 1997; de la Cruz and Sanchez, 1998). For example, the anatomic right and left ventricles are derived from the following segments: the trabeculated region from original segments one and two, respectively; the inlet and outlet (up to the level of the valves) of each ventricle are derived from the AV junction (segment 3) and the conus and truncus of the outlet segment, respectively (Markwald et al, 1998). Thus, mature ventricular chambers are anatomic but not embryologic units. While the variation in trabeculation can be used to confirm the identity of a ventricle (e.g., as in persistent AV canal or situs inversus), most congenital defects of the ventricle involve either the inlet segment (e.g., tricuspid atresia) or the outlet segment (double outlet right ventricle). For this and other reasons, Anderson (1989) developed the sequential segmental analysis approach (Shinbourne et al, 1976) as a means of more precisely characterizing and identifying congenital heart defects. It is important to recognize that the majority of heart defects are problems in the integration or alignment of segments or the septa that form within or between segments (Lamers et al, 1992; Webb et al, 1998; Mjaatvedt et al, 1999; Markwald et al, 1997).

Origin of the Outlet Segment

The terms *outlet segment* and *outflow tract*, which are used to describe the vascular conduit between the embryonic right ventricular segment and the aortic arches, include the conus, truncus, and aortic sac (Figure 1.2) (Pexieder et al, 1989). Only the conus and truncus have a myocardial epithelial mantle (Thompson and Fitzharris, 1979). The conus forms cushions that fuse to form the conal or outlet septum that divides the conus into an outlet for both the left and right ventricle (the infundibulum and aortic vestibule, respectively) (Barteling and Gittenberger-deGroot, 1989). Cushions also develop within the truncus and are known as the truncal cushions. These cushions become "impaled" by the leading edge of the aorticopulmonary septum that divides the aortic sac into the root of the aorta and the pulmonary trunk (Ya et al, 1998). Truncal cushions differentiate into the valve leaflets that guard the entrances to the aorta and pulmonary trunk (Wunsch et al, 1994; Noden et al, 1995). The transcription factors *NFAT-C* and *Sox-4* are expressed in the endocardium of the AV canal and truncus. However, targeted deletions of each only affect formation of the valves derived from the truncal cushions (Schilham et al, 1996; Ranger et al, 1998), which raises the question as to why genes expressed by the endocardium of both cushion-forming segments affect development of only one. Since transformation of endocardium into cushion mesenchyme (see below) is not affected in either knockout, one might predict there are proteins expressed differentially between segments that can enhance or inhibit expression of AV vs. outlet endocardial transcription factors. Such a possibility is consistent with the hypothesis that the two segments may originate from different heart fields as discussed below.

Despite its enormous relevance to congenital heart disease, little is known about the origin of the conotruncal segment. One possibility is that precursors could form by an extension (outward growth) of myocardial cells from the existing right ventricular segment (Castro-Quezada et al, 1972) or, alternatively, they could be derived de novo from mesoderm located at or rostral to the distal rim or the heart tube. The latter possibility would suggest the existence of a third heart forming field (a rostral or anterior heart-forming field), which to our knowledge has never been described. However, such a concept was clearly inferred by the experimental findings of de la Cruz et al (1977), who, using in vivo labeling approaches, demonstrated that no in vivo marker placed into the original paired (classical) heart fields of chick embryos ever traced, over time, to the outlet segment. Rather, as shown by the dark circles in Figure 1.1, the conotruncal segment was labeled only when markers are placed into mesodermal cells located rostal or anterior to the distal end of the heart tube at stages 9, and incubation continued until stages 12 to 17 (window of conus formation). Note in Figure 1.1, the conus (S5) is developing from tissue located behind (or rostral) to the marker (this experiment is reviewed in Markwald et al, 1998). The conclusion from these studies was that the conus and later the truncus are added as new structures to the distal end of the tubular heart. A similar interpretation was made by Fisher and Watanabe (1996) using adenoviral/lacZ vectors to infect, in situ, all myocardial cells present at stage 14 and then continuing incubation until stage 17 to 24 (window of truncus formation). The outcome was that part of the conus and all the truncus were unlabeled, i.e., they did not express β-galactosidase, a finding consistent with the hypothesis of an anterior heart field. Had they been "blue" (lacZ+) like the rest of the tubular heart, it would have suggested an origin from the myocardium at the distal rim of the tubular heart that existed at the time of infection. These findings predict the existence of an anterior or rostral heart-forming field, probably located at or near the junction of the truncal end of the tubular heart and the posterior wall of the aortic sac. If true, one might expect to observe reexpression of genes or protein first observed in the classic heart-forming fields, e.g., bmp 2 or 4, tinman homologues, and wnt-11. Such spatial and temporal expression studies have yet to be reported and, unfortunately, the knockouts of either bmp 2, bmp 4, tinman, or *dHAND* are early lethals (Lyons et al, 1995; Winnier et al, 1995; Zhang and Bradley, 1996; Srivastava et al, 1997), which preclude concluding anything about the role of these morphogens or transcription factors in the formation of the outlet segment. One potential future solution may be to place early lethal mouse models onto different genetic backgrounds to attempt to prolong development.

CUSHION FORMATION VS. TRABECULATION

Each segment of the heart tube participates in forming septal tissues that collectively integrate over time to divide the primary heart tube into four chambers. In the AV and outlet segments, the septal tissues are mesenchymal swellings called cushions, which fuse to form septal structures from which the valve leaflets arise (de la Cruz et al, 1998; Webb et al, 1998). In the ventricles, clonal expansions of myocardial cells project into the lumen to form trabeculae (Mikawa et al, 1992). Historically, the coalescence of trabeculae at the interventricular junction has been the mechanism proposed for the formation of the primary ventricular muscular septum (Ben-Shachar et al, 1985). However, in vivo labeling studies point to a specific site of origin for the ventricular septum from cells derived from the original ventral fusion line of the heart fields (de la Cruz et al, 1997), a finding supported by the expression of the mammalian tolloid-like 1 gene (Clark et al, 1999). In the sinoatrium, proliferation of atrial myocardium form the muscular portion of the primary atrial septum. Formation of this septum is associated with extracardiac mesenchymal tissue that penetrates the posterior wall of the atrium to form a midline structure, the spina vestibuli (or dorsal mesenchymal protrusion) (Figure 1.2) (Tasaka et al, 1996; Webb et al, 1998; Wessels et al, 2000), a structure that also expresses tolloid-like-1. The mesenchyme of the fused AV cushions serves to knit together and anchor the primary atrial and ventricular muscular septa (de la Cruz et al, 1998; Webb et al, 1998; Wessels et al, 2000). In the outlet segment, the fusion of the conal cushions forms a septum that when brought into alignment with the primary ventricular septum creates a separate outlet for each ventricle. Thus, the formation of cushions is critical for cardiac morphogenesis.

A central tenet of cushion tissue formation is that the myocardium of the AV and outlet segments induces a competent endocardium to transform into mesenchyme (Runyan and Markwald, 1983; Runyan et al, 1990). Myocardial cells of these segments secrete multicomponent complexes we have called adherons (Mjaatvedt and Markwald, 1989) which promote some but not all endothelial cells to transform into mesenchyme (reviewed in Eisenberg and Markwald, 1995, and Mjaatvedt et al, 1999). Endothelial cells competent to transform express a marker, JB3, which may reflect their origin from the heart fields (Wunsch et al, 1994) vs. an origin outside the heart fields (Sugi and Markwald, 1996). Overall, some 50+ genes or proteins have been identified that affect endothelial transformation or whose expression is modified during transformation and migration. These include *ES-130* (Krug et al, 1995), *hLAMP* (Sinning et al, 1995; Sinning, 1997), transferrin (Isokawa et al, 1994), serine proteases (McGuire and Alexander, 1993), growth factors, particularly of the TGF-β family (Potts et al, 1991; Nakajima et al, 1997a,b; Boyer et al, 1999), receptors to TGF-β growth factors (Brown et al, 1999), and extracellular proteins that bind TGF-β (Nakajima et al, 1997a), and cell and extracellular adhesion molecules (Crossin and Hoffman, 1991; Little and Rongish, 1995; Zhang et al, 1995; Bouchey et al, 1996; Capehart et al, 1999). Thus, multiple levels of myocardial-endocardial interaction appear to be required to initiate, sustain, or amplify the formation of cushion tissue or to "fine tune" its final form or shape. (A more detailed analysis of the complex process of endocardial transformation to cushion mesenchyme is presented in Chapter 11.)

Despite intense investigation, to date, no consensus myocardial signal has been confirmed as the one "master" upstream signal that starts the whole process of cushion formation. That such a signal exists was reaffirmed in a recent, high stringency, temporal refinement of the culture assay for myocardial-endocardial interaction (Ramsdell and Markwald, 1997; Ramsdell et al, 1998). Using early activation markers, these studies indicated that myocardial induction actually began during the formation of the actual AV or outlet segments (much earlier than previously thought). Thus, the search for the myocardial inducer continues and appears to be heading toward the future consideration and/or identification of proteins like bmp2/4 or wnt-11 (Eisenberg and Eisenberg, 1999) that are restricted to cushion forming segments, particularly during the early morphogenesis of the segment itself. Surprisingly, although their temporal and spatial expression makes bmp2/4 "perfect" candidates for being the upstream myocardial signal that initiates cushion formation, neither bmp2 or bmp4 has been tested in culture assays and, unfortunately, both bmp2/4 knockouts were lethal too early to determine if they had a role in cushion formation.

Similar links between segmental morphogenesis and trabeculation may occur in the ventricles. Results from gene knockouts indicate that it is unlikely that myocardial cells invade the cardiac jelly to form trabeculae as a "default" response to the failure of ventricular endocardial endothelial cells to first fill the extracellular space with cushion tissues. Rather, specific endocardial-myocardial interactions appear to drive trabeculation as shown by the profound disruption of trabeculae formation in mice null for neuregulin receptors (Meyer and Birchmeier, 1995) or for either Tie 2 or angiopoietin 1. Tie 2 is a receptor tyrosine kinase on endocardial cells whose ligand, angiopoietin 1, is secreted by the myocardium (Sato et al, 1995; Suri et al, 1996). In this instance, it is proposed that signaling from the myocardium through secretion of angiopoietin 1 is received by the endocardium via the tie 2 receptor, which in turn feeds back to affect myocardial development, in some yet unknown manner (see Hanahan, 1997, for a review). Thus, future investigations regarding ventricular segmental development appear to be directed toward an understanding of how proteins like neuregulin or angiopoietin change cell behavior to promote formation of focalized myocardial evaginations (trabeculae).

TRANSFORMING THE PRIMARY HEART TUBE INTO A FOUR-CHAMBERED ORGAN

Prior to the onset of cushion formation and trabeculation (about stage 16 in the chick and day 9 in the mouse), the heart tube is exclusively derived from precardiac mesoderm of the heart fields (including a possible anterior candidate heart field). Thereafter, multiple populations of extracardiac mesenchymal cells enter the heart tube. As shown in Figure 1.2, neural crest enter the heart as two prongs or columns of cells that extend from the aorticopulmonary septum of the aortic sac (Figure 2.2) (Poelmann et al, 1998; Waldo et al, 1998). "Vent" cells are thought to be a second wave of neural crest-like cells that are derived from the ventral wall of the neural tube and that enter the venous pole through a conduit of dorsal mesocardial tissue (i.e., the spina vestibuli) that forms in association with the primary atrial septum (R. Poelmann, unpublished data presented at the Weinstein meeting, May 20, 1999, in Tuscon, Arizona). Yet a third mesenchymal population also penetrates the myocardial wall to enter the AV cushions as well as those of the conus and ventricular trabeculae. This mesenchymal population is derived from the epicardial epithelium and is more characterized than are VENT (ventral neural tube-derived cells) (Gittenberger-deGroot et al, 1998). The invasion of the epicardial, neural crest, and VENT cells correlates spatially and temporally with the remodeling of the inner curvature of the looped heart by a process called "myocardialization." The latter is the morphogenetic event proposed as the mechanism for completing looping (i.e., the convergence of the two limbs of the U-shaped heart tube), which serves to align segments and their internal septa to form an inlet and outlet for each ventricle (Markwald et al, 1998).

Epicardialization of the Primary Heart Tube

The epicardium, despite the implication of the historical term *epimyocardium*, is not a derivative of the myocardial mantle formed by the fusion of the original heart fields. Rather, it is formed from a villous (grape-like) outgrowth from the coelomic mesothelium associated with the liver and venous pole termed the proepicardial organ (Viragh and Challice 1973; Viragh et al, 1994; Dettman et al, 1998). Virtually nothing is known, however, about the origin of proepicardial organ or how or why it develops where it does. The importance of answering these questions is that the proepicardial organ gives origin to an epithelial monolayer that migrates to progressively cover the surface of all segments of the heart tube expect for the conotruncal segment (Robert Poelmann, unpublished data). The origin of the conotruncal epicardium has not been established but appears likely to be another derivative of the hypothetical anterior heart field associated with the arterial pole. After migration to cover the surface of the heart, some but not all epicardial epithelial cells transform into mesenchyme or epicardial-derived cells (EPDCs) that accumulate in the extracellular space between the myocardium and the epicardium (Figures 1.2 and 1.3; see color insert) (Perez-Pomares et al, 1997). The location and subsequent fate of the EPDCs can be multiple. In the subepicardium, many persist as fibroblastic cells, or they incorporate as endothelium or smooth muscle into the coronary vasculature (Dettman et al, 1998). Inhibiting formation of the epicardium completely prevents formation of coronary vessels over much of the heart tube (Gittenberger-deGroot et al, 1998). (For more details, see Chapter 8).

Surprisingly, chimera or retroviral marking experiments also revealed that EPDC can penetrate the myocardial mantle and enter the AV cushions and occasionally the endocardium (Gittenberger-deGroot et al, 1998) (Figures 1.2 and 1.3). The exact role of the EPDC in cushion development remains to be elucidated; however, it has been reported that hearts with impaired epicardial development show abnormal development of the AV cushions (Poelmann and Gittenberger-deGroot, unpublished data). Genetic models for pertubation of epicardial development are found in the vascular cell adhesion model (VCAM)-1 (expressed by myocardial cells) knockout (Kwee et al, 1995), the α_4-integrin (expressed on epicardial cells and mesenchymal cells found in the AV cushions) knockout (Yang et al, 1995), and the Wilms' tumor-1 gene (*Wt1*) knockout mice (Moore et al, 1999). These mice show a lack of proper epicardial formation and like the epicardial ablation model in the chick (Gittenberger-deGroot et al, 1998), show a lack of coronary vessels. Together these model systems also should provide future opportunities to determine how EPDCs manage to traverse the myocardium and whether they play a regulatory role in determining if cushion mesenchyme tissue differentiates into valve tissue or becomes "myocardialized," i.e., invaded by myocardial tissue (Figure 1.2). Given their presence in the AV cushions and subendocardially in the ventricles, there is also speculation that they might have a potential regulatory role in the formation of the central or peripheral conduction systems (Gourdie et al, 1999). It might be particularly informative to determine if the EPDCs express growth factors such as TGF-β, platelet-derived growth factors, endothelin, or ligands for receptor tyrosine kinases. For example, if EPDCs do express endothelin, they could be the source of signals that direct myocardial cells into a conduction phenotype (Gourdie et al, 1998).

Morphologically, the formation of the subepicardial mesenchyme resembles the formation of cushion mesenchyme. A testable hypothesis might be that the inductive processes that play a role in endocardial transformation (Mjaatvedt et al, 1991, 1998; Krug et al, 1995) similarly play a role in epicardial transformation. If so, a first step might be to determine if epicardial cells respond to myocardial signals. ES130, a protein upregulated in activated endocardial cells (Krug et al, 1995) and required for transformation (Ramsdell and Markwald, 1997) is also expressed by epicardial cells (Perez-Pomares et al, 1997). In addition, the expression of the JB3 antigen (fibrillin-2) (Wunsch et al, 1994; Rongish et al, 1998) correlates temporally and spatially with the transformation of both endocardial cells and epicardium-to-mesenchymal transformation (Perez-Pomares et al, 1998). These and other observations indicate that the differentiation programs for epicardial and endocardial development may have some steps in common. Whether shared mechanism include TGF-β family members will be an "acid test" as these growth factors play rate-limiting roles in cushion formation (Potts et al, 1991; Nakajima et al, 1994, 1997a; Ramsdell and Markwald, 1997). The characterization of a novel basic helix-loop-helix (bHLH) transcription factor, epicardin, by Robb and co-workers (1998) would suggest that the two processes may have differences. The reported similarities between epicardin and myocardial bHLH proteins, *eHAND* and *dHAND*, will undoubtedly lead to more extensive studies on the molecular and cellular pathways involved in epicardial transformation and differentiation, if for no other reason than its relationship to coronary vasculogenesis.

Myocardialization

The notion that myocardial invasion of the conal cushions plays a role in conotruncal development is not entirely new (Okamoto et al, 1981). However, this process has only recently gained considerable attention (Mjaatvedt et al, 1999; Moorman et al, 1999; van den Hoff et al, 1999), partly because the process affects the conal septum, a structure frequently associated with defects in the outflow tract. The term *myocardialization* was first introduced in 1995 (Lamers et al, 1995) to describe the active ingrowth of existing myocardium into mesenchymalized tissues of the heart. It is important to understand that the term does *not* refer to the differentiation of mesenchyme into cardiac muscle but rather to a replacement of mesenchyme with existing muscle cells. In both the avian and mammalian heart, myocardialization is associated with the cushions (dorsodextral conal cushion and superior AV cushion) that lie along inner curvature of the U-shaped heart, where it results in the formation of the muscular outlet (conal) septum and the supraventricular crest (de la Cruz, 1998). However, in the chick heart, myocardialization of the right lateral and superior AV cushions also occurs (as does the membranous AV septum). Whether this is an evolutionary offshoot or some functional adaptation in birds remains unclear. (As described below, if neural crest cells play a role in triggering myocardialization, then myocardialization of the AV cushions may reflect that in birds, unlike mammals, neural crest cells migrate beyond the level of the conus to reach the AV segment.)

The importance of myocardialization with respect to outflow tract septation is clearly suggested by studies in which the absence or inhibition of myocardialization correlated with congenital conotruncal defects (Hogers et al, 1997; Sanford et al, 1997). Most of these malformations involved alignment of the outlet septum, with the muscular interventricular septum, resulting in a spectrum of defects correlated with the degree of misalignment, e.g., simple ventricular septal defect to double outlet right ventricle (DORV) (Markwald et al, 1998). As described by Mjaatvedt et al (1999), in DORV, the embryonic pattern of blood flow is retained, i.e., blood flows in series from the atrium into the left ventricle, then into the right ventricle, and finally into the outlet segment (Figure 1.2). In other words, just as in the early embryo, in DORV neither the left nor right ventricle has a separate inlet and outlet. For this reason, it has been proposed that myocardialization serves to remodel the inner curvature of the heart such that the inlet and outlet limbs of the U-shaped heart are able to converge at the level of the AV junction to create an outlet for the left ventricle while simultaneously allowing for expansion of the right side of the AV junction to create an inlet for the right ventricle (Markwald et al, 1998). What, if any, cascade of cellular events is engendered by myocardialization of the conus that could affect septal or segmental alignments remains to be elucidated. One prediction is that there may exist some type of extracellular or structural link between the conal and AV cushions that lies along the inner curvature. Such a structure could serve to transmit physical forces created by an inward growth of myocardium into the conal cushions to the right side of the AV junction or could serve as a cell-to-cell conduit for communicating changes in cell activities.

If myocardialization is the driving force for aligning segments and their internal septa in order to create an inlet and outlet for each ventricle, then a key question becomes why only certain cushions, like the conal cushions, undergo this remodeling process in mammals (van den Hoff et al, 1999). Circumstantial evidence would suggest neural crest as a possible answer. As described above and shown in Figure 1.2, crest cells enter the conus region of the outlet segment as two prongs or columns of cells extending from the aorticopulmonary septum shortly before the onset of myocardialization. Experimental or genetically induced perturbations that affect neural crest formation or migration (Nishibatake et al, 1987; Kirby, 1993; Conway et al, 1997) result in a similar spectrum of defects to those that occur from inhibiting myocardialization, e.g., DORV (Hogers et al, 1997; Sanford et al, 1997). This interesting correlation may hold answers not only for the process of myocardialization but also for one of the major mysteries of heart development: the role, if any, of neural crest cells that invade the heart below the level of the outlet valves.

In the same context, it may also be potentially important to know that a concentration of apoptotic cells is uniquely found in the conal cushions during the active phase of myocardialization in developing chick and rat hearts (Ya et al, 1998). Some studies suggest that the apoptotic cells in this region are of neural crest origin (Poelmann et al, 1998) but other studies indicate that the majority of neural crest cells do not die but instead contribute to the developing outlet septum as part of the "condensed mesenchyme" found on either side of the conal cushion fusion line (Waldo et al, 1998). The foci of apoptotic cells appear to locally recruit macrophages into the conal cushions and the outlet septum formed by their fusion (Ya et al, 1998). Whether the purpose of the apoptotic clusters of cushion mesenchyme and their subsequent phagocytosis by macrophages is solely to eliminate cells that migrated too far into the heart or to open up space for myocardialization may be an oversimplification in either case. Because the matrix of conal cushions is enriched in the latent form of TGF-β (Ghosh and Brauer, 1996), the time may be ripe to determine if the apoptotic cells (whether they be of crest, EPDC, or cushion origin) release proteases that could activate latent TGF-β. One significance of focally activating this growth factor is that TGF-β is a candidate signal for initiating myocardialization. First, in the TGF-β_2 knockout mouse, myocardialization was impaired. Second, there are preliminary data that TGF-β_2 stimulated myocardialization in culture assays (Wessels et al, Weinstein Cardiovascular Development Meeting, Nashville TN, 1998). Finally, ligation of the vitelline vein in the developing chick downregulated expression of TFG-β receptors, resulting in impaired myocardialization and abnormalities of the conotrucal outlet segment and AV junction (Hogers et al, 1997; Adriana Gittenberger-deGroot, personal communication). If true, myocardialization would be yet another example of a centrally defining morphogenetic event in heart development that is regulated by members of the TGF-β gene family (see Chapter 11). Understanding the process of myocardialization and its molecular regulation might also provide candidate genes for a single gene defect in the Keeshond dog that causes conotruncal defects in which the primary defect appears to be in the myocardium (Patterson et al, 1993).

Finally, consistent with the proposed role of neural crest in regulating myocardialization, a newly developed culture assay confirmed that it is a nonmyocardial cell type that initiates myocardialization (van den Hoff et al, 1999). However, the culture data also indicate that myocardial cells competent to myocardialize do so, only after differentiating to a specific developmental stage, and that no matter what the stage, some myocardial cells (e.g., ventricular myocardium) remain incompetent to myocardialize. This suggests considerable levels of regulation and that the molecular basis for myocardialization may be segmentally based. The elucidation of the mechanisms regulating myocardialization would provide invaluable clues to how the segments and septa of the heart are aligned to complete formation of the four-chambered heart.

The Conduction System

The development of a functional four-chambered heart could not occur without a conduction system to regulate blood flow and contraction. The developmental aspects, morphologic characteristics, as well as the functional features of the conduction system have remained controversial since workers at the turn of the 20th century proposed that a population of specialized cells existed in the heart to regulate the impulse over the respective chambers of the heart (Keith and Flack, 1907). However, new aspects of the conduction system have been more recently revealed by recognition that conduction cells had special staining characteristics that could be used to map their three-dimensional networks (Thompson et al, 1995). For instance, it was demonstrated that parts of the atrioventricular ("central") conduction system coexpressed both atrial and ventricular myosin heavy chains (Wessels et al, 1991), that the slow-conducting proximal portions of the central conduction system (AV node and His-bundle) were characterized by the extremely low expression of the M isoform of creatine kinase (Wessels et al, 1990) and virtual absence of connexin 43 proteins (van Kempen et al, 1991), the expression of the homeobox gene *msx-2* (Chan-Thomas et al, 1993) and the expression of min-K (Kupershmidt et al, 1999). Interestingly, the myocytes of the conduction system in the mammalian heart and neural tissues share a series of antigens. Elevated levels of cholinesterase were found in the mammalian heart (Anderson and Taylor, 1972; Lamers et al, 1987), the central and sinoatrial conduction system of the rabbit is characterized by the expression of neurofilaments (Gorza and Vitadello, 1989) and, in rats and humans, the tissues of the central conduction system are specifically recognized by antibodies (GlN2/Leu7/HNK1) that also recognize neural antigens (Wessels et al, 1992; Gourdie et al, 1999). A functional role for these proteins has yet to be described. The expression of the neuromuscular markers has led some to propose that the conduction system of the heart is a neural crest derivative (Gorza et al, 1988). However, none of the cell-fate experiments performed to date provide any evidence for this hypothesis, neither the quail-chick chimeric system (Kirby, 1993), nor retroviral labeling of neural crest cells (Poelmann et al, 1998). In contrast, Studies by Gourdie et al (1995) have demonstrated using viral approaches that the specialized cells of the conduction system share the same progenitor as the working myocytes of the heart and are hence of myocardial origin, which was already suggested is several earlier studies (e.g., Wessels et al, 1992). The myocytes of the conduction system leave the mitotic cell cycle and differentiate much earlier than other myocardial cells, possibly under the regulatory influence of endothelin (Thompson et al, 1995; Gourdie et al, 1998). Obviously, many question remain regarding the function and development of the conduction system: What is the basis for the large variation in molecular phenotype between working myocytes and conduction cells when they have a common origin and, in particular, if there is a role or function for the neuromuscular antigens? Above all, what are the molecular cascades that direct a myocardial precursor cell into a conduction cell lineage? For example, are there transcription factors uniquely expressed in cells committed to a conduction cell differentiation pathway? (The conduction system is discussed in detail in Chapter 7.)

SUMMARY

Many current studies in heart development appropriately emphasize identifying new genes or genetic "wiring." In the last 5 years, dozens, if not a hundred or more, potential regulatory genes have been identified as being expressed in the heart and many of these have been or are being used to create animal models in which to study the function of these genes. Thus, what we have today is a rapidly growing number of genes and genetically created animals. While the identification of new genes expressed in the heart is exciting and always with anticipation of being an upstream master regulator (e.g., a "myo-H"; Miner et al, 1992), the fact remains that the heart is the outcome of many cellular processes operating at different time points and at different levels of structural complexity. Moreover, as described above, not all cells in the heart are derived from the heart-forming fields and as many as three heart fields may be required to sustain the progressive formation of segments. Thus, to fully exploit the wealth of new genetic information, what is needed are studies that can provide an understanding of how fundamental, cellular interactions are integrated over time to engender changes in gene expression that direct the morphogenetic processes required for the formation of a tubular heart and its subsequent remodeling into a four-chambered organ. The chapters presented in this volume present new data and/or review studies that provide insight for mechanistically interpreting (or hypothesizing) structural or functional consequences of those gene activities that can continue to move the field of cardiac morphogenesis.

ACKNOWLEDGMENT

The authors wish to express the gratitude to Dr. Thomas Trusk for his invaluable assistance in preparing this chapter, particularly for skillful help in developing the computer-generated figures.

REFERENCES

Anderson, R. (1989). The present-day place of correlations between embryology and anatomy in the understanding of congenitally malformed hearts. *Correlations between experimental cardiac embryology and teratology and congenital cardiac defects*. A. A. a. P. T, University of Granada Press: 265–295.

Anderson, R.H., and Taylor, I.M. (1972). Development of atrioventricular specialized tissue in human heart. *British Heart Journal* 34(12):1205–1214.
Antin, P.B., Taylor, R.G., and Yatskievych, T. (1994). Precardiac mesoderm is specified during gastrulation in quail. *Dev Dyn* 200:144–154.
Bao, Z.-Z., Bruneau, B.G., Seidman, J.G., Seideman, C.E., and Cepko, C.L. (1999). Regulation of chamber-specific gene expression in the developing heart by *Irx4*. *Science* 283:1161–1164.
Barteling, M.M., and Gittenberger-deGroot, A.C. (1989). The outflow tract of the heart—embryologic and morphologic correlations. *Int J Cardiol* 22:289–300.
Ben-Shachar, G., Arcilla, R.A., Lucas, R.V., and Manasek, F.J. (1985). Ventricular trabeculations in the chick embryo heart and their contribution to ventricular and muscular development. *Circ Res* 57:759–766.
Bodmer, R. (1993). The gene tinman is required for specification of the heart and visceral muscles in *Drosophila*. *Development* 118:719–729.
Bouchey, D., Argraves, W.S., and Little, C.D. (1996). Fibulin-1, vitronectin and fibronectin expression during avian valve and septal development. *Anat Rec* 244:540–551.
Bouman, H.G.A., Broekhuizen, M.L.A., Baasten, M.J., Gittenberger-deGroot, A.C., and Wenink, A.C.G. (1995). A spectrum of looping disturbances in stage 34 chicken hearts after retinoic acid treatment. *Anat Rec* 243:101–108.
Boyer, A.S., Ayerinskas, I.I., Vincent, E.B., McKinney, L.A., Weeks, D.L., and Runyan, R.B. (1999). TGF beta2 and TGF beta3 have separate and sequential activities during epithelial-mesenchymal cell transformation in the embryonic heart. *Dev Biol* 208:530–545.
Brand, T., Andree, B., Schneider, A., Buchberger, A., and Arnold, H.-H. (1997). Chicken *NKx2.8*, a novel homeobox gene expressed during early heart and foregut development. *Mech Dev* 64:53–59.
Brown, C.B., Boyer, A.S., Runyan, R.B., and Barnett, J.V. (1999). Requirement of the type III TGF-beta receptor for endocardial cell transformation in the heart. *Science* 283:2080–2082.
Brown, S., Biben, C., Ooms, L.M., et al. (1999). The cardiac expression of striated muscle LIM protein 1 (SLIM1) is restricted to the outflow tract of the developing heart. *J Mol Cell Cardiol* 31:837–841.
Brueckner, M.D., and Horwich, A.L. (1989). Linkage mapping of a mouse gene, iv, that controls left-right asymmetry of the heart and viscera. *Proc Natl Acad Sci USA* 86:5035.
Capehart, A.A., Mjaatvedt, C.H., Hoffman, S., and Krug, E.L. (1999). Dynamic expression of a native chrondroitin sulfate epitope reveals microheterogeneity of extracellular matrix organization in the embryonic chick heart. *Anat Rec* 1:1–15.
Castro-Quezada, A., Nidal-Ginard, B., and de la Cruz, M.V. (1972). Experimental study of the formation of the bulboventricular loop in the chick. *J Embryol Exp Morphol* 27:623–637.
Chan-Thomas, P., Thompson, R.P., Robert, B., Yacoub, M.H., and Barton, P.J. (1993). Expression of homeobox genes *Msx-1* (*Hox-7*) and *Msx-2* (*Hox-8*) during cardiac development in the chick. *Dev Dyn* 197:203–216.
Clark, T.G., Conway, S.J., Scott, I.C., et al. (1999). The mammalian tolloid-like 1 gene, *Tl1*, is necessary for normal septation and positioning of the heart. *Development* 126:2631–2642.
Cohen-Gould, L., and Mikawa, T. (1996). The fate diversity of mesodermal cells within the heart field during early embryogenesis. *Dev Biol* 177:265–273.
Conway, S., Henderson, D., and Copp, A. (1997). Pax3 is required for cardiac neural crest migration in the mouse: evidence from the splotch (Sp2H) mutant. *Development* 124:505–514.
Crossin, K., and Hoffman, S. (1991). Expression of adhesion molecules during the formation and differentiation of the avian endocardial cushion tissue. *Dev Biol* 145:277–286.

de la Cruz, M.V. (1998). Embryological development of the outlet of each ventricle. In: de la Cruz, M.V., and Markwald, R.R., eds. *Living Morphogenesis of the Heart*. Birkhauser (Springer/Verlag), Boston, pp. 157–168.

de la Cruz, M.V., Castillo, M., Villavicencio, L., Valencia, G.A., and Moreno-Rodriguez, R.A. (1997). Primitive interventricular septum, its primordium, and its contribution in the definitive interventricular septum: in vivo labelling study in the chick embryo. *Anat Rec* 247:512–520.

de la Cruz, M.V., Moreno-Rodriguez, R., and Markwald, R.R. (1998). Embryological development of the ventricular inlets. Septation and atrioventricular valve apparatus. In: de la Cruz, M.V., and Markwald, R.R., eds. *Living Morphogenesis of the Heart*. Birkhauser (Springer-Verlag), Boston, pp. 1131–1156.

de La Cruz, M.V., Robota, G.M., Saravalli, O., and Cayre, R. (1983). The contribution of the inferior cushion of the atrioventricular valves: study in the chick embryo. *Am J Anat* 166:63–72.

de la Cruz, M.V., and Sanchez-Gomez, C. (1998). Straight tube heart: primitive cardiac cavities vs. primitive cardiac segments. In: de la Cruz, M.V., and Markwald, R.R., eds. *Living Morphogenesis of the Heart*. Birkhauser (Springer Verlag), Boston, pp. 85–99.

de la Cruz, M.V., Sanchez-Gomez, C., Arteaga, M.M., and Arguello, C. (1977). Experimental study of the development of the truncus and the conus in the chick embryo. *J Anat* 123:661–686.

de la Cruz, M.V., and Sanchez Gomez, C., Cayre, R. (1991). The developmental components of the ventricles: their significance in congenital cardiac malformations. *Cardiol Young* 1:123–128.

de La Cruz, M.V., Sanchez-Gomez, C., and Palomino, M.A. (1989). The primitive cardiac regions in the straight tube heart (stage 9-) and their anatomical expression in the mature heart: an experimental study in the chick heart. *J Anat* 165:121–131.

DeRuiter, M.C., Poelmann, R.E., Vanderplass-deVries, I., Mentink, M.M.T., and Gittenberger-de Groot, A.C. (1992). The development of the myocardium and endocardium in mouse embryos: fusion of two heart tubes? *Anat Embryol* 185:461–473.

Dettman, R.W., Denetclaw, W. Jr., Ordahl, C.P., and Bristow, J. (1998). Common epicardial origin of coronary smooth muscle, perivascular fibroblasts and intermyocardial fibroblasts in the avian heart. *Dev Biol* 193:169–181.

Dumont, D.J., Jussila, L., Taipale, J., et al. (1999). Cardiovascular failure in mouse embryos deficient in VEGF receptor-3. *Science* 282:946–949.

Eisenberg, C.A., Gourdie, R.G., and Eisenberg, L.M. (1997). Wnt-11 is expressed in early avian mesoderm and required for the differentiation of the quail mesoderm cell line QCE-6. *Development* 124:525–536.

Eisenberg, C.A., and Markwald, R.R. (1997). Mixed cultures of avian blastoderm cells and the quail mesoderm cell line QCE-6 provide evidence for the pluripotentiality of early mesoderm. *Dev Biol* 191:167–181.

Eisenberg, L.M., and Markwald, R.R. (1995). Molecular regulation of atrioventricular valvuloseptal morphogenesis. *Circ Res* 77:1–6.

Fisher, S.A., and Watanabe, M. (1996). Expression of exogenous protein and analysis of morphogenesis in the developing chicken heart using an adenoviral vector. *Cardiovasc Res* 31:E86–E95.

Franco, D., Kelly, R., Buckingham, M., and Mooman, A.F.M. (1997). Regionalized transcriptional domains of myosin light chain 3f transgenes in the embryonic mouse heart: morphogenetic implications. *Dev Biol* 187:17–33.

Garcia-Martinez, V., Schoenwolf, G.C. (1993). Primitive steak origin of the cardiovascular system in avian embryos. *Dev Biol* 159:706–719.

Ghosh, S., and Brauer, R.P. (1996). Latent transforming growth factor-beta is present in the extracellular matrix of embryonic hearts in situ. *Dev Dyn* 205:126–134.

Gittenberger-deGroot, A.C., Bartelings, M.M., and Poelmann, R.E. (1995). Cardiac morphogenesis. In: Clark, E.B., Markwald, R.R., Takao, A., eds. *Developmental Mechanisms of Heart Disease*, Futura, Armonk, NY, pp. 157–168.
Gittenberger-deGroot, A.C., Vrancken Peeters, M.P., Mentink, M.M., Gourdie, R.G., and Poelmann, R.E. (1998). Epicardium-derived cells contribute a novel population to the myocardial wall and atrioventricular cushion. *Circ Res* 82:1043–1052.
Gorza, L., Schiaffino, S., and Vitadella, M. (1988). Heart conduction system: a neural crest derivative? *Brain Res* 457:360–366.
Gorza, L., and Vitadello, M. (1989). Distribution of conduction system fibers in the developing and adult rabbit heart revealed by an antineurofilament antibody. *Circ Res* 65:360–369.
Gourdie, R.G., Kubalak, S., and Mikawa, T. (1999). Conducting the embryonic heart: orchestrating development of the specialized cardiac tissues. *Trends Cardiovasc Med* 9:18–26.
Gourdie, R.G., Mima, T., Thompson, R.P., and Mikawa, T. (1995). Terminal diversification of the myocyte lineage generates Purkinje fibers of the cardiac conduction system. *Development* 121:1423–1431.
Gourdie, R.G., Wei, Y., Kim, D., Klatt, S.C., and Mikawa, T. (1998). Endothelin-induced conversion of embryonic heart muscle cells into impulse-conducting Purkinje fibers. *Proc Natl Acad Sci USA* 95:6815–6818.
Han, Y., Dennis, J.E., Cohen-Gould, L., Bader, D.M., Fischman, D.A. (1992). Expression of sarcomeric myosin in the presumptive myocardium of chicken embryos occurs within six hours of myocyte commitment. *Dev Dyn* 193:257–265.
Hanahan, D. (1997). Signaling vascular morphogenesis and maintenance. *Science* 277(5322):48–50.
Hogers, B., DeRuiter, M.C., Gittenberger-deGroot, A.C., and Poelmann, R.E. (1997). Unilateral vitelline vein ligation alters intracardiac blood flow patterns and morphogenesis in the chick embryo. *Circ Res* 80:473–481.
Hoyle, C., Brown, N.A., Wolpert, L. (1992). Development of left/right handedness in the chick heart. *Development* 115:1071–1078.
Icardo, J.M., and Sanchez, M.J. (1991). Spectrum of heart malformations in mice with situs solitus, situs inversus, and associated visceral heterotaxy. *Circulation* 84:2547–2558.
Isokawa, K., Rezaee, M., Wunsch, A., Markwald, R., and Krug, E.L. (1994). Identification of transferrin as one of multiple EDTA-extractable extracellular proteins involved in early chick heart morphogenesis. *J Cell Biochem* 54:207–218.
Jones, C.M., Lyons, K.M., and Hogan, B.L.M. (1991). Involvement of bone morphogenetic protein-4 (BMP-4) and Vgr-1 in morphogenesis and neurogenesis in the mouse. *Development* 111:531–542.
Keith, A., and Flack, M.W. (1907). The form and nature of the muscular connections between the primary divisions of the vertebrate heart. *J Anat Physiol* 41:172–189.
Kirby, M.L. (1993). Cellular and molecular contributions of the cardiac neural crest to cardiovascular development. *Trends Cardiovasc Med* 3:18–23.
Krug, E.L., Mjaatvedt, C.H., and Markwald, R.R. (1987). Extracellular matrix from embryonic myocardium elicits an early morphogenetic event in cardiac endothelial differentiation. *Dev Biol* 120:348–355.
Krug, E.L., Rezaee, M., Isokawa, K., et al. (1995). Transformation of cardiac endothelium into cushion mesenchyme is dependent of *ES/130*: temporal, spatial, and functional studies in the early chick embryo. *Cell Mol Biol Res* 41:263–277.
Kupershmidt, S., Yang, T., Wessels, A., Niswender, K.D., Magnuson, M.A., and Roden, D.M. (1999). Replacement by homologous recombination of the minK gene with LacZ reveals restriction of minK expression to the mouse cardiac conduction system. *Circ Res* 84:146–152.

Kwee, L., Baldwin, H.S., Shen, H.M., et al. (1995). Defective development of the embryonic and extraembryonic circulatory systems in vascular cell adhesion molecule (VCAM-1) deficient mice. *Development* 121:489–503.

Ladd, A.N., Yatskievych, T.A., and Antin, P.B. (1998). Regulation of avian cardiac myogenesis by actin/TGF beta and bone morphogenetic proteins. *Dev Biol* 204:407–419.

Lamers, W.H., te Korschot, A., Los, J.A., and Mooman, A.F. (1987). Acetylcholinesterase in prenatal rat heart: a marker for the early development of the cardiac conductive tissue? *Anat Rec* 217:361–370.

Lamers, W.H., Viragh, S.Z., Wessels, A., Moorman, A.F.M., and Anderson, R.H. (1995). Formation of the tricuspid valve in the human heart. *Circ Res* 91:111–121.

Lamers, W.H., Wessels, A., Verbeek, F.J., et al. (1992). New findings concerning ventricular septation in the human heart. *Circulation* 86:1194–1205.

Laverriere, A.C., MacNeill, C., Mueller, C., Poelmann, R.E., Burch, J.B., and Evans, T. (1994). *GATA-4/5/6*, a subfamily of three transcription factors transcribed in developing heart and gut. *J Biol Chem* 269:2377–2384.

Lin, Q., Schwartz, J.A., and Olsen, E.N. (1997). Control of cardiac morphogenesis and myogenesis by the myogenic transcription factor *MEF-2C. Science* 276:1404–1407.

Linask, K.K. (1992). N-cadherin localization in early heart development and polar expression of Na+,K+ ATPase and integrin during pericardial coelom formation and epithelialization of the differentiating myocardium. *Dev Biol* 151:213–224.

Linask, K.K., and Lash, J.W. (1988). A role for fibronectin in the migration of avian precardiac cells. I. Dose-dependent effects of fibronectin antibody. *Dev Biol* 114:87–101.

Linask, K.K., and Lash, J.W. (1993). Early heart development: dynamics of the endocardial cell sorting suggests a common origin with cardiocytes. *Dev Dyn* 195:62–69.

Linask, K.K., and Lash, J.W. (1998). Morphoregulatory mechanisms underlying early heart development. In: de la Cruz, M.V., and Markwald, R.R., eds. *Living Morphogenesis of the Heart.* Birkhauser (Springer/Verlag), Boston, pp. 1–41.

Lints, T.J., Parsons, L.M., Hartley, L., Lyons, I., and Harvey, R.P. (1993). NKX-2.5: a novel murine homeobox gene expressed in early hearts progenitor cells and their myogenic descendants. *Development* 119:419–431.

Little, C.D., and Rongish, B.J. (1995). Extracellular matrix in heart development. *Experientia* 51:873–882.

Litvin, J., Montgomery, M., Gonzalez-Sanchez, A., Bisaha, J.G., and Bader, D. (1992). Commitment and differentiation of cardiac myocytes. *Trends Cardiovasc Med* 2:27–32.

Lough, J., Barron, M., Brogley, M., Sugi, Y., Bolender, D.L., and Zhu, X. (1996). Combined BMP-2 and FGF-4 but neither factor alone, induce cardiogenesis in non-precardiac embryonic mesoderm. *Dev Biol* 178:198–202.

Lyons, I., Parsons, L.M., Hartley, L., et al. (1995). Myogenic and morphogenetic defects in the heart tubes of murine embryos lacking the homeobox gene *Nkx2-5. Genes Dev* 9:1654–1666.

Markwald, R., Eisenberg, C., Eisenberg, L., Trusk, T., and Sugi, Y. (1996). Epithelial-mesenchymal transformations in early heart development. *Acta Anat* 156:173–186.

Markwald, R.R., Trusk, T., Gittenberger-deGroot, A.C., and Poelman, R. (1997). Cardiac morphogenesis: formation and septation of the primary heart tube. In: Kavlock, R., and Datson, G., eds. *Handbook of Experimental Pharmacology*, vol 124/I. Springer-Verlag, Berlin, pp. 11–40.

Markwald, R.R., Trusk, T., and Moreno-Rodriguez, R. (1998). Formation and septation of the tubular heart: integrating the dynamics of morphology with emerging molecular concepts. In: de la Cruz, M.V., and Markwald, R.R. eds. *Living Morphogenesis of the Heart.* Birkhauser (Springer-Verlag), Boston, pp. 43–84.

McGuire, P.G., and Alexander, S.M. (1993). Inhibition of urokinase synthetase and cell surface binding alters the motile behavior of embryonic endocardial derived mesenchymal cells in vitro. *Development* 118:931–939.

Meyer, D., and Birchmeier, C. (1995). Multiple essential functions of neuregulin in development [see comments]. *Nature* 378(6555):386–390.

Mikawa, T., Borisov, A., Brown, A.M.C., and Fischman, D.A. (1992). Clonal analysis of cardiac morphogenesis in the chicken embryo using a replication-defective retrovirus: I. Formation of the ventricular myocardium. *Dev Dyn* 195:133–141.

Miner, J.H., Miller, J.B., and Wold, B.J. (1992). Skeletal muscle phenotypes initiated by ectopic MyoD in transgenic mouse heart. *Development* 114:853–860.

Mjaatvedt, C.H., Krug, E.L., and Markwald, R.R. (1991). An antiserum (ES1) against a particulate form of extracellular matrix blocks the transformation of cardiac endothelium into mesenchyme in culture. *Dev Biol* 145:219–230.

Mjaatvedt, C.H., and Markwald, R.R. (1989). Induction of epithelial-mesenchymal transition by an in vivo adheron-like complex. *Dev Biol* 136:118–128.

Mjaatvedt, C.H., Yamamura, H., Capehart, A.A., Turner, D., and Markwald, R.R. (1998). The Cspg2 gene, disrupted in the hdf mutant is required for right cardiac chamber and endocardial cushion formation. *Dev Biol* 202:56–66.

Mjaatvedt, C.H., Yamamura, H., Ramsdell, A., Turner, D., and Markwald, R.R. (1999). Mechanisms of segmentation and remodeling of the tubular heart: endocardial cushion fate and cardiac looping. In: Harvey, R.P., and Rosenthal, N., eds. *Heart Development.* Academic Press, New York, pp. 159–177.

Molkentin, J.D., Kalvakolanu, D.V., and Markham, B.E. (1994). Transcription factor GATA-4 regulates cardiac muscle-specific expression of the alpha-myosin heavy chain gene. *Mol Cell Biol* 14:4947–4957.

Montgomery, M.O., Litvin, J., Gonzalez-Sanchez, A., and Bader, D. (1994). Staging of commitment and differentiation of avian cardiac myocytes. *Dev Biol* 164:63–71.

Moore, A.W., McInnes, L., Kreidberg, J., Hastie, N.D., and Schedl, A. (1999). YAC complementation shows a requirement for Wt1 in the development of epicardium, adrenal gland and throughout nephrogenesis. *Development* 126:1845–1857.

Moorman, A.F.M., van den Hoff, M.J.B., deJong, F., et al. (2000). Myocardialization: a novel mechanism of cardiac septation. In: Clark, E.B., Takao, A., and Nakazawa, M., eds. *Etiology of Congenital Heart Disease*, vol 5. (In press.)

Nakajima, Y., Krug, E.L., and Markwald, R.R. (1994). Myocardial regulation of transforming growth factor-beta expression by outflow tract endothelium in the early embryonic chick heart. *Dev Biol* 165:615–626.

Nakajima, Y., Miyazono, K., Kato, M., Takase, M., Yamagishi, T., and Nakamura, H. (1997a). Extracellular fibrillar structure of latent TGF beta binding protein-1: role in TGF beta-dependent endothelial-mesenchymal transformation during endocardial cushion tissue formation in mouse embryonic heart. *J Cell Biol* 136:193–204.

Nakajima, Y., Yamagishi, T., Nakamura, H., Markwald, R.R., and Krug, E.L. (1997b). An autocrine function for transforming growth factor (TGF) beta3 in the transformation of atrioventricular canal endocardium into mesenchyme during chick heart development. *Dev Biol* 194:58–72.

Nishibatake, M., Kirby, M.L., and Van Mierop, H.L. (1987). Pathogenesis or persistent truncus arteriosus and dextroposed aorta in the chick embryo after neural crest ablation. *Circulation* 75:255–264.

Noden, D.M., Poelmann, R.E., and Gittenberger-deGroot, A.C. (1995). Cell origins and tissue boundaries during outflow tract development. *Trends Cardiovasc Med* 5:69–75.

Nusse, R., and Varmus, H.E. (1992). *Wnt* genes. *Cell* 69:1073–1087.

Okamoto, N., Akimoto, N., et al. (1981). Role of cell death in conal ridges of developing human heart. *Perspectives in Cardiovascular Research. Vol. 5. Mechanisms of cardiac morphogenesis and teratogenesis.* T. Pexieder. New York, Raven Press: 127–137.

Okamoto, N., Akimoto, N., Satow, Y., Hidaka, N., and Miyabara, S. (1981). Role of cell death in conal ridges of developing human heart. In: Pexieder, T., ed. *Perspectives in Car-*

diovascular Research, Vol 5: Mechanisms of Cardiac Morphogenesis and Teratogenesis. Raven Press, New York, pp. 127–137.

Olsen, E.N., and Srivastava, D. (1996). Molecular pathways controlling heart development. *Science* 272:671–676.

Patterson, D.F., Pexieder, T., Schnarr, W.R., Navratil, T., and Alaili, R. (1993). A single major-gene defect underlying cardiac conotruncal malformations interferes with myocadial growth during embryonic development: studies in the CTD line of Keeshond dogs. *Am J Hum Genet* 52:388–397.

Perez-Pomares, J.M., Macias, D., Garcia-Garrido, L., and Munoz-Chapuli, R. (1997). Contribution of the primitive epicardium to the subepicardial mesenchyme. *Dev Dyn* 210:96–105.

Perez-Pomares, J.M., Macias, D., Garcia-Garrido, L., and Munoz-Chapuli, R. (1998). The origin of the subepicardial mesenchyme in the avian embryo: an immunohistochemical and quail-chick chimera study. *Dev Biol* 200:57–68.

Pexieder, T., Wenink, A.C., et al. (1989). A suggested nomenclature for the developing heart. Working Group for Embryology and Teratology of the European Society of Cardiology. *International Journal of Cardiology* 25(3):255–263.

Poelmann, R.E., Mikawa, T., and Gittenberger-deGroot, A.C. (1998). Neural crest cells in outflow tract septation of the embryonic chicken heart: differentiation and apoptosis. *Dev Dyn* 212:373–384.

Potts, J.D., Dagle, J.M., Walder, J.A., Weeks, D.L., and Runyan, R.B. (1991). Epithelial mesenchymal transformation of embryonic cardiac endothelial cells is inhibited by a modified antisense oligodeoxynucleotide to transforming growth factor b3. *Proc Natl Acad Sci USA* 88:1516–1520.

Ramsdell, A., and Markwald, R. (1997). Induction of endocardial cushion tissue in the avian heart is regulated, in part, by TGFbeta-3-mediated autocrine signalling. *Dev Biol* 187:64–74.

Ramsdell, A.F., Moreno-Rodriguez, R.A., Weinecke, M.M., et al. (1998). Identification of an autocrine signaling pathway that amplifies induction of endocardial cushion tissue in the avian heart. *Acta Anat* 162:1–15.

Ranger, A., Grusby, M., Hodge, M., et al. (1998). The transcription factor NFATc is essential for cardiac valve formation. *Nature* 392:186–190.

Rawles, M.E. (1943). The heart-forming areas of the early chick blastoderm. *Physiol Zool* 16:22–42.

Robb, L., Mifsud, L., Hartley, L., et al. (1998). Epicardin, a novel basic helix-loop-helix transcription factor gene expressed in epicardium, branchial arch myoblasts and mesenchyme of developing lung, gut, kidney and gonads. *Dev Dyn* 213:105–113.

Rongish, B.J., Drake, C.J., Argraves, W.S., and Little, C.D. (1998). Identification of the developmentally expressed JB3 antigen as avian fibrillin-2. *Dev Dyn* 212:461–471.

Rosenquist, G.C., and DeHaan, R.L. (1966). Migration of precardiac cells in the chick embryo: a radiographic study, Carnegie Inst Washington. *Contrib Embryol* 38:111–121.

Ross, R.S., Navankasattusas, S., Harvey, R.P., and Chien, K.R. (1996). An HF-1a/HF-1b/MEF-2 combinatorial element confers cardiac ventricular specificity and establishes an anterior posterior gradient of expression. *Development* 122:1799–1809.

Runyan, R.B., and Markwald, R.R. (1983). Invasion of mesenchyme into three-dimensional gels: a regional and temporal analysis of interaction in embryonic heart tissue. *Dev Biol* 95:108–114.

Runyan, R.B., Potts, J.D., Sharma, R.V., Loeber, C.P., Chiang, J.J., and Bhalla, R.C. (1990). Signal transduction of a tissue interaction during embryonic heart development. *Cell Regul* 1:301–313.

Ruzicka, D.L., and Schwartz, R.J. (1988). Sequential activation of alpha-actin genes during avian cardiogenesis: vascular smooth muscle alpha-actin gene transcripts mark the onset of cardiomyocyte differentiation. *J Cell Biol* 107:2755–2586.

Sato, T.N., Tozawa, Y., et al. (1995). Distinct roles of the receptor tyrosine kinases Tie-1 and Tie-2 in blood vessel formation. *Nature* 376(6535):70–74.

Sanford, L.P., Ormsby, I., Gittenberger-deGroot, A.C., et al. (1997). TGF beta-2 knockout mice have multple developmental defects that are non-overlapping with other TGF beta knockout phenotypes. *Development* 124:2645–2657.

Schilham, M.W., Oosterwegel, M.A., Moerer, P., et al. (1996). *Sox-4* gene is required for cardiac outflow tract formation and pro-B lymphocyte expansion. *Nature* 380:711–714.

Schultheiss, T.M., Burch, J.B.E., and Lassar, A.B. (1997). A role for bone morphogenetic proteins in the induction of cardiac myocytes. *Genes Dev* 11:451–462.

Shinbourne, E.A., Macartney, F.J., and Anderson, R.H. (1976). Sequential chamber localization—a logical approach to diagnosis in congenital heart disease. *Br Heart J* 38:327–340.

Sinning, A.R. (1997). Partial purification of HLAMP-1 provides direct evidence for the multicomponent nature of the particulate matrix associated with cardiac mesenchyme formation. *J Cell Biochem* 66:112–122.

Sinning, A.R., Hewitt, C.C., and Markwald, R.R. (1995). A subset of SBA lectin-binding proteins isolated from myocardial-conditioned media transforms cardiac endothelium into mesenchyme. *Acta Anat* 154:111–119.

Smith, S.M., Dickman, E.D., Thompson, R.P., Sinning, A.R., Wunsch, A.M., and Markwald, R.R. (1997). Retinoic acid directs cardiac laterality and expression of early markers of cardiogenesis. *Dev Biol* 182:162–181.

Soler, A.P., and Knudsen, K.A. (1994). N-cadherin involvement in cardiac myocyte interaction and myofibrillogenesis. *Dev Biol* 162:9–17.

Spence, S.G., Argraves, W.S., Walters, L., Hungerford, J.E., and Little, C.D. (1992). Fibulin is localized at sites of epithelial-mesenchymal transitions in the early embryo. *Dev Biol* 151:473–484.

Srivastava, D., Thomas, T., Lin, Q., Kirby, M.L., Brown, D., and Olson, E.N. (1997). Regulation of cardiac mesodermal and neural crest development by the bHLH transcription factor dHAND. *Nature Genet* 16:154–160.

Stahlsberg, H. (1969). The origin of heart asymmetry: right and left contributions to the early chick embryo heart. *Dev Biol* 19:109–127.

Stainier, D.Y., and Fishman, M.C. (1992). Patterning the zebrafish heart tube: acquisition of anteroposterior polarity. *Dev Biol* 153(1):91–101.

Sugi, Y., and Markwald, R.R. (1996). Formation and early morphogenesis of endocardial precursor cells and the role of endoderm. *Dev Biol* 175:66–83.

Supp, D.M., Witte, D.P., Potter, S.S., and Brueckner, M. (1997). Mutation of an axonemal dynein affects left-right symmetry in inversus viserum mice. *Nature* 389:963–966.

Suri, C., Jones, P., et al. (1996). Requisite role of angiopoietin-1, a ligand for the TIE2 receptor, during embryonic angiogenesis. *Cell* 87:1171–1180.

Tasaka, H., Krug, E.L., and Markwald, R.R. (1996). Origin of the pulmonary venosus orifice in the mouse and its relationship to the morphogenesis of the sinus venosus, extracardiac mesenchyme (spina vestibuli) and atrium. *Anat Rec* 246:107–113.

Thompson, R.P., and Fitzharris, T.P. (1979). Morphogenesis of the truncus arteriosus of the chick embryo heart: tissue reorganization during septation. *Am J Anat* 156:251–264.

Thompson, R.P., Kanai, T., Germroth, P.G., et al. (1995). Organization and function of early specialized myocardium. In: Clark, E.B., Markwald, R.R., and Takao, A., eds. *Developmental Mechanisms of Congenital Heart Disease*. Futura, Armonk, NY, pp. 269–279.

van den Hoff, M.J.B., Bennington, R.W., et al. (1999). Myocardialization of the cardiac outflow tract. *Dev Biol* 212:477–490.

van Kempen, M.J., Fromaget, C., et al. (1991). Spatial distribution of connexin43, the major cardiac gap junction protein, in the developing and adult rat heart. *Circulation Research* 68(6):1638–1651.

Viragh, S., and Challice, C.E. (1973). The origin of the epicardium and the embryonic myocardial circulation in the mouse. *Anat Rec* 201:157–168.

Viragh, S., Gittenberger-deGroot, A.C., Poelmann, R.E., and Kalman, F. (1994). Early development of quail heart epicardium and associated vascular and glandular structures. *Anat Embryol* 188:381–393.

Viragh, S., Szabo, E., and Challice, C.E. (1989). Formation of the primitive myo- and endocardial tubes in the chicken embryo. *J Mol Cell Cardiol* 21:123–137.

Waldo, K., Miyagawa-Tomita, S., Kumiski, D., and Kirby, M.L. (1998). Cardiac neural crest cells provide new insight into septation of the cardiac outflow tract: aortic sac to ventricular septal closure. *Dev Biol* 196:129–144.

Waldo, K., Zdanowicz, M., Burch, J., et al. (1999). A novel role for cardiac neural crest in heart development. *J Clin Inuest* 103:1499–1507.

Wall, N.A., and Hogan, B. (1994). TGF-beta related genes in development. *Curr Opin Genet Dev* 4:517–522.

Webb, S., Brown, N.A., and Anderson, R.H. (1998). Formation of the atrioventricular septal structures in the normal mouse. *Circ Res* 82:645–656.

Wessels, A., Anderson, R.H., Markwald, R.R., Webb, S., Brown, N.A., Viragh, S.Z., Moorman, A.F.M., and Lamers, W.H. (2000). Atrial Development in the Human Heart: An immunohistochemical study with emphasis on the role of mesenchymal tissues. *The Anatomical Record* 259:288–300.

Wessels, A., Vermeulen, J.L.M., Virágh, S.Z., Kálmán, F., Morris, G.E., Nguyen, T.M., Lamers, W.H., and Moorman, A.F.M. (1990). Spatial distribution of "tissue-specific" antigens in the developing human heart and skeletal muscle. I. An immunohistochemical analysis of creatine kinase isoenzyme expression patterns. *Anat Rec* 228:163–176.

Wessels, A., Vermeulen, J.L.M., Verbeek, F.J., et al. (1992). Spatial distribution of "tissue-specific" antigens in the developing human heart and skeletal muscle: III. An immunohistochemical analysis of the distribution of the neural tissue antigen G1N2 in the embryonic heart; implications for the development of the atrioventricular conduction system. *Anat Rec* 231:97–111.

Wessels, A., Vermeulen, J.L.M., Viragh, S.Z., Lamers, W.H., and Moorman, A.F.M. (1991). Spatial distribution of "tissue-specific" antigens in the developing human heart and skeletal muscle: II. An immunohistochemical analysis of myosin heavy chain isoform expression patterns in the embryonic heart. *Anat Rec* 229:355–368.

Winnier, G., Blessing, M., Labosky, P.A., and Hogan, B.L.M. (1995). Bone morphogenetic protein-4 is required for mesoderm formation and patterning in the mouse. *Genes Dev* 9:2105–2116.

Wunsch, A., Markwald, R.R., and Little, C.D. (1994). Cardiac endothelial heterogeneity defines valvular development as demonstrated by the diverse expression of JB3 antigen, a fibrillin-like protein of the endocardial cushion tissue. *Dev Biol* 165:585–601.

Ya, J., van den Hoff, M.J.B., de Boer, P.A.J., et al. (1998). The normal development of the outflow tract in the rat. *Circ Res* 82:464–472.

Yamamura, H., Zhang, M., Mjaatvedt, C.H., and Markwald, R.R. (1997). A heart segmental defect in the anterior/posterior axis of a transgenic mutant mouse. *Dev Biol* 186:58–72.

Yang, J.T., Rayburn, H., and Hynes, R.O. (1995). Cell adhesion events mediated by alpha-4 integrins are essential in placental and cardiac development. *Development* 121:549–560.

Yutzey, K.E., Rhee, J.T., and Bader, D.M. (1994). Expression of the atrial-specific myosin heavy chain AMHC1 and the establishment of anteroposterior polarity in the developing chicken heart. *Development* 120:871–883.

Zhang, H., and Bradley, A. (1996). Mice deficient for BMP2 are nonviable and have defects in amnion/chorion and cardiac development. *Development* 122(10):2977–2986.

Zhang, H.-Y., Chu, M.-L., Te-Cheng, P., Sasaki, T., Timple, R., and Ekblom, R. (1995). Extracellular matrix protein fibulin-2 is expressed in the embryonic endocardial cushion tissue and is a prominent component of valves in adult heart. *Dev Biol* 167:18–26.

CHAPTER 2

Myofibrillogenesis in the Heart

Carol C. Gregorio and Parker B. Antin

The formation of perfectly aligned myofibrils in cardiac muscle represents the assembly of one of the most highly ordered macromolecular structures found in eukaryotic cells. As a result of this precise assembly, many different classes of proteins function together to efficiently convert the molecular-level movements of actin and myosin into macroscopic movements of efficient contractile activity.

Historically, cardiac myofibrillogenesis was difficult to study since the heart is the first functional organ in the developing vertebrate, and suitable cell culture model systems that faithfully recapitulate the in vivo assembly process observed have been lacking. As a result, delineation of regulatory pathways responsible for muscle-specific expression of myofibrillar proteins and events contributing to their assembly into contractile sarcomeres were largely derived from studies using skeletal muscle. Although skeletal and cardiac muscle share a similar sarcomeric organization and function, and express many closely related contractile proteins, they arise from spatially distinct populations of mesodermal progenitors and have different physiologic mechanical properties. McGrew et al (1999) provide an excellent comparative analysis of cardiac versus skeletal muscle development. This chapter focuses on the present state of knowledge concerning myofibril assembly in the heart and speculates on the promising future that lies ahead in this field.

ORIGIN OF HEART MUSCLE CELLS

A full understanding of myofibril assembly in cardiac myocytes requires that this process be placed in the context of cell lineage diversification and the morphogenetic movements leading to formation of the linear heart tube. The pregastrula avian embryo and mammalian embryo are composed of two cell layers, the epiblast and the hypoblast (called the visceral or primitive endoderm in mouse). The entire embryo proper arises from the epiblast, while the hypoblast is ultimately displaced anteriorly during gastrulation to extraembryonic regions. At the onset of gastrulation in chick and mouse, cells destined to form cardiac myocytes are found in the lateral epiblast (Lawson et al, 1991; Garcia-Martinez and Schoenwolf, 1993; Tam et al, 1997). In chick, heart-forming cells involute during early to midgastrula stages (HH stage 3–4, 12–19 hours; Hamburger and Hamilton, 1951) through a broad area of the primitive streak just caudal to Hensen's node, and migrate anterolaterally at the leading (anterior) edge of the mesoderm (Figure 2.1; see color insert). By the late gastrula stage (stage 6, 23–25 hours), cardiogenic regions of the mesoderm occupy anterior and lateral positions in the embryo (Figure 2.1). Soon thereafter, the lateral mesoderm splits into the dorsal somatic and ventral splanchnic mesoderm. Anterior regions of the splanchnic mesoderm contain the cardiac progenitors. This layer remains in intimate contact with the endoderm and gives rise to both cardiac myocytes and the endothelial lining of the heart, the endocardium. Many studies have defined an important role of the anterior lateral (cardiogenic) endoderm in inducing and supporting heart muscle cell development (reviewed in Jacobson and Sater, 1988; Schultheiss and Lassar, 1999). In chick, anterior lateral endoderm exerts its cardiogenic effects at least partly by producing bone morphogenetic protein 2 (BMP-2) beginning at stages 5 to 6 (19–25 hours). BMP-2 upregulates expression of cardiogenic regulatory transcription factors such as *Nkx2.5* and *GATA-4* within adjacent splanchnic mesoderm (Schultheiss et al, 1997; Andrée et al, 1998). At approximately the same time, premyocardial cells in the splanchnic mesoderm form an epithelium and are joined by tight junctions, a process that is required for differentiation and expression of muscle-specific contractile protein genes (Linask and Gui, 1995). Interactions with the extracellular matrix are necessary for conversion to an epithelium; the change in cell shape requires reorganization of the cytoskeleton that is also likely to be essential for the proper assembly of contractile myofibrils. Continued ventral folding of the splanchnic mesoderm and endoderm brings the two cardiogenic layers toward the ventral midline (Figure 2.1). These layers then fuse in a rostral to caudal progression to form the primitive heart tube, composed of an outer myocardial layer and an inner endocardial tube separated by a thick extracellular matrix layer called the cardiac jelly. Expression of muscle-specific contractile protein genes can first be detected at stages 7 to 8 (23–29 hours), shortly following formation of the epithelial myocardial layer and just prior to fusion of the heart primordia (Han et al, 1992; Yutzey et al, 1995). Myofibril assembly occurs during the next few hours, and by the time of linear heart tube formation (stage 10, 10 somites, 33–38 hours), myocytes have begun to rhythmically contract. Myofibrils therefore form rapidly and simultaneously arrange into a highly organized network that traverses through the myocytes such that synchronized contraction produces unidirectional blood flow. See Figure 2.1 for an overview of heart formation in chick.

Although it is known that the four basic helix-loop-helix transcription factors MyoD, myogenin, Myf5, and MRF4 are largely responsible for induction of skeletal myogenesis, none of these genes is expressed in cardiac myocytes. Specification of premyocardial cells and activation of contractile protein genes appears to require the coordinated activity of several types of transcription factors, including *Nkx2.5*, *GATAs 4*, *5*, and *6*, *MEF-2c*, *dHAND*, and *eHAND*, among others (reviewed in Olson and Srivastava, 1996; Patterson et al, 1998). None of these factors were reported to activate a myocardial phenotype in noncardiac cells, and most contractile protein genes contain *cis* elements that bind several of these factors, suggesting combinatorial activation of gene transcription (for example see, Durocher et al, 1997; Reecy et al, 1999). Interestingly, atrial and ventricular myocytes arise from distinct cardiac myocyte lineages, and express distinct subsets of contractile protein genes. Although all of the known cardiac transcription factors are initially expressed throughout the primitive heart, in mouse *dHAND* and *eHAND* become restricted to the right and left ventricles, respectively (Srivastava et al, 1997), and *Nkx2.8* becomes restricted to the atria in chick (Brand et al, 1997). These chamber-restricted localization patterns correlate with chamber-specific expression of contractile proteins that likely have important implications for myofibril assembly and contractile function. For a review of mechanisms regulating chamber-specific gene expression, see Wang and Stockdale (1999).

SARCOMERE STRUCTURE

Because the heart forms earlier than skeletal muscles in birds and mammals, the first sarcomeres in the developing embryo are assembled in cardiac myocytes. The assembly of contractile proteins into myofibrils is a complex process that requires coordinate synthesis of the constituent proteins, polymerization of actin and myosin (and many associated proteins) into thin and thick filaments, respectively, and association of the two filament systems into highly organized sarcomeres. Newly assembled sarcomeres consist of parallel arrays of approximately 1.0-μm-long thin (actin) filaments that interdigitate with laterally aligned 1.6-μm-long thick (myosin) filaments. Rod-like tropomyosin molecules are associated with each other head to tail, forming two polymers per thin filament. Each tropomyosin molecule binds one troponin complex (composed of troponins T, I, and C), and together they mediate the calcium regulation of myosin adenosine triphosphatase (ATPase). Thin filaments are polarized in muscle sarcomeres with the barbed ends crosslinked by α-actinin, capped by CapZ, and anchored at the Z disk. The thin filament pointed ends terminate in the A band, are capped by tropomodulin, and overlap with the bipolar myosin filaments. ("Barbed" and "pointed" ends of actin filaments refer to the orientation of arrowheads generated by myosin subfragment 1 binding to actin filaments, and are the fast-growing and the slow-growing ends of the thin filaments, respectively.)

A third filament system in striated muscle is formed by titin (also known as connectin), the largest protein yet discovered in humans at 3,000 to 3,700 kd. Single molecules of titin span from Z- to M-lines of the sarcomere. This molecule centers thick filaments in the middle of the sarcomere and acts as an efficient elastic spring element during muscle contraction. In vitro binding assays have demonstrated that titin interacts directly with several thick filament-associated proteins, including myosin-binding protein C (MyBP-C), myosin-binding protein H (MyBP-H), M-protein, myomesin, and the rod region of myosin heavy chain. Titin also interacts directly with actin filaments near the Z line, and with the Z line components α-actinin and titin-cap (T-cap/telethonin). Through its interaction with titin, α-actinin functions to crosslink the third filament system to the thin filaments; these interconnections likely contribute to force transmission between adjacent sarcomeres. The N-terminal regions of titin molecules from opposite sarcomeres overlap in Z-lines, while the C-terminal regions of titin molecules from opposite half-sarcomeres overlap in M-lines. Therefore, in myofibrils composed of many sarcomeric units, the titin filaments form a contiguous filament system through interaction with various components of the myofibril.

A fourth filament system thus far found only in skeletal muscle is composed of nebulin, a protein of approximately 800 kd that spans the length of thin filaments.

Although a precise function for nebulin has not been defined, evidence suggests that it plays a role in thin filament length specification. While a large nebulin-like molecule has not been detected in cardiac muscle, cardiac myocytes contain nebulette, a 100-kd protein that shares extensive structural homology with the Z disk region of nebulin. Since nebulette is predicted to extend only 25% along the thin filament, it is unlikely on its own to have a function similar to nebulin in skeletal muscle. While actin and titin filaments span the entire Z-line, nebulette (or nebulin in skeletal muscle) appears to insert only within the periphery of the Z-line. Similarly, polymers of tropomyosin molecules that extend along and stabilize actin filaments are absent from the central Z-disk region of the thin filament. Figure 2.2 is a schematic of the known distribution patterns of cardiac muscle sarcomeric components (see color insert). For reviews on sarcomere organization see Huxley (1960), Vigoreaux (1994), Schiaffino and Reggiani (1996), and Squire (1997).

SARCOMERE ASSEMBLY—IN VITRO STUDIES

For more than 30 years, a major focus of studies investigating myofibril assembly in heart muscle cells has been to identify the sequence of events involved in the assembly of contractile proteins into myofibrils. To date, most studies of cardiac myofibril assembly have focused on the localization of sarcomeric proteins in isolated cardiac myocytes from embryonic chicks or fetal, newborn, or adult rats. Although plagued by relatively few electron microscopic analyses, these studies have nevertheless provided insight into the process of myofibril assembly and have also contributed to our understanding of the influences of cell substrate, cell–cell interactions, and signaling pathways under defined conditions. Due to their flat, spread out morphology and the ability to control the timing of myofibril assembly (occurring over several days) by modulating medium conditions, isolated myocytes are readily observable by light microscopy and accessible to experimental manipulation.

Two prominent models of myofibrillogenesis have been proposed from studies using primary cultures of cardiac myocytes. In one model, microfilament bundles resembling the stress fibers of nonmuscle cells (stress fiber-like structures, SFLSs) are proposed to act as a scaffold during sarcomere assembly (Dlugosz et al, 1984). I-Z-I complexes, structures containing α-actinin, sarcomeric actin, and titin are hypothesized to be organized in register on these filamentous structures to form nonstriated (stress-fiber–like) myofibrils (NSMFs) in the absence of muscle-myosin–containing thick filaments. Thick filaments assemble and incorporate independently into these preformed structures (Wang et al, 1988; Schultheiss et al, 1990; Epstein and Fischman, 1991; Holtzer et al, 1997).

A different model proposes the importance of "premyofibrils" as precursor structures during myofibril assembly (Rhee et al, 1994; Dabiri et al, 1997). Premyofibrils are composed of minisarcomeres (with the distance between Z bodies ranging from 0.3 to 1.4 μm), containing short thin filaments that grow in length during maturation. Premyofibrils contain nonmuscle myosin as "acting positional spacers" for other sarcomeric proteins and do not contain titin. In this model, late in assembly the distance between individual Z disks increases, thin filaments grow in length, titin assembles, and nonmuscle myosin is replaced by its muscle isoforms to form mature sarcomeres (with Z-line spacings ranging from 1.8 to 2.5 μm).

Despite the accessibility of cultured myocytes to direct observation and manipulation, several factors suggest that some aspects of myofibrillogenesis in cultured cells may not faithfully reflect myofibril assembly in vivo. One concern is that primary cultures of cardiac myocytes are obtained by isolating fully differentiated heart tissue from animals at a time when the cardiomyocytes already contain functional, fully assembled sarcomeres. Therefore, cultured cells appear to assemble their myofibrils by reutilizing preexisting material; hence, it is difficult to determine whether myofibrillar structures are undergoing "assembly" or "disassembly" (Lin et al, 1989). Additionally, the reappearance of myofibrils, whose timing is highly dependent on the plating density, takes place over the course of several days, whereas in vivo fully assembled myofibrils are present in the heart only a few hours following the appearance of muscle-specific contractile proteins. Furthermore, while it is known that proper cell–cell contact and shape are crucially important for myofibril assembly and alignment, cells in culture also modify their cytoarchitecture in response to isolation procedures and the artificial tissue culture environment. Cardiac myocytes in culture spread on the substrate, are largely two-dimensional (often less than 2 µm thick), and express cytoskeletal isoforms not found in myocytes in vivo. In contrast, myocytes in vivo have a three-dimensional cytoarchitecture, and are interconnected on all surfaces through intercellular junctions and intercalated disks. Consequently, myofibrils course through the cytoplasm in complex patterns reflecting the orientation of external forces that exist on the myocyte in vivo. Thus, myofibrillogenesis in cultured myocytes may not reflect the de novo assembly of myofibrils in developing embryonic cardiac myocytes.

SARCOMERE ASSEMBLY—IN VIVO STUDIES

Although few studies have investigated myofibril assembly during development in vivo, ultrastructural observations combined with high-resolution immunofluorescence experiments have provided a fairly comprehensive view of cardiac myofibril assembly in the developing heart. Beginning more than 30 years ago, many superb electron microscopic observations revealed that myofibrils form along the cell periphery in close association with the plasma membrane, and defined the timing of assembly with respect to the onset of rhythmic contractions (e.g., Manasek, 1968; Legato, 1972; Markwald, 1973; Brook et al, 1983; Hill and Lemanski, 1985). Models of assembly based on these ultrastructural observations have proven to be quite accurate, even though many myofibril protein constituents were not yet identified.

Initial immunolocalization studies in vivo were hindered by the presence of cardiac jelly that covers the inner cardial wall surface, blocking antibody access. This problem was overcome in the late 1980s by treating intact myocardial walls from 4 to 12 somite stage chick embryos with hyaluronidase, which degrades a key component of the cardiac jelly and allows antibodies to penetrate (Tokuyasu and Maher, 1987a,b). Although whole mount specimens of myocardium are relatively thick, the utilization of confocal microscopy allows for resolution that is comparable to conventional microscopy on cultured cells (e.g., Shiraishi et al, 1992; Ehler et al, 1999).

The elegant and detailed immunolocalization studies in the chick heart by Tokuyasu and Maher (1987a,b) and recently by Perriard, Ehler, and colleagues

(Ehler et al, 1999) have now provided a comprehensive description of myofibril assembly in situ. At the earliest stages of myocyte differentiation, all sarcomeric proteins studied are accumulated diffusely within the cytoplasm as early as the 4-somite stage (stage 8, 26–29 hours; Tokuyasu and Maher, 1987a), with the exception of titin, which is present both diffusely and in a punctate pattern in the cytoplasm. (The dots likely represent aggregates of several titin molecules). Combined immunofluorescence and electron microscopic studies on the same specimens failed to identify any discernible cytoskeletal structures associated with the titin dots (Tokuyasu and Maher, 1987b). Within a few hours, the titin dots appear at the sites of α-actinin localization at the plasma membrane, forming nascent Z bodies. These accumulations are components of the "dense Z bodies/plaques" observed by electron microscopic analysis (Manasek, 1968; Legato, 1972; Markwald, 1973; Hill and Lemanski, 1985; Tokuyasu and Maher, 1987a,b). Thus, the early appearance of Z bodies suggest that they may function as fundamental organizing structures during myofibrillogenesis, furnishing the initial cue for the polarized organization of filaments in nascent sarcomeres. Filamentous actin accumulates near the cell membrane and rapidly becomes associated with Z bodies, forming I-Z-I complexes consisting of actin filaments, the N terminus of titin, and α-actinin (Figure 2.3a) (Tokuyasu and Maher, 1987a; Ehler et al, 1999).

Like other myofibrillar constituents, myosin first appears diffusely in the cytoplasm by immunofluorescence at the 4-somite stage, and slightly later filaments that are not associated with I-Z-I complexes can be observed in the cytoplasm (Figure 2.3a: Manasek, 1968; Hiruma and Hirakow, 1985; Tokuyasu and Maher, 1987a). Between 8 and 9 somites (stage 9–10; ~30 hours), a pattern for M-line epitopes of titin, are discerned (Ehler et al, 1999), and myosin thick filaments are observed in a sarcomeric arrangement organized together with I-Z-I structures. The assembly of the myosin-binding protein myomesin correlates with this event, suggesting that this molecule may anchor thick filaments to titin filaments, analogous to the crosslinking role of α-actinin within Z lines (Ehler et al, 1999). Interestingly, myosin thick filaments are aligned in their mature sarcomeric pattern before actin thin filaments are organized into their mature pattern (Tokyasu and Maher, 1987a; E. Ehler, S.P. Hämmerle, B.M. Rothen, and J.-C. Perriard, unpublished results). Ultrastructural analysis revealed that the shortest sarcomere lengths (Z-band spacings) seen in nascent myofibrils are approximately 1.5 μm with no visible I bands, suggesting that the A band width determines the initial sarcomeric length of the first myofibrils (Markwald, 1973; Tokuyasu and Maher, 1987b; Bishop et al, 1990). At approximately this stage, developing myofibrils become disassociated from the cell membrane and extend throughout the cytoplasm (Figure 2.3b) (Markwald, 1973; Tokuyasu, 1989).

The late maturation of actin filament distribution patterns may be explained by the delayed alignment or attainment of uniform filament lengths (~1 μm), and/or by the distinct localization of different actin isoforms during myofibril assembly in vivo. For example, α-vascular actin is transiently expressed during the earliest stages of cardiac myocyte development. In contrast, α-cardiac actin is expressed continuously from the earliest stages of development through adulthood. Indirect immunofluorescence reveals that all thin filaments have attained their mature lengths by 12 to 13 somites (stage 11, 40–45 hours), and this correlates with the downregulation of α-vascular actin and assembly of the actin filament pointed end

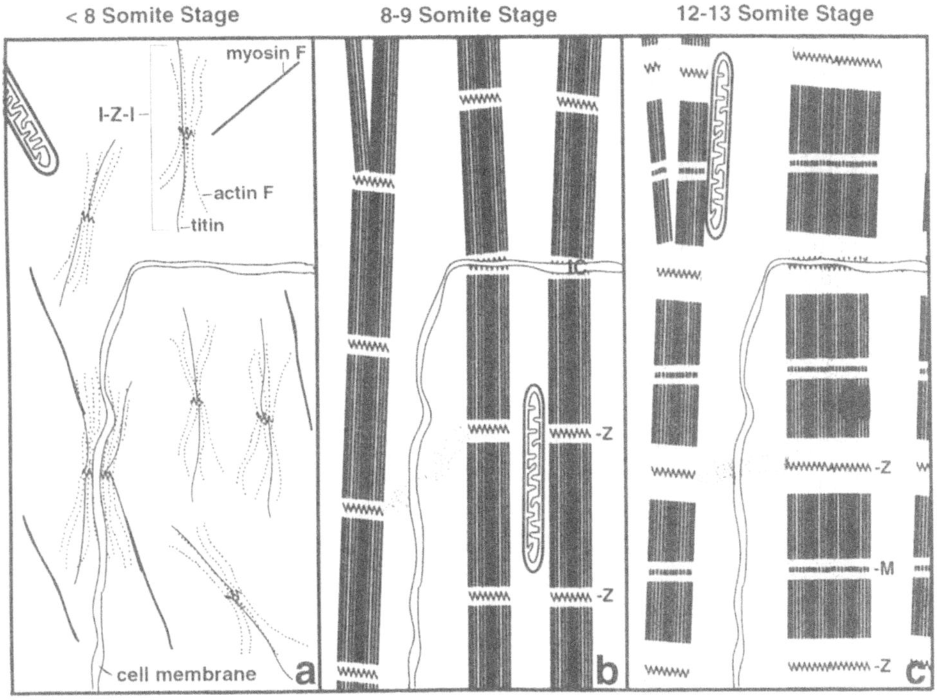

FIGURE 2.3. Model of proposed temporal steps involved in cardiac myofibrillogenesis in chick heart in vivo. (Modified from Figure 1 in Markwald, 1973.) (a) At the <8 somite stage, I-Z-I complexes are composed of Z bodies that contain α-actinin and titin linked to filaments composed of actin (and other thin filament proteins), often in close proximity to the membrane. Note in this model that the actin filaments are at variable lengths at this stage. Myosin thick filaments are scattered diffusely in the cytoplasm and become associated with I-Z-I complexes, perhaps via interaction with the giant molecule titin and alternatively, via their interaction with the thin filaments. (b) At the 8 to 9 somite stage, myosin thick filaments are organized together with I-Z-I structures forming nascent sarcomeres. The shortest sarcomere lengths (Z-band spacings) seen in nascent myofibrils are approximately 1.5 μm with no visible I bands, suggesting that the A band width determines the initial sarcomeric length of the first myofibrils. (c) At the 12 to 13 somite stage, thin filaments can be resolved at their mature lengths. Sarcomere lengths (distance between Z lines) range from 1.8 to 2.5 μm and individual myofibrils are properly aligned in register, forming a three-dimensional network. Note, continuity of myofibrils of neighboring cells at intercalated disks.

capping protein, tropomodulin (E. Ehler, S.P. Hämmerle, B.M. Rothen, and J.-C. Perriard, unpublished results). The late assembly of tropomodulin is consistent with its essential role in maintenance of actin filament length (Gregorio et al, 1995; Sussman et al, 1998). The late assembly of tropomodulin and resolution (nonoverlap) of the actin filament pointed ends correlate with electron microscopy studies showing that the organization of thin filaments into a mature striated pattern is a relatively late event in myofibril assembly. Restriction of filament length and separation of thin filaments into two half-sarcomeres (as evidenced by narrower Z lines and defined H zones and M lines) occurs late in myofibril assembly, after interdigitation of thick and thin filaments (Figure 2.3c) (Legato, 1972; Markwald, 1973; Brook et al, 1983; Bishop et al, 1990). The final alignment of the thick and thin filaments in the mature sarcomere is likely to be regulated, in part, by cell adhesion and mechanical forces (i.e., contractile activity) (e.g., Bishop et al, 1990; Simpson et al, 1993; Littlefield and Fowler, 1998).

By the 12- or 13-somite stage, extensive intercalated disks and branching of myofibrils are apparent, and individual myofibrils are properly aligned in register, forming a three-dimensional network (Tokuyasu, 1989; Shiraishi et al, 1993, 1995, 1997). The circumferential alignment and stability of myofibrils is the last step during myofibrillogenesis and seems to be important for optimal load management.

Although there is some discrepancy in the field due to limitations in extrapolating dynamic assembly events from fixed tissues and the asynchronous development of myofibrils within cells, the in vivo observations detailed above define a series of events during cardiac myofibril assembly that can be summarized as follows: (1) appearance of all sarcomeric proteins diffusely distributed in the cytoplasm with titin also organized into discrete dots; (2) association of titin dots with α-actinin dots at the cell membrane forming dense Z bodies; (3) association of α-actinin and the N terminus of titin with actin filaments to form nascent I-Z-I complexes; (4) intercalation of full-length 1.6-μm muscle myosin filaments with I-Z-I complexes to form immature sarcomeres with >1.5 μm sarcomeric periodicity; (5) organization of ~1 μm long actin filaments and association of tropomodulin with the pointed ends of thin filaments, forming mature sarcomeres; and finally (6) alignment of a fully registered mature myofibrils with one another. Figure 2.3 is a schematic of the stages of sarcomere assembly during cardiac muscle development in vivo.

It is clear that many aspects of myofibril assembly observed utilizing primary cultures of cardiac myocytes resemble those seen during the process in the developing heart. Some important examples of similarities between the two systems include (1) the early detection of titin; (2) I-Z-I complexes comprising the first organized structures that initally form independently of myosin thick filaments; (3) the presence of different actin isoforms; (4) sequential detection of titin epitopes (from N terminus to C terminus); and (5) the late assembly of the pointed end capping protein tropomodulin (e.g., Wang et al, 1988; Furst et al, 1989; Schultheiss et al, 1990; Handel et al, 1991; Epstein and Fischman, 1991; Komiyama et al, 1993; Gregorio and Fowler, 1995). Other aspects of myofibrillogenesis observed in cultured cardiac myocytes, however, do not appear to reflect events occurring in vivo. Minisarcomeres containing short actin filaments (with distances between α-actinin dots/dense bodies of less than 1.5 μm), and "stress-fiber–like" myofibrils are not evident (these structures have been reported in most, if not all, cell culture studies). In particular, the early appearance of titin, its ability to span the half-sarcomere, and its propensity to interact with a large number of myofibrillar constituents argue for a central role for this molecule in regulating and directing myofibril assembly. This possibility is discussed at greater detail below.

MOLECULAR TEMPLATES FOR ASSEMBLY

While the model described above provides a framework for describing events that occur during myofibril assembly in differentiating cardiac myocytes, many questions remain unresolved. Of particular interest are the mechanisms by which specific myofibrillar proteins, and even different isoforms of the same protein, become selectively localized within the cell. If myofibril assembly is to occur in a rapid and ordered fashion, every protein must sort correctly to its proper place within the growing myofibril. Data suggest that information for targeting of individual sarcomeric proteins is embedded within discrete protein domains. Domains that are responsible for sorting have already been identified in several sarcomeric proteins including myosin light chains, MyBP-C, M-protein, and myomesin (Soldati and Perriard, 1991; Gilbert et al, 1996; Komiyama et al, 1996; Obermann et al, 1998; Auerbach et al, 1999). It is also predicted that a molecular template (blueprint of the sarcomere) must be present for the assembly of sarcomeric components that specifies the precise position of proteins within each half-sarcomere. The titin third filament system has emerged as an excellent candidate for providing this template. This theory is based on the following observations: first, that titin interacts directly with various thick and thin filament components; second, that it is one of the first myofibrillar proteins to assemble into nascent sarcomeres (see above); and third, that many distinct isoforms are generated by alternative splicing that show differences in titin's Z disk, M line, or I band region (Labeit and Kolmerer, 1995; Gautel et al, 1996; Kolmerer et al, 1996; Sorimachi et al, 1997; Freiburg et al, unpublished results). Thus, sarcomeres could be created with different ultrastructures through expression of distinct titin isoforms. The ability of distinct titin isoforms to interact with subsets of myofibrillar constituents would be expected to confer unique elastic and mechanical properties on the different muscle types. Interestingly, titin isoforms with different structural elements within their central I-band region are expressed within human myocardium, for example, most likely reflecting the physiologic demands of the morphologically distinct regions of the heart (Helmes et al, 1999). If titin is indeed the template for sarcomere assembly, one would predict that muscle type-specific titin splice pathways would emerge early during myogenesis. At present it is unclear if this finely tuned regulation of myofibril assembly is coordinated with muscle fiber-type specificity, which is dictated predominantly by the myosin isoforms expressed, or if both levels of complexity are regulated independently. For reviews on the properties of titin see Labeit et al (1997) and Gregorio et al (1999).

FILAMENT LENGTH SPECIFICATION

The precise regulation of thick and thin filament length is an essential property of all striated muscles. Although little is known about the complex mechanisms required for cardiac muscle cells to control the assembly, length determination, and organization of myosin and actin into thick and thin filaments, respectively, a diverse repertoire of interacting components appears to be involved. It is clear from extensive in vitro experiments (where wide variations in filament lengths are observed) that the properties of actin alone cannot account for the thin filament length distributions found in striated muscle (e.g., Oosawa and Asakura, 1975). Although the distribution of filament lengths in cardiac muscle is broader (0.8 μm to about 1.3 μm; Robinson and Winegrad, 1979) than in skeletal muscle (1.11 ± 0.03 μm; Sosa et al, 1994), a regulatory mechanism must be present since the thin filament lengths resemble a gaussian and not the exponential distribution seen with pure actin filaments in vitro.

In vivo, actin filament length distributions are maintained partly by actin-binding proteins that have profound effects on many aspects of actin filament polymerization and organization within cells. For example, there are actin-binding proteins responsible for regulating the rate and site of assembly of the actin filaments either by binding to the ends of the filaments (CapZ at the barbed end, tropomodulin at the pointed end), by sequestering soluble subunits (e.g., profilin), or by depolymerizing filaments (actin depolymerizing factor/cofilin family). For reviews see Schafer and Cooper (1995), Carlier (1998), and Weber (1999). Also, other proteins such as tropomyosin stabilize actin filaments, and α-actinin links actin filaments to one another or to other cellular components (e.g., titin filaments). Interestingly, while actin monomers polymerize to strikingly uniform lengths in the thin filaments of striated muscle, they also exchange dynamically, indicating that actin monomer association and dissociation at filament ends must be highly regulated to maintain the uniformity of filament lengths (this phenomenon is referred to as "dynamic stability"; for a detailed discussion see Littlefield and Fowler, 1998). It has also been proposed that template molecules exist to regulate the lengths of actin filaments by providing stability to filaments that are the length of the template (e.g., nebulin in skeletal muscle, its equivalent in cardiac muscle?; for discussion see Trinick, 1994; Fowler, 1996; Littlefield and Fowler, 1998). Recently, it was also suggested that acquisition of correct thin filament length may be dependent on the correct assembly of a myosin thick filament-titin scaffold (Littlefield and Fowler, 1998).

One model to explain the observation (described above) that mature actin filament length distributions are not visible by immunofluorescence and electron microscopy until late in myofibril assembly is the variable length model (Gregorio, 1997). This model predicts that in the absence of the capping protein tropomodulin, actin filaments lengths are heterogeneous, with particular lengths perhaps correlating with the actin isoform expressed. Thus, early in assembly, prior to the appearance of tropomodulin and when multiple actin isoforms are expressed, the absolute lengths of the individual polarized thin filaments are heterogeneous, being both shorter and longer than their mature ~1-μm lengths. During subsequent stages of assembly, by as yet unknown mechanisms, elongation or shortening of filaments at the pointed ends occurs to generate the 1-μm-long mature thin filament; these processes give the actin filaments their mature striated appearance. In this model, the acquisition of mature thin filament length may therefore be contingent on the absence of tropomodulin, and thus uncapped actin filament pointed ends early in the assembly process. In this capacity, the assembly of tropomodulin into sarcomeres would be required to maintain the final length of thin filaments, as well as perhaps being the rate-limiting step required to orchestrate muscle contraction. Additional experiments are required to test the validity of this proposed model during in vivo embryonic development.

The exact mechanisms by which vertebrate sarcomeric myosin assembles into thick filaments is also unknown. The ability of myosin to form ordered structures is an essential property of the molecule. Purified muscle myosin can assemble spontaneously to form filaments in vitro and when introduced into nonmuscle cell types; however, these are always heterogeneous in length to some degree (Huxley, 1963; Moos et al, 1975; Vikstrom et al, 1993; Seiler et al, 1996). Light meromyosin (LMM), a proteolytic fragment of the myosin heavy chain that corresponds to about two thirds of the myosin rod, is the primary determinant responsible for the assembly of myosin into filaments (Szent-Gyorgyi et al, 1960; Lowey et al, 1969). Specifically, a conserved 29 amino acid domain in the C-terminal region for the myosin heavy chain rod appears to be required for thick filament assembly

(Sohn et al, 1997). Increasing evidence indicates that MyBP-C and MyBP-H, two of the large number of thick filament-associated proteins, contribute significantly to thick filament length regulation and organization (e.g., Epstein and Fischman, 1991; Seiler et al, 1996). In addition, it has been proposed that titin acts as a template for the precise assembly of myosin molecules into thick filaments in both cardiac and skeletal muscle (Trinick, 1996). The mechanisms contributing to the precise assembly of thick filaments are under investigation. For a review see Vikstrom et al (1997).

GENETICS OF MYOFIBRIL ASSEMBLY

Animal Models

Naturally occurring and targeted mutations in genes coding for myofibrillar proteins have provided considerable insight into the function of sarcomeric proteins in myofibril assembly and stability. In *Drosophila melanogaster* and *Caenorhabditis elegans*, null mutations in sarcomeric proteins such as the myosins, actins, α-actinins, and associated proteins, cell adhesion proteins and extracellular matrix components have identified roles in sarcomere assembly and delineated functional differences between related protein isoforms. Point mutations resulting in amino acids substitutions have defined domains within individual proteins that are important for interaction with other sarcomeric constituents or for contractile function. For review see Epstein and Bernstein (1992).

Few mutations affecting myofibril assembly have been identified in vertebrates, perhaps because proper heart function is crucial for embryo survival and so only mutations with modest effects would be present in a population. A homozygous recessive mutation identified in the Mexican axolotl results in greatly reduced amounts of α-tropomyosin and absence of organized myofibrils. Coculture of developing myocardium from mutant embryos with cardiogenic endoderm from wild-type embryos rescues the phenotype (Davis and Lemanski, 1987), indicating that the lack of α-tropomyosin protein is secondary to a defect in the endoderm. Myofibril assembly and beating is also reportedly rescued by incubation with RNA from wild-type endoderm (LaFrance et al, 1993), or by introduction of tropomyosin protein into mutant myocytes (Zajdel et al, 1998). Accumulating evidence in chick indicates that although precardiac mesoderm explanted from stage 5 embryos can differentiate in culture, endoderm is required until a somewhat later stage for proper myofibrillogenesis to occur. One possibility is that the sequential activation of genes encoding contractile proteins, mediated at least partly by signals from the endoderm, is an important mechanism for regulating the timing of myofibrillogenesis and contractile activity in developing myocytes (Antin et al, 1994; Gannon and Bader, 1995).

Relatively few genes coding for cardiac myofibrillar proteins have been targeted for mutation in mouse, and perhaps not surprisingly the resulting phenotypes shed little light on mechanisms of myofibril assembly. Most mice homozygous for mutations in the cardiac α-actin gene die prior to birth, although some live for up to 2 weeks (Kumar et al, 1997). While upregulation of other actin genes partially compensates for loss of cardiac α-actin, myofibrillar disarray is observed, reflecting reduced levels of actin protein. Lethality can be rescued by targeted overexpression of enteric smooth muscle γ-actin, though cardiac function remains abnormal. Ablation of the α-tropomyosin gene leads to lethality between embryonic days 9.5 and 13.5 due to myofibrillar abnormalities (Blanchard et al, 1997; Rethinasamy et al, 1998). Mice carrying a null mutation in the α-myosin heavy chain gene die between embryonic days 11 and 12 of gross heart defects (Jones et al, 1996), as might be expected in the absence of this essential component of the myofibril. Mice carrying a null mutation in cardiac troponin I are born apparently normal due to compensation by a fetal troponin I isoform. Upon downregulation of the fetal isoform around day 15 after birth, however, mice rapidly die of acute heart failure (Huang et al, 1999). In general, ablation of a gene coding for myofibrillar component is lethal, unless compensated for by a closely related isoform, as is the case for the actins. In these instances, knockouts can tell us something about the functions of individual isoforms, though relatively little about myofibril assembly. Several reports have described replacement of genes coding for one isoform with another (Fewell et al, 1998; Pawloski-Dahm et al, 1998). This approach, combined with more subtle mutations to alter specific amino acids or protein domains within target genes, may be more informative for studies of myofibril assembly. Nevertheless, studies in mice are limited by the early requirement of a beating heart for embryo survival. In this context, cell culture model systems may provide a more advantageous approach for genetic analysis of myofibrillogenesis (see below).

Human Myopathies

Research on myofibril assembly and maintenance took on clinical relevance with the discovery that mutations within certain contractile proteins are responsible for genetic cardiac muscle diseases (reviewed in Vikstrom and Leinwand, 1996). One member of this class is familial hypertrophic cardiomyopathy (FHC) (Clark et al, 1973), an autosomal-dominant inherited disease characterized by unexplained left ventricular hypertrophy, myofibril disarray, and a wide range of clinical symptoms. It is a disease of the sarcomere (Thierfelder et al, 1994). Thus far, at least seven different genes and multiple alleles encoding highly abundant structural proteins are associated with FHC: β-myosin heavy chain (β-MyHC), cardiac myosin essential light chain, cardiac myosin regulatory light chain, cardiac troponin T and I (cTnT, and cTnI, respectively), myosin-binding protein C (MyBP-C), and α-tropomyosin (for a recent review on FHC see Bonne et al, 1998). Although it is not yet known how mutations in structural sarcomeric proteins cause this disease, it is thought that they operate through a common pathway, presumably by causing either a structural and/or functional alteration in the cardiac sarcomere (Cuda et al, 1993; Sweeney et al, 1994; Lankford et al, 1995).

Despite the fact that all FHC alleles discovered to date occur in structural proteins, the disease is clinically heterogeneous. Patients with mutations in myosin heavy chain, for example, show consistent cardiac hypertrophy but variable rates of sudden death, while patients with cardiac troponin T mutations exhibit little hypertrophy but have a high propensity for sudden death (Solomon et al, 1993; Watkins et al, 1995; Moolman et al, 1997). Understanding the molecular basis for these differences is likely to shed light on the function and mechanism of interactions of specific myofibrillar constituents. Although little is presently known about how defined mutations lead to different disease states, multiple cellular mechanisms appear to result in human FHC depending on the specific mutation, including primary contractile dysfunctions, dominant-negative mechanisms of action, deficiencies in motor activity of the myosin molecules, increased energetic load on the heart, diminshed power production, and primary hypercontactile states (e.g., Sweeney et al, 1994, 1998; Lin et al, 1996; Poetter et al, 1996; Sata and Ikebe, 1996; Vikstrom and Leinwand, 1996; Watkins et al, 1996; Bottinelli et al, 1998; Tardiff et al, 1998). Targeted mutations in mice have been generated that mimic several human mutations leading to FHC that now provide a valuable tool for understanding how subtle perturbations in contractile proteins affect sarcomeric structure and function (Geisterfer-Lowrance, 1996; Vikstrom et al, 1996; Tardiff et al, 1998; Yang et al, 1998). These mouse models of FHC, along with approaches to create more subtle mutations (e.g., point mutations) in specific domains of contractile proteins, are likely to be useful for investigating cardiac myofibrillogenesis. This is an active area of research at the present time.

NEW APPROACHES FOR INVESTIGATING MYOFIBRIL ASSEMBLY IN HEART

For the most part, studies of myofibril assembly to date have focused on histologic analysis of the normal assembly process in developing myocytes. As detailed above, this approach has provided a fairly comprehensive view of the temporal aspects of myofibril assembly. While some investigators have tested the roles of various myofibrillar constituents in fully differentiated embryonic or fetal cardiac myocytes by microinjection of truncated proteins or function-perturbing antibodies, by anti-sense strategies, or by transfection of cDNA expression vectors (e.g., Schultheiss et al, 1992; Gregorio et al, 1995, 1998; Schafer et al, 1995; Peckham et al, 1997; Turnacioglu et al, 1997; Sussman et al, 1998; Linke et al, 1999), virtually no work has been directed toward functional analysis of myofibril assembly in differentiating heart muscles. Such experiments would require manipulation and/or introduction of cDNAs or proteins into precardiac cells, a process that presents numerous technical difficulties in vivo. One culture system that may prove useful for investigating the function of various proteins in cardiac myofibril assembly is explants of precardiac regions from avian embryos. Precardiac regions from late gastrula (stages 5–6) avian embryos, consisting of mesoderm plus endoderm or mesoderm alone, can be explanted and maintained in culture. Within 24 hours, premyocardial cells differentiate, assemble contractile myofibrils, and commence beating in a temporal sequence that corresponds closely with the timing of myocyte differentiation and myofibril assembly in vivo (DeHaan, 1963; Lough et al, 1990; Antin et al, 1994; Imanaka-Yoshida, 1997; Yatskievych et al, 1997). Premyocardial explants have the added advantage of being more accessible to direct manipulation than intact embryos, either by utilizing microinjection or transfection techniques. Despite the potential for directly manipulating the assembly process, to date few studies have exploited this culture system.

Murine embryonic stem (ES) cells provide another cell culture system that holds much promise for investigating myofibril assembly in differentiating cardiac myocytes. ES cells isolated from the inner cell mass of blastocyst stage mouse embryos retain their totipotency, both when grown in culture or when reintroduced into the inner cell mass of recipient embryos. When 500 to 4,000 ES cells are cultured in hanging droplets, under appropriate culture conditions embryoid bodies form that within 7 to 10 days readily give rise to foci of beating cardiac myocytes. Although ES cell cultures have been used extensively to study many aspects of cardiac myocyte development, including contractile protein gene expression (Robbins et al, 1990; Ng et al, 1997), the role of signaling molecules in the regulation of heart muscle cell development (Wobus et al, 1997) and various aspects of myocyte ultrastructure and cell–cell coupling (Westfall et al, 1997), little if any work has addressed the issue of myofibril assembly. In this regard, the ability to genetically modify ES cells provides a powerful tool for addressing questions of macromolecular assembly. Genes whose function would almost certainly be required for proper heart function, and whose inactivation would likely lead to embryo lethality in vivo, can be ablated in ES cells and the effects on myofibril assembly examined in culture. The ability to substitute genes encoding defined mutations for wild-type genes ("knock ins"), should allow detailed analysis of defined protein domains in the function of myofibrillar constituents. Finally, it should also be possible to assess the effects of expressing truncated or dominant negative forms of myofibrillar proteins in ES cell-derived differentiating cardiac myocytes, once again bypassing the potential problem of early lethality in vivo. ES cells exhibiting a potentially lethal phenotype could be reintroduced into early developing embryos to generate chimeras. If introduced cells represented only a minority of heart muscle cells, it may be feasible to study the effects of these modifications in the whole animal (for review see Robbins et al, 1992).

SUMMARY

Cardiac muscle sarcomeres are complex structures composed of numerous proteins organized in an exquisitely precise manner. Studies using cultured myocytes combined with analysis of myofibril assembly in the developing heart in vivo have yielded a model for assembly that provides a framework for future experiments. The stage is now set for mechanistic and functional analyses in differentiating cardiac myocytes. These types of experiments will require developmental systems in which precardiac cells are targeted with function altering proteins or antibodies, or in which expression of specific myofibrillar constituents is ablated or foreign genes introduced. Explants from precardiac regions of avian embryos are accessible to experimental manipulation and may offer one avenue for approaching these questions. Mouse embryonic stem cells can be induced to differentiate into beating cardiac myocytes in culture, and this system provides another potentially powerful approach for defining the function of proteins and protein domains during myofibrillogenesis. Importantly, the rapid technologic advances in imaging techniques at the light microscopic level, including confocal, deconvolution, and two (multi-) photon microscopy, is predicted to have a significant impact on this field, allowing for the generation of three-dimensional images of the spatial relationships of myofibrillar constituents during assembly. In particular, the ability to monitor the expression and assembly of sarcomeric proteins [e.g., green fluorescent protein (GFP)-tagged] in individual differentiating cardiac myocytes in real time will allow us to address questions that cannot be answered from static images. The combination of improved imaging methods, new developmental models, and gene ablation and modification technologies promise to enable detailed analyses of the mechanisms involved in myofibril assembly including important protein interactions regulating thick and thin filament length specification, and the potential role of titin as a molecular scaffold in the assembly process.

ACKNOWLEDGMENTS

The authors wish to thank Elisabeth Ehler and Jean-Claude Perriard (Institute of Cell Biology, Zurich, Switzerland) for sharing unpublished data; Velia M. Fowler (Scripps Research Institute), Elisabeth Ehler, Abigail McElhinny, Catherine McLellan, Ryan Mudry, and Diane Rudy (University of Arizona) for helpful discussions and critical reading of the chapter; and Tania Yatskievych and David Carroll for preparing the figures. This work was supported by National Institutes of Health grants HL57461, HL03985 (to C.C.G.) and HL54133 (to P.B.A.).

REFERENCES

Andrée, B., Duprez, D., Vorbusch, B., Arnold, H.-H., and Brand, T. (1998). BMP-2 induces ectopic expression of cardiac lineage markers and interferes with somite formation in chicken embryos. *Mech Dev* 70:119–131.

Antin, P.B., Taylor, R.G., and Yatskievych, T.A. (1994). Precardiac mesoderm is specified during gastrulation in quail. *Dev Dyn* 200:144–153.

Auerbach, D., Bantle, S., Keller, S., et al. (1999). Different domains of the M-band protein myomesin are involved in myosin binding and M-band targeting. *Mol Biol Cell* 10:1297–1308.

Bishop S.P., Anderson, P.G., and Tucker, D.C. (1990). Morphological development of the rat heart growing in oculo in the absence of hemodynamic work load. *Circ Res* 66:84–102.

Blanchard, E.M., Lizuka, K., Christe, M., et al. (1997). Targeted ablation of the murine alpha-tropomyosin gene. *Circ Res* 81:1005–1010.

Bonne, G., Carrier, L., Richard, P., Hainque, B., and Schwartz, K. (1998). Familial hypertrophic cardiomyopathy: from mutations to functional defects. *Circ Res* 83:580–593.

Bottinelli, R., Coviello, D.A., Redwood, C.S., et al. (1998). A mutant tropomyosin that causes hypertrophic cardiomyopathy is expressed in vivo and associated with an increased calcium sensitivity. *Circ Res* 82:106–115.

Brand, T., Andree, B., Schneider, A., Buchberger, A., and Arnold, H.H. (1997). Chicken Nkx2-8, a novel homeobox gene expressed during early heart and foregut development. *Mech Dev* 64:53–59.

Brook, W.H., Connell, S., Cannata, J., Maloney, J.E., and Walker, A.M. (1983). Ultrastructure of the myocardium during development from early fetal life to adult life in sheep. *J Anat* 137:729–741.

Carlier, M.-F. (1998). Control of actin dynamics. *Curr Opin Cell Biol* 10:45–51.

Clark, C.E., Henry, W.L., and Epstein, S.E. (1973). Familial prevalence and genetic transmission of idiopathic hypertrophic subaortic stenosis. *N Eng J Med* 289:709–714.

Cuda, G., Fananapazir, L., Zhu, W.S., Sellers, J.R., and Epstein, N.D. (1993). Skeletal muscle expression and abnormal function of beta-myosin in hypertrophic cardiomyopathy. *Clin Invest* 91:2861–2865.

Dabiri, G.A., Turnacioglu, K.K., Sanger, J.M., and Sanger, J.W. (1997). Myofibrillogenesis visualized in living embryonic cardiomyocytes. *Proc Natl Acad Sci USA* 19:9493–9498.

Davis, L.A., and Lemanski, L.F. (1987). Induction of myofibrillogenesis in cardiac lethal mutant axolotl hearts rescued by RNA derived from normal endoderm. *Development* 99:145–154.

DeHaan, R.L. (1963). Migration patterns of the precardiac mesoderm in the early chick embryo. *Exp Cell Res* 29:544–560.

Dlugosz, A.A., Antin, P.B., Nachmias, V.T., and Holtzer, H. (1984). The relation between stress fiber-like structures and nascent myofibrils in cultured cardiac myocytes. *J Cell Biol* 99:2268–2278.

Durocher, D., Charron, F.R.W., Schwartz, R., and Nemer, M. (1997). The cardiac transcription factors Nkx2-5 and GATA-4 are mutual cofactors. *EMBO J* 16:5687–5696.
Ehler, E., Rothen, B.M., Hämmerle, S.P., Komiyama, M., and Perriard, J.-C. (1999). Myofibrillogenesis in the developing chicken heart: assembly of Z-disk, M-line and the thick filaments. *J Cell Sci* 112:1529–1539.
Epstein, H.F., and Bernstein, S.I. (1992). Genetic approaches to understanding muscle development. *Dev Biol* 154:231–344.
Epstein, H.F., and Fischman, D.A. (1991). Molecular analysis of protein assembly in muscle development. *Science* 251:1039–1044.
Fewell, J.G., Hewett, T.E., Sanbe, A., et al. (1998). Functional signficance of cardiac myosin essential light chain isoform switching in transgenic mice. *J Clin Invest* 101:2630–2639.
Fowler, V.M. (1996). Regulation of actin filament length in erythrocytes and striated muscle. *Curr Opin Cell Biol* 8:86–96.
Furst, D.O., Osborn, M., and Weber, K. (1989). Myogenesis in the mouse embryo: differential onset of expression of myogenic proteins and the involvement of titin in myofibril assembly. *J Cell Biol* 109:517–527.
Gannon, M., and Bader, D. (1995). Initiation of cardiac differentiation occurs in the absence of anterior endoderm. *Development* 121:2439–2450.
Garcia-Martinez, V., and Schoenwolf, G.C. (1993). Primitive streak origin of the cardiovascular system in avian embryos. *Dev Biol* 159:706–719.
Gautel, M., Goulding, D., Bullard, B., Weber, K., and Fürst, D.O. (1996). The central Z-disk region of titin is assembled from a novel repeat in variable copy numbers. *J Cell Sci* 109:2747–2754.
Geisterfer-Lowrance, A.A., Christe, M., Conner, D.A., et al. (1996). A mouse model of familial hypertrophic cardiomyopathy. *Science* 272:731–734.
Gilbert, R., Kelly, M.G., Mikawa, T., and Fischman, D.A. (1996). The carboxyl terminus of myosin binding protein C (MyBP-C, C-protein) specifies incorporation into the A-band of striated muscle. *J Cell Sci* 109:101–111.
Gregorio, C.C. (1997). Models of thin filament assembly in cardiac and skeletal muscle. *Cell Struct Funct* 22:191–195.
Gregorio, C.C., and Fowler, V.M. (1995). Mechanisms of thin filament assembly in embryonic chick cardiac myocytes: tropomodulin requires tropomyosin for assembly. *J Cell Biol* 129:683–695.
Gregorio, C.C., Granzier, H., Sorimachi, H., and Labeit S. (1999). Muscle assembly: a titanic achievement? *Curr Opin Cell Biol* 11:18–25.
Gregorio, C.C., Trombitás, K., Centner, T., et al. (1998). The NH_2 terminus of titin spans the Z disc: its interaction with a novel 19kD ligand (T-cap) is required for sarcomeric integrity. *J Cell Biol* 143:1013–1027.
Gregorio, C.C., Weber, A., Bondad, M., Pennise, C.R., and Fowler, V.M. (1995). Requirement of pointed-end capping by tropomodulin to maintain actin filament length in embryonic chick cardiac myocytes. *Nature* 377:83–86.
Hamburger, V., and Hamilton, H.L. (1951). A series of normal stages in the development of the chick embryo. *J Morphol* 88:49–92.
Han, Y., Dennis, J.E., Cohen-Gould, L., Bader, D.M., and Fischman, D.A. (1992). Expression of sarcomeric myosin in the presumptive myocardium of chicken embryos occurs within six hours of commitment. *Dev Dyn* 193:257–265.
Handel, S.E., Greaser, M.L., Schultz, E., et al. (1991). Chicken cardiac myofibrillogenesis studied with antibodies specific for titin and the muscle and nonmuscle isoforms of actin and tropomyosin. *Cell Tissue Res* 263:419–430.
Helmes, M, Trombitas, K., Centner, T., et al. (1999). Mechanically driven contour-length adjustment in rat cardiac titin's unique N2B sequence: titin is an adjustable spring. *Circ Res* 84:1339–1352.

Hill, C.S., and Lemanski, L.F. (1985). Immunoelectron microscopic localization of alpha actinin and actin in embryonic hamster heart cells. *Eur J Cell Biol* 39:300–312.

Hiruma, T., and Hirakow, R. (1985). An ultrastructural topographical study on myofibrillogenesis in the heart of the chick embryo during pulsation onset period. *Dev Dyn* 196:291–299.

Holtzer, H., Hijikata, T., Lin, Z.X., et al. (1997). Independent assembly of 1.6μm long bipolar MHC filaments and I-Z-I bodies. *Cell Struct Funct* 22:83–93.

Huang, X., Pi, Y., Lee, K.J., et al. (1999). Cardiac troponin I gene knockout: a mouse model of myocardial troponin I deficiency. *Circ Res* 84:1–8.

Huxley, H.E. (1960). Muscle cells. In: Brachet, J., and Mirsky, A.E., eds. *The Cell (Biochemistry, Physiology, Morphology)*. Academic, New York, pp. 365–481.

Huxley, H.E. (1963). Electron microscope studies on the strucure of natural and synthetic proteins from striated muscle. *J Mol Biol* 7:281–308.

Imanaka-Yoshida, K. (1997). Myofibrillogensis in precardiac mesoderm explant culture. *Cell Struct Funct* 22:45–49.

Jacobson, A.G., and Sater, A.K. (1988). Features of embryonic induction. *Development* 104:341–359.

Jones, W.K., Grupp, I.L., Doetschman, T., et al. (1996). Ablation of the murine alpha myosin heavy chain gene leads to dosage effects and functional deficits in the heart. *J Clin Invest* 98:1906–1917.

Kolmerer, B., Olivieri, N., Witt, C.C., Herrmann, B.G., and Labeit, S. (1996). Genomic organization of the M-line titin and its tissue-specific expression in two distinct isoforms. *J Mol Biol* 256:556–563.

Komiyama, M., Kouchi, K., Maruyama, K., and Shimada, Y. (1993). Dynamics of actin and assembly of connectin (titin) during myofibrillogenesis in embryonic chick cardiac muscle cells in vitro. *Dev Dyn* 196:291–299.

Komiyama, M., Soldati, T., von Arx, P., and Perriard, J.-C. (1996). The intracompartmenal sorting of myosin alkali light chain isoproteins reflects the sequence of developmental expression as determined by double epitope-tagging competition. *J Cell Sci* 109:2089–2099.

Kumar, A., Crawford, K., Close, L., et al. (1997). Rescue of cardiac alpha-actin deficient mice by enteric smooth muscle gamma-actin. *Proc Natl Acad Sci USA* 94:4406–4411.

Labeit, S., and Kolmerer, B. (1995). Titins, giant proteins in charge of muscle ultrastructure and elasticity. *Science* 270:293–296.

Labeit, S., Kolmerer, B., and Linke, W.A. (1997). The giant protein titin. Emerging roles in physiology and pathophysiology. *Circ Res* 80:290–294.

LaFrance, S.M., Fransen, M.D., Ergenel-Unaltuna, N., et al. (1993). RNA from normal anterior endoderm/mesoderm-conditioned medium stimulates myofibrillogenesis in developing mutant axolotl hearts. *Cell Mol Biol Res* 39:547–560.

Lankford, E.B., Epstein, N.D., Fananapazi, R.L., and Sweeney, H.L. (1995). Abnormal contractile properties of muscle fibers expressing beta-myosin heavy chain gene mutations in patients with hypertrophic cardiomyopathy. *Clin Invest* 95:1409–1144.

Lawson, K.A., Meneses, J.J., and Pedersen, R.A. (1991). Clonal analysis of epiblast fate during germ layer formaton in the mouse embryo. *Development* 113:891–911.

Legato, M.J. (1972). Ultrastructural characteristics of the rat ventricular cell grown in tissue culture with special reference to sarcomerogenesis. *J Mol Cell Cardiol* 4:299–317.

Lin, D., Bobkova, A., Homsher, E., and Tobacman, L.S. (1996). Altered cardiac troponin T in vitro function in the presence of a mutation implicated in familial hypertrophic cardiomyopathy. *J Clin Invest* 97:2842–2848.

Lin, Z., Holtzer, S., Schultheiss, T., et al. (1989). Polygons and adhesion plaques and the disassembly and assembly of myofibrils in cardiac myocytes. *J Cell Biol* 10:2355–2367.

Linask, K.K., and Gui, Y.-H. (1995). Inhibitory effects of ouabain on early heart development and cardiomyogenesis in the chick embryo. *Dev Dyn* 203:93–105.

Linke, W.A., Rudy, D.R., Centner, T., et al. (1999). I-band titin in cardiac muscle is a three-element molecular spring and is critical for maintaining thin filament structure. *J Cell Biol* 146:631–644.
Littlefield, R., and Fowler, V.M. (1998). Defining actin filament length in striated muscle: rulers and caps or dynamic stability? *Annu Rev Dev Biol* 14:487–525.
Lough, J.W., Bolender, D.L., and Markwald, R.R. (1990). A culture model for cardiac morphogenesis. *Ann NY Acad Sci* 588:421–424.
Lowey, S., Slayter, H.S., Weeds, A.G., and Baker, H. (1969). Substructure of the myosin molecule. I. Subfragments of myosin by enzymic degradation. *J Mol Biol* 42:1–29.
Manasek, F.J. (1968). Embryonic development of the heart. I. A light and electron microscopic study of myocardial development in the early chick embryo. *J Morphol* 125:329–365.
Markwald, R.R. (1973). Distribution and relationship of precursor Z material to organizing myofibrillar bundles in embryonic rat and hamster ventricular myocytes. *J Mol Cell Cardiol* 5:341–350.
McGrew, M.J., Xavier-Neto, J., Pourquie, O., and Rosenthal, N. (1999). Molecular genetics of skeletal muscle development. In: Harvey, R.P., and Rosenthal, N., eds. *Heart Development*. Academic Press, New York, pp. 493–512.
Moolman, J.C., Corfield, V.A., Posen, B., et al. (1997). Sudden death due to troponin T mutations. *J Am Coll Cardiol* 29:49–55.
Moos, C., Offer, G., Starr, R., and Bennett, P. (1975). Interaction of C-protein with myosin, myosin rod and light meromyosin. *J Mol Biol* 97:1–9.
Ng, W.A., Doetschman, T.J.R., and Lessard, J.L. (1997). Muscle isoactin expression during in vitro differentiation of murine embryonic stem cells. *Pediatr Res* 41:285–292.
Obermann, W.M., van der Ven, P.F.M., Steiner, F., Weber, K., and Fürst, D.O. (1998). Mapping of a myosin-binding doman and a regulatory phosphorylation site in M-protein, a structural protein of the sarcomeric M band. *Mol Biol Cell* 9:829–840.
Olson, E.N., and Srivastava, D. (1996). Molecular pathways controlling heart development. *Science* 272:671–676.
Oosawa, F., and Asakura, S. (1975). Thermodynamics of the polymerization of protein. In: Horecker, B., Kaplan, N.O., Marmur, J., and Scheraga, H.A., eds. Academic Press, New York, pp. 1–194.
Patterson, K.D., Cleaver, O., Gerber, W.V., Grow, M.W., Newman, C.S., and Krieg, P.A. (1998). Homeobox genes in cardiovascular development. *Curr Top Dev Biol* 40:1–44.
Pawloski-Dahm, C.M., Song, G., Kirkpatrick, D.L., et al. (1998). Effects of total replacement of atrial myosin light chain-2 with the ventricular isoform in atrial myocytes of transgenic mice. *Circulation* 97:1508–1513.
Peckham, M., Young, P., and Gautel, M. (1997). Constitutive and variable regions of Z-disk titin/connectin in myofibril formation: a dominant-negative screen. *Cell Struct Funct* 22:95–101.
Poetter, K., Jiang, H., Hassanzadeh, S., et al. (1996). Mutations in either the essential or regulatory light chains of myosin are associated with a rare myopathy in human heart and skeletal muscle. *Nat Genet* 13:63–69.
Reecy, J.M., Xuyang, L., Yamada, M., et al. (1999). Identification of upstream regulatory regions in the heart-expressed homeobox gene Nkx2-5. *Development* 126:839–849.
Rethinasamy, P., Muthuchamy, M., Hewett, T., et al. (1998). Molecular and physiological effects of alpha-tropomyosin ablation in the mouse. *Circ Res* 82:116–123.
Rhee, D., Sanger, J.M., and Sanger, J.W. (1994). The premyofibril: evidence for its role in myofibrillogenesis. *Cell Motil Cytol* 28:1–24.
Robbins, J., Doetschman, T., Jones, W.K., and Sanchez, A. (1992). Embryonic stem cells as a model for cardiogenesis. *Trends Cardiovasc Med* 2:44–50.
Robbins, J., Gulick, J., Sanchez, A., Howles, P., and Doetschman, T. (1990). Mouse embryonic stem cells express the cardiac myosin heavy chain genes during development in vitro. *J Biol Chem* 265:11905–11909.

Robinson, T.F., and Winegrad, S. (1979). The measurement and dynamic implications of thin filament lengths in heart muscle. *J Physiol* 286:617–619.
Sata, M., and Ikebe, M. (1996). Functional analysis of the mutations in the human cardiac beta-myosin that are responsible for familial hypertrophic cardiomyopathy. Implication for the clinical outcome. *J Clin Invest* 98:2866–2873.
Schafer, D.A., and Cooper, J.A. (1995). Control of actin assembly at filament ends. *Annu Rev Cell Dev Biol* 11:497–518.
Schafer, D.A., Hug, C., and Cooper, J.A. (1995). Inhibition of CapZ during myofibrillogenesis alters assembly of actin filaments. *J Cell Biol* 128:61–70.
Schiaffino, S., and Reggiani, C. (1996). Molecular diversity of myofibrillar proteins: gene regulation and functional signficance. *Physiol Rev* 76:371–423.
Schultheiss, T.M., Burch, J., and Lassar, A. (1997). A role for bone morphogenetic proteins in the induction of cardiac myogenesis. *Genes Dev* 11:451–462.
Schultheiss, T., Choi, J., Lin, Z., Cohen-Gould, L., Fischman, D., and Holtzer, H. (1992). A sarcomeric α-actinin truncated at the carboxyl end induces the breakdown of stress fibers in PtK2 cell and the formation of nemaline-like bodies and breakdown of myofibrils in myotubes. *Proc Natl Acad Sci USA* 89:9282–9286.
Schultheiss, T.M., and Lassar, A.B. (1999). Heart formation and the heart field in amphibian embryos. In: Harvey, R.P., and Rosenthal, N., eds. *Heart Development*. Academic Press, San Diego, pp. 37–47.
Schultheiss, T., Lin, Z., Lu, M.-H., et al. (1990). Differential distribution of subsets of myofibrillar proteins in cardiac nonstriated and striated myofibrils. *J Cell Biol* 110:1159–1172.
Seiler, S.H., Fischman, D.A., and Leinwand, L.A. (1996). Modulation of myosin filament organization by C-protein family members. *Mol Biol Cell* 7:113–127.
Shiraishi, I., Simpson, D.G., Carver, W., et al. (1997). Vinculin is an essential component for normal myofibrillar arrangment in fetal mouse cardiac myocytes. *J Mol Cell Cardiol* 29:2041–2052.
Shiraishi, I., Takamatsu, T., and Fujita, S. (1993). 3-D observation of N-cadherin expression during cardiac myofibrillogenesis of the chick embryo using a confocal laser scanning microscope. *Anat Embryol* 187:115–120.
Shiraishi, I., Takamatsu, T., and Fujita, S. (1995). Three-dimensional observation with a confocal scanning laser microscope of fibronectin immunolabeling during cardiac looping in the chick embryo. *Anat Embryol* 191:183–189.
Shiraishi, I., Takamatsu, T., Minamikawa, T., and Fujita, S. (1992). 3-D observation of actin filaments during cardiac myofibrillogenesis in chick embryo using a confocal laser scanning microscope. *Anat Embryol* 185:401–408.
Simpson, D.G., Decker, M.L., Clark, W.A., and Decker, R.S. (1993). Contractile activity and cell-cell contact regulate myofibrillar organization in cultured cardiac myocytes. *J Cell Biol* 123:323–336.
Sohn, R.L., Vikstrom, K.L., Strauss, M., Cohen, C., Szent-Gyorgyi, A.G., and Leinwand, L.A. (1997). A 29 residue region of the sarcomeric myosin rod is necessary for filament formation. *J Mol Biol* 21:317–330.
Soldati, T., and Perriard, J.-C. (1991). Intracompartmental sorting of essential myosin light chains: molecular dissection and in vitro monitoring by epitope tagging. *Cell* 66:277–289.
Solomon, S.D., Wolff, S., Watkins, H., et al. (1993). Left ventricular hypertrophy and morphology in familial hypertrophic cardiomyopathy associated with mutations of the beta-myosin heavy chain gene. *Am Coll Cardiol* 22:498–505.
Sorimachi, H., Freiburg, A., Kolmerer, B., et al. (1997). Tissue-specific expression and α-actinin binding properties of the Z-disc titin. Implications for the nature of vertebrate Z-discs. *J Mol Biol* 270:688–695.
Sosa, H., Popp, D., Ouyang, G., and Huxley, H.E. (1994). Ultrastructure of skeletal muscle fibers studied by a plunge quick freezing method: myofilament lengths. *Biophys J* 67:283–292.

Squire, J.M. (1997). Architecture and function in the muscle sarcomere. *Curr Opin Struct Biol* 7:247–257.
Srivastava, D., Thomsas, T., Lin, Q., Kirby, M.L., Brown, D., and Olson, E.N. (1997). Regulation of cardiac mesodermal and nerual crest development by the bHLH trnscription factor, dHAND. *Nat Genet* 16:154–160.
Sussman, M.A., Baquè, U.C.-S., Daniels, M.P., et al. (1998). Altered expression of tropomodulin in cardiomyocytes disrupts the sarcomeric structure of myofibrils. *Circ Res* 82:94–105.
Sweeney, H.L., Feng, H.S., Yang, Z., and Watkins, H. (1998). Functional analyses of troponin T mutations that cause hypertrophic cardiomyopathy: insights into disease pathogenesis and troponin function. *Proc Natl Acad Sci USA* 95:14406–14410.
Sweeney, H.L., Straceski, A.J., Leinwand, L.A., Tikunov, B.A., and Faust, L. (1994). Heterologous expression of a cardiomyopathic myosin that is defective in its actin interaction. *J Biol Chem* 269:1603–1605.
Szent-Gyorgyi, A.G., Cohen, C., and Philpott, D.E. (1960). Light meromyosin fraction I: a helical molecule from myosin. *J Mol Biol* 2:133–142.
Tam, P.P.L., Parameswaran, M., Kinder, S.J., and Weinberger, R.P. (1997). The allocation of epiblast cells to the embryonic heart and other mesodermal lineages: the role of ingression and tissue movement during gastrulation. *Development* 124:1631–1642.
Tardiff, J.C., Factor, S.M., Tompkins, B.D., et al. (1998). A truncated cardiac troponin T molecule in transgenic mice suggests multiple cellular mechanisms for familial hypertrophic cardiomyopathy. *J Clin Invest* 101:2800–2811.
Thierfelder, L., Watkins, H., MacRae, C., et al. (1994). Beta-tropomyosin and cardiac troponin T mutations cause familial hypertrophic cardiomyopathy: a disease of the sarcomere. *Cell* 77:701–712.
Tokuyasu, K.T. (1989). Immunocytochemical studies of cardiac myofibrillogenesis in early chick embryos. III. Generation of fasciae adherents and costameres. *J Cell Biol* 108:43–53.
Tokuyasu, K.T., and Maher, P.A. (1987a). Immunocytochemical studies of cardiac myofibrillogenesis in early chick embryos. I. Presence of immunofluorescent titin spots in premyofibrillar stages. *J Cell Biol* 105:2781–2793.
Tokuyasu, K.T., and Maher, P.A. (1987b). Immunocytochemical studies of cardiac myofibrillogenesis in early chick embryos. II. Generation of alpha-actinin dots within titin spots at the time of the first myofibril formation. *J Cell Biol* 105:2795–2801.
Trinick, J. (1994). Titin and nebulin protein rulers in muscle? *Trends Biochem Sci* 19:405–408.
Trinick, J. (1996). Cytoskeleton: titin as a scaffold and spring. *Curr Biol* 6:258–260.
Turnacioglu, K.K., Mittal, B., Dabiri, G.A., Sanger, J.M., and Sanger, J.W. (1997). An N-terminal fragment of titin coupled to green fluorescent protein localizes to the Z-bands in living muscle cells: overexpression leads to myofibril disassembly. *Mol Biol Cell* 8:705–717.
Vigoreaux, J.O. (1994). The muscle Z-band: lessons in stress management. *J Muscle Res Cell Motil* 15:237–255.
Vikstrom, K.L., Factor, S.M., and Leinwand, L.A. (1996). Mice expressing mutant myosin heavy chains are a model for familial hypertrophic cardiomyopathy. *Mol Med* 5:556–567.
Vikstrom, K.L., and Leinwand, L.A. (1996). Contractile protein mutations and heart disease. *Curr Opin Cell Biol* 8:97–105.
Vikstrom, K.L., Rovner, A.S., Bravo-Zehnder, M., Saez, C.G., Straceski, A.J., and Leinwand, L.A. (1993). Sarcomeric myosin heavy chain expressed in nonmuscle cells forms thick filaments in the presence of stoichiometric amounts of light chains. *Cell Motil Cytoskeleton* 26:192–204.
Vikstrom, K.L., Seiler, S.H., Sohn, R.L., et al. (1997). The vertebrate myosin heavy chain: genetics and assembly properties. *Cell Struct Funct* 22:123–129.

Wang, S.-M., Greaser, M.L., Schultz, E., Bulinski, J.C., Lin, J.J.-C., and Lessard, J.L. (1988). Studies on cardiac myofibrillogenesis with antibodies to titin, actin, tropomyosin and myosin. *J Cell Biol* 107:1075–1083.

Wang, G.F., and Stockdale, F.E. (1999). Chamber-specific gene expression and regulation during heart development. In: Harvey, R.P., and Rosenthal, N., eds. *Heart Development*. Academic Press, San Diego, pp. 357–372.

Watkins, H., McKenna, W.J., Thierfelder, L., et al. (1995). Mutations in the genes for cardiac troponin T and alpha-tropomyosin in hypertrophic cardiomyopathy. *N Engl J Med* 332:1058–1064.

Watkins, H., Seidman, C.E., Seidman, J.G., Feng, H.S., and Sweeney, H.L. (1996). Expression and functional assessment of a truncated cardiac troponin T that causes hypertrophic cardiomyopathy. Evidence for a dominant negative action. *J Clin Invest* 98:2456–2461.

Weber, A. (1999). Actin binding proteins that change extent and rate of actin monomer-polymer distribution by different mechanisms. *Mol Cell Biochem* 190:67–74.

Westfall, M.V., Pasyk, K.A., Yule, D.I., Samuelson, L.C., and Metzger, J.M. (1997). Ultrastructure and cell-cell coupling of cardiac myocytes differentiating in embryonic stem cell cultures. *Cell Motil Cytoskeleton* 36:43–54.

Wobus, A.M., Kaomei, G., Shan, J., et al. (1997). Retinoic acid accelerates embryonic stem cell-derived cardiac differentiation and enhances development of ventricular cardiomyocytes. *J Mol Cell Cardiol* 29:1525–1539.

Yang, Q., Sanbe, A., Osinska, H., Hewett, T.E., Klevitsky, R., and Robbins, J. (1998). A mouse model of myosin binding protein C human familial hypertrophic cardiomyopathy. *J Clin Invest* 102:1292–1300.

Yatskievych, T.A., Ladd, A.N., and Antin, P.B. (1997). Induction of cardiac myogenesis in avian pregastrula epiblast: the role of the hypoblast and activin. *Development* 124:2561–2570.

Yutzey, K.E., Gannon, M., and Bader, D. (1995). Diversification of cardiomyogenic cell lineages in vitro. *Dev Biol* 170:531–541.

Zajdel, R.W., McLean, M.D., Lemanski, S.L., et al. (1998). Ectopic expression of tropomyosin promotes myofibrillogenesis in mutant axolotl hearts. *Dev Dyn* 213:412–420.

CHAPTER 3

Cardiac Looping and the Left-Right Axis: Integrating Morphologic, Molecular, and Genetic Analyses of Vertebrate Left-Right Asymmetry

Ann F. Ramsdell and H. Joseph Yost

OVERVIEW OF CARDIAC LOOPING

Comparative vertebrate anatomy reveals a strikingly diverse complexity of heart structures, ranging from the more familiar four-chambered hearts of mammals and avians to the three-chambered forms of reptiles, amphibians, and fish. Yet, despite variations in anatomic form, the functionality of all vertebrate hearts is the same: to support and maintain separation of pulmonary and systemic blood flow. To this end, one of the highly conserved morphologic features of vertebrate embryonic heart development is the process of cardiac looping. Simply defined, cardiac looping is the rightward bending and rotation of primary heart tube that occurs over several developmental stages and that culminates in the contact of the venous and arterial poles of the heart. The significant outcome of cardiac looping is that the initially adjacent regions of the heart tube become repositioned with respect to one another and with the great vessels such that "remodeling" of the tube into a multichambered organ with distinct inlet and outlet components can occur.

As emphasized elsewhere (Markwald et al, 1997), the major portion of the primary heart tube arises from the gradual fusion of cells derived from the paired, lateral plate mesodermal cardiac fields. The primary heart tube moreover is composed of primitive "segments"—right ventricle, left ventricle (or, for some vertebrates, common ventricle), atrioventricular (AV) canal, and sinoatrium—that are initially formed in an anteroposterior progression in the cardiac fields, followed by sequential incorporation into the fusing heart tube (Markwald et al, 1997, 1998) (Figure 3.1a). The last primitive segment to be formed is the conotruncus. This segment differs from the others in that it appears to be derived from mesoderm located anterior to the heart fields and is joined to the anterior aspect of the extended heart tube, connecting the primitive right ventricle with the aortic sac (Markwald et al, 1998; Mjaatvedt et al, 1999). The first indication of cardiac looping during the formation of four-chambered hearts is the ventricular infundibular fold, which results from a rightward movement of the right ventricular segment imme-

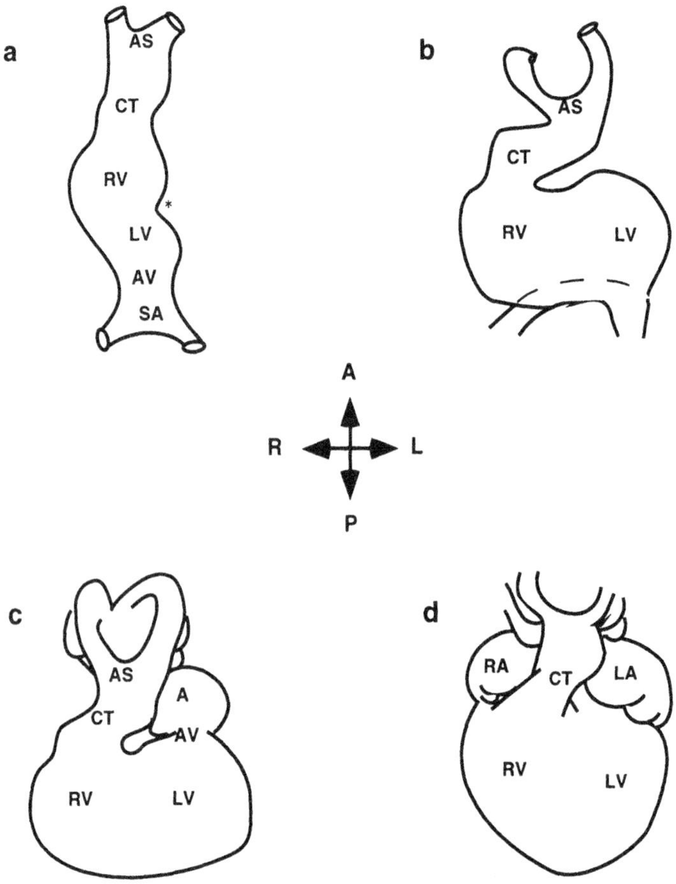

FIGURE 3.1. Schematic illustration (ventral view) of progressive looping during formation of the four-chambered heart. (a) The primary heart tube is composed of primitive segments (labeled) that are positioned adjacent to each other prior to the onset of overt looping morphogenesis. The slight rightward shift of the primitive right ventricle marks the interventricular fold (*). (b) Early looping results in the characteristic S-shaped heart tube as it bends and rotates to the right. (c) As looping continues, the ventricles become positioned ventral to the common atrioventricular (AV) canal and the atrium. (d) The final aspect of looping "wedges" the conotrucus in between the left and right atria and allows continuity of the conotruncus with both ventricles. A, common atrium; AS, aortic sac; AV, atrioventricular canal; CT, conotruncus; LA, left atrium; LV, left ventricle; RA, right atrium; RV, right ventricle; SA, sinoatrium. (Panels b–d were redrawn and modified from Kirby and Waldo, 1995.)

diately following the addition of the left ventricular segment. As remaining primitive segments become incorporated, the heart tube continues its bending and rightward displacement until the AV canal and the conotruncus become aligned with both the left and right ventricles (Figure 3.1b,c). Shortly thereafter, the final aspect of looping (termed "wedging"; Kirby and Waldo, 1995) places the conotruncus between the mitral and tricuspid valves so that inflow and outflow septation can be completed (Figure 3.1d). Similar basic movements of looping and convergence must occur in formation of three-chambered hearts to allow functional separation of oxygenated and deoxygenated blood flow, e.g., as facilitated by the slightly more rightward positioning of the conotruncus over the common ventricle in the amphibian heart (Burggren, 1988). Thus, given the role of cardiac looping in overall vertebrate heart morphogenesis, it is not unexpected that aberrations in the looping process can disrupt other processes such as septation or alignment of chambers with the great vessels, thereby resulting in a broad spectrum of inflow, outflow, and/or septal defects.

LATERALITY DISORDERS AND ASSOCIATED CARDIAC DEFECTS

Situs Solitus and Situs Inversus

The importance of cardiac looping with respect to heart morphogenesis is particularly evident when considering the types of heart defects that are causally linked with most human laterality disorders. In individuals with normal situs (situs solitus), hearts are positioned to the left of the thorax, with the apex pointed in a leftward direction. The left lung in such individuals is bilobed, the right lung is trilobed, and the visceral organs also are predictably positioned relative to the midline, with the stomach and spleen to the left, and the liver and pancreas to the right (Figure 3.2a). By contrast, full mirror-image reversal of the normal orientation of the heart and visceral organs is observed in individuals afflicted with situs inversus (Figure 3.2d), a condition that also is experimentally observed in the *inv/inv* mouse (Yokoyama et al, 1993) and in *Xenopus* embryos targeted for ectopic expression of the Vg1 ligand (Hyatt and Yost, 1998). Situs inversus differs from other classes of laterality disorders in that no other clinically adverse effects are detectable upon diagnosis, presumably because a complete left-right inversion of the body plan allows properly ordered and seamless connections to be made in developing organ systems during embryogenesis. For this reason, clinical presentation is rare, and the exact incidence is uncertain, with estimates ranging from 1/2,500 to 1/20,000 (Bowers et al, 1996). The apparent lack of physiologic defects associated with situs inversus moreover has raised the intriguing (and yet to be answered) question of why left-right asymmetries in any given vertebrate popu-

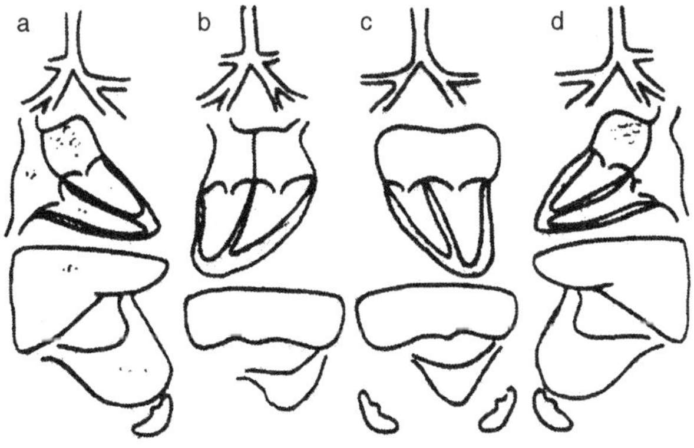

FIGURE 3.2. Schematic illustration (ventral view) of some possible arrangements of the heart and visceral organs. (a) *Situs solitus*. (b) Right isomerism. (c) Left isomerism. (d) *Situs inversus*. Heterotaxia (not shown) could include any possible combination of thoracic and visceral organs that exhibit disconcordant alignment. (From Bowers et al, 1996, with permission.)

lation are invariantly aligned relative to the dorsoventral and anteroposterior axes, as opposed to an equal mix of individuals showing situs solitus and situs inversus. For discussion of this issue, see (Wood and Kershaw, 1991; Wood, 1997; Levin and Mercola, 1998a).

Heterotaxia

By contrast to situs inversus, heterotaxia is a condition in which the heart and/or visceral organs are disconcordantly aligned relative to the left-right body axis, such that some organs show normal situs and others show left-right reversals. Thus, individuals afflicted with heterotaxia do not necessarily exhibit identical positioning of the heart or viscera (Burn, 1991). In instances where the heart is the only organ that is directionally reversed ("isolated dextrocardia"), associated cardiac defects are frequently observed, including AV septal defects, ventricular septal defects, and transposition of the great arteries (Winer-Muram, 1995; Bowers et al, 1996). The incidence of heterotaxia is believed to be similar to that of situs inversus, and the incidence of cardiac malformations accompanying heterotaxia has been reported to exceed 50% of diagnosed individuals (Bowers et al, 1996). As detailed in following sections, the mouse mutation *iv* (Hummel and Chapman, 1959) is a classic animal model of heterotaxia, and a rapidly growing list of other experimental conditions also result in this phenotype. It should be noted that the term *randomized* is often used interchangeably with *heterotaxia* to refer to the disconcordant alignment of the heart and viscera of a given individual with respect to the midline. However, randomization can also describe heart looping that occurs by chance (i.e., 50% of individuals in a population exhibit reversals, and the remaining develop normally). Thus, the term *randomization* takes on different meanings, according to the context in which it is used (i.e., individual vs. population), and to avoid confusion, its use will be limited in this chapter.

Isomerism

A third type of laterality disorder associated with a significant frequency of cardiac defects is isomerism. Isomerism can occur regardless of the direction of cardiac looping, and is thought to represent a condition in which bilateral symmetry occurs in place of left-right asymmetry (Bowers et al, 1996). For example, individuals with complete right isomerism exhibit hearts with two morphologically right atrial appendages, two trilobed lungs, and an absence of the spleen (Figure 3.2b). Conversely, complete left isomerism presents as two morphologically left atrial appendages, two bilobed lungs, and polysplenia (Figure 3.2c). It is not uncommon for only the cardiopulmonary system to be affected in individuals afflicted with isomerism, and this type of condition is prevalent in *lefty-1* (Meno et al, 1998) and *ActRIIb* (Oh and Li, 1997) null mice, which exhibit left and right thoracic isomerism, respectively. In general, the frequency and severity of cardiac defects that accompany isomerism is greater than that of heterotaxia, and virtually always includes atrial septal defects (Bowers et al, 1996). Transposition of the great arteries, single ventricle, anomalous venous pulmonary return, and persisting left superior vena cava are some of the other defects typically found in affected individuals. Mortality rates of 60% and 80% for left and right isomerism, respectively, have been reported (Bowers et al, 1996).

Monozygotic Twinning

Although not classified as a laterality disorder, it has long been recognized that monozygotic twinning is accompanied by a statistically higher frequency of disturbances in left-right development than that predicted for individuals, a phenomenon that occurs almost exclusively in one sibling of the pair (Burn, 1991). Defects detected in monozygotic twins include primarily the same types of cardiopulmonary malformations associated with the laterality disorders described above. In instances of conjoined twinning, the general observation is that the left twin is normal for left-right development and the right twin is not (Burn, 1991). Complete left-right reversal (situs inversus) of the right twin does not always occur, and the heart appears to be the most frequently affected organ, typically exhibiting either dextrocardia or isomerism. Recent progress in our understanding of left-right development has led to hypothetical models of how laterality might become altered during the process of twinning, and experimental approaches to testing these models are presented below (see Left-Right Development in Conjoined Twins).

OVERVIEW OF VERTEBRATE LEFT-RIGHT PATTERNING

It is inferred from the various defects elicited by human laterality disorders that not only is the establishment of left-right asymmetries per se important for normal organ form and function, but also that a consistent alignment of such asymmetries must occur relative to the orthogonal body axes. Thus, with respect to the heart, it is both the generation of asymmetry plus the rightward direction of looping that must occur to establish a seamlessly connected cardiopulmonary system.

Left-right orientation of the heart is critically dependent on the underlying mechanisms that generate asymmetry from the initially bilaterally symmetric vertebrate body plan. Several studies of left-right development in the African clawed frog, *Xenopus laevis*, have identified developmental periods in which left-right patterning is susceptible to experimental manipulation. Although these types of experiments cannot definitively establish when left-right patterning is initiated in the embryo, they do illustrate that perturbations quite early in development can lead to profound effects on laterality once organogenesis begins. For example, disturbances of the ectodermal extracellular matrix in the late blastula to gastrula stages elicit heterotaxia in tadpoles (Yost, 1992), and treating embryos at neurula stages with inhibitors of heparan sulfate proteoglycan synthesis results in a failure of the heart to undergo looping morphogenesis (Yost, 1990). Moreover, disruption of microtubule-dependent events in the first cell cycle can result in complete left-right reversal (situs inversus) of the heart and viscera (Yost, 1991).

Because experimental conditions leading to altered cardiac left-right development typically result in one of four outcomes—heterotaxia, situs inversus, isomerism, or failure to loop—it is thought that such experimental manipulations represent perturbations of specific parts of the left-right developmental pathway. Thus, a failure to establish global left-right asymmetry would be predicted to result in heterotaxia, whereby the heart still undergoes looping, albeit in a stochastic fashion, as occurs in response to most types of experimental treatments that are discussed in the following sections. By contrast, generation of global left-right asymmetry that is in reversed alignment with the other body axes would result in situs inversus, e.g., as observed when perturbations are made during the first cell cycle (Yost, 1991) or early blastula stages in *Xenopus* (Hyatt et al, 1996; Hyatt and Yost, 1998) (see Early Left-Right Patterning Events, below). Whether the processes of generating global left-right asymmetry and alignment of this asymmetry with the dorsoventral and anteroposterior axes occur concomitantly is not known, although it is certain that both processes are necessary to establish the left-right body axis. Experimental outcomes in which the heart fails to loop are probably not related to formation of the left-right axis, but rather are thought to result either from a lack of reception or response to laterality cues by the cardiac primordia, or alternatively, from a failure of the biomechanical process of looping that is intrinsic to the heart tube (see Interpretation of Left-Right Patterning Signals by the Cardiac Primordia). With respect to isomerism, the underlying problem is more obscure, and it could be envisioned that aberrations in any part of the left-right pathway might result in this type of patterning defect.

CONSERVED EVENTS IN LEFT-RIGHT PATTERNING

Asymmetric Gene Expression in the Left Lateral Plate Mesoderm

The common link between formation of the left-right axis and directed, rightward looping of the heart in all vertebrates thus far examined appears to be the conserved, asymmetrical expression of transforming growth factor-β (TGF-β)–related family members by the lateral plate mesoderm (LPM). One such family member is encoded by *nodal*, which is normally expressed within the left LPM, including the posterior-most aspect of the left cardiac field, of mouse (Collignon et al, 1996; Lowe et al, 1996), chick (Levin et al, 1995, 1997; Pagan-Westphal and Tabin, 1998), *Xenopus* (Hyatt et al, 1996; Hyatt and Yost, 1998; Lohr et al, 1997), and zebrafish (Rebagliati et al, 1998; Sampath et al, 1998). Bilateral or absent expression of *nodal* in the LPM is correlatively associated with disturbances in laterality, as indicated by the complex *nodal* expression patterns detected in *inv/inv* (Collignon et al, 1996; Lowe et al, 1996) and *iv/iv* mice (Lowe et al, 1996). Moreover, mice doubly heterozygous for mutations of *nodal* and *HNF3β*, a transcription factor required for normal midline development, exhibit left-right heart reversals (Collignon et al, 1996). In addition to *nodal*, a second TGF-β–related molecule, encoded by *lefty-2* (Meno et al, 1996, 1997), is expressed by the left LPM, and to a much lesser extent, by the left half of the ventral neural tube (Meno et al, 1996). In mouse (Meno et al, 1996), *Xenopus* (J. Essner and H.J. Yost, unpublished), and zebrafish (Bisgrove et al, 1999), the normal spatiotemporal expression of *lefty-2* in the LPM is similar to that of *nodal*, and also as observed for *nodal*, *lefty-2* expression is similarly altered in *inv/inv* and *iv/iv* mice (Meno et al, 1996, 1997). Ectopic, right-sided misexpression of *nodal* in the LPM of *Xenopus* (Sampath et al, 1997) or chick (Levin et al, 1997) results in stochastic heart looping, and misexpression of *lefty-2* (via null mutation of a related gene, *lefty-1*, discussed in Role of the Midline, below) results in cardiopulmonary left isomerism, despite a normal orientation of heart looping (Meno et al, 1998). Moreover, experimental perturbations of early events in left-right axis formation predictably lead to aberrant *nodal* and *lefty-2* expression patterns, as described below.

Whereas direct functional roles for both *nodal* and *lefty-2* in cardiac left-right development are suggested by misexpression studies, it should be noted that neither *nodal* nor *lefty-2* gene expression is detected within the heart tube itself. The connection between LPM *nodal* and *lefty-2* gene expression and directed heart looping appears to be, at least in part, the regulation of downstream genes involved in looping morphogenesis. One such gene that is subject to regulation by *nodal* and *lefty-2* and that is required for heart looping is the *bicoid*-related homeobox gene, *Pitx2*. By contrast to *nodal* and *lefty-2* expression, *Pitx2* expression is detected in the left LPM and as well as during stages of heart tube formation when it is detected exclusively in the left, but not right, side of the developing heart (Logan et al, 1998; Ryan et al, 1998; St Amand et al, 1998; Campione et al, 1999). In mouse (Piedra et al, 1998; Yoshioka et al, 1998; Campione et al, 1999) chick (Logan et al, 1998; St Amand et al, 1998), *Xenopus* (Ryan et al, 1998; Campione et al, 1999), and zebrafish (Campione et al, 1999), asymmetric *Pitx2* expression in the LPM is detected just following the onset of *nodal* and *lefty-2* expression. As observed for *nodal* and *lefty-2*, complex expression patterns of *Pitx2* occur in *iv/iv* mice (Piedra et al, 1998; Ryan et al, 1998; Yoshioka et al, 1998; Campione et al, 1999). Regulation of *Pitx2* expression by *nodal* and *lefty-2* is suggested by experiments in which ectopic, right-sided expression of *nodal* leads to bilateral misexpression of *Pitx2* (Logan et al, 1998; Piedra et al, 1998; Yoshioka et al, 1998). In turn, right-sided misexpression of *Pitx2* in chick (Logan et al, 1998; Ryan et al, 1998) or *Xenopus* (Ryan et al, 1998; Campione et al, 1999) results in cardiac laterality defects, including reversed heart looping and/or cardiac isomerism. These findings, coupled with the additional observation of *Pitx2* expression in the muscular components of other organs that exhibit left-right asymmetry (e.g., gut, lung bud), have led to the suggestion that *Pitx2* is involved in processes of bending and looping that are required for execution of laterality decisions during organogenesis, perhaps through induction of genes that regulate contractile proteins (Logan et al, 1998; Ryan et al, 1998; Campione et al, 1999).

Asymmetric Gene Expression in the Right Lateral Plate Mesoderm

Given the apparent lack of TGF-β–related gene expression in the right LPM, it might be thought that right-sided identities of developing organs occur via a default patterning pathway. However, two lines of evidence suggest otherwise. First, a snail-related zinc finger transcription factor, *cSnR*, is expressed by the right, but not left, LPM, suggesting that right-sided signaling pathways also are operative in establishing left-right asymmetry (Isaac et al, 1997). Similar to *nodal* and *lefty-2*, *cSnR* is expressed also in the posterior aspect of the (right) cardiac field, but not the fused heart tube itself. Nevertheless, treatment of chick embryos with *cSnR* antisense oligonucleotides at stages preceding formation of the heart tube results in stochastic heart looping (Isaac et al, 1997). Intriguingly, the expression of *cSnR* and *nodal* appears to be mutually exclusive. Signals that inhibit *nodal* expression in chick have been found to induce *cSnR* expression, and conversely, those signals that induce *nodal* have been found to inhibit *cSnR* (Isaac et al, 1997). Although it has yet to be examined, one prediction of these findings is that *cSnR* and *Pitx2* (which is downstream of *nodal*) also might have mutually exclusive roles in left-right development.

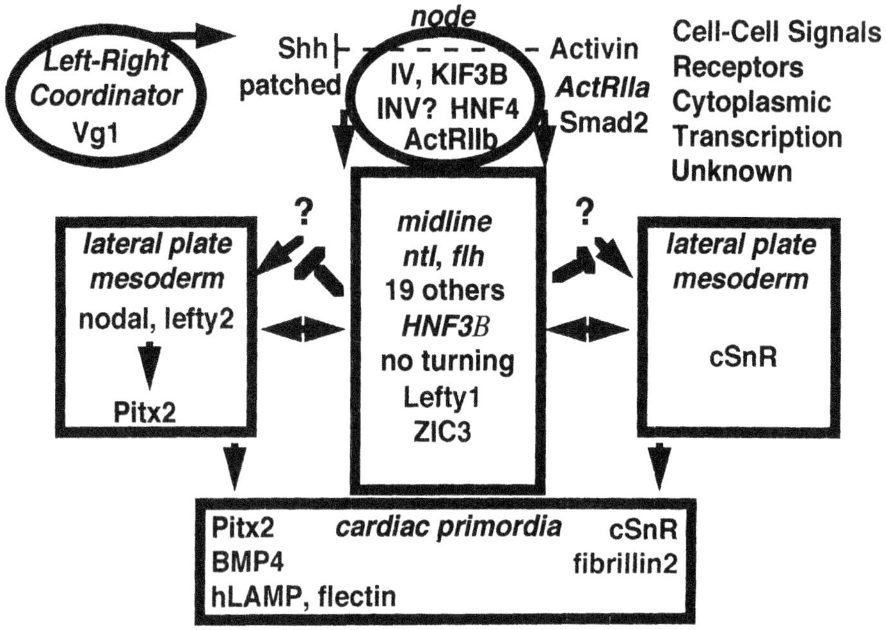

FIGURE 3.3. Summary of inductive interactions in the left-right developmental pathway. Left-right coordinator activity induces left-right identity in tissues lateral to the organizer/node. By an undefined mechanism, these tissues in turn elicit a cascade of asymmetrical gene expression in or near the organizer/node. Once specified for left-right identity, the node signals back to its surrounding tissues to elicit asymmetrical gene expression in paraxial and lateral plate mesoderm. Following regression of the organizer/node, the notochord, and floor plate function to maintain asymmetry of gene expression to either side of the midline and to prevent diffusion of molecules from one side of the embryo to the other. Transmission of left-right patterning cues from the lateral plate mesoderm results in asymmetrical gene expression by the heart primordia. Genes that are expressed at stages coincident with these inductive processes include cell–cell signaling molecules, cell surface receptors, transcription factors, cytoplasmic proteins, or unknown factors. Arrows and bars represent positive and negative regulatory interactions between genes, but do not necessarily imply direct interactions. Unidentified signals are indicated by "?". See text for details and references. (Modified from Ramsdell and Yost, 1998.)

In concurrence with the discovery of restricted *cSnR* expression in the right LPM, a more recent study provides additional evidence of right-sided signaling activity during left-right development. In a series of grafting experiments in *Xenopus* embryos (Lohr et al, 1998), it was found that replacing either left or right LPM tissue of a host embryo with LPM tissue derived from the contralateral side of a donor embryo results in normal, unilateral left-sided *nodal* expression in the host. The ability of the host to suppress *nodal* expression in right-sided grafts of LPM tissue derived from the left side of the donor indicates that a right-sided signaling pathway in the host can repattern the grafted tissue to acquire a normal, right-sided identity. Undoubtedly, there are many other genes expressed by the LPM that are requisitely involved in left-right development, and it is anticipated that future endeavors to identify these other players will reveal members of both left- and right-sided signaling pathways. Those already known or implicated to function in the left-right pathway, including genes discussed in the following sections, are summarized in Figure 3.3.

INTERPRETATION OF LEFT-RIGHT PATTERNING SIGNALS BY THE CARDIAC PRIMORDIA

Perturbations of Cardiac Anteroposterior Patterning Also Result in Altered Heart Looping

As focus in the field of left-right development has expanded to include consideration of how the cardiac primordia receive and respond to left-right patterning cues, studies hint at the emerging concept that anteroposterior patterning of the primitive segments and their subsequent execution of left-right morphogenesis might be intimately linked during formation and looping of the primary heart tube. Identification of the the helix-loop-helix transcription factors *dHAND* and *eHAND* is one such example (Srivastava et al, 1995). Based on initial observations that *eHAND* is expressed in the cardiac fields and later in an asymmetric pattern in the looped heart tube in mouse, it was postulated that restricted *eHAND* expression might result from regulation by left-right patterning signals (Biben and Harvey, 1997). However, a subsequent study of *HAND* genes has shown that the restricted expression of *eHAND* in the left ventricle, and of *dHAND* in the right ventricle, is regulated by anteroposterior patterning of the heart tube and is indicative of a role for these genes in chamber specification (Thomas et al, 1998). Thus, the restricted, asymmetric expression of the *HAND* genes appears to be a consequence of alignment of the ventricular chambers following cardiac looping, but not from the direct influence of left-right patterning per se. Yet despite the clear role of anteroposterior patterning in regulating *HAND* gene expression, *HAND* genes also are required for the execution of cardiac left-right morphogenesis. Treatment of cultured chick embryos with *HAND* antisense oligonucleotides (Srivastava et al, 1995) or null mutations of *dHAND* in mouse (Srivastava et al, 1997) result in hearts that appear to initiate normal rightward looping prior to a complete arrest of the looping process. In addition to the example provided by the *HAND* genes, experimental perturbations of anteroposterior patterning by retinoic acid treatment also can result in perturbations of cardiac left-right asymmetry (Smith, 1997; Smith et al, 1997; Yasui et al, 1998), raising the possibility that other (retinoid-responsive) cardiac genes (e.g., *GATA-4*, *Nkx2.5*) might serve multiple, overlapping functional roles in heart development (Lyons et al, 1995; Kostetskii et al, 1999).

Formation of the Primary Heart Tube Involves Genes That Are Implicated in Cardiac Left-Right Development

If anteroposterior patterning and left-right morphogenesis are driven, at least to some extent, by genes that can participate in both pathways, the question then becomes at what point in heart development do these two pathways intersect? Analyses of other genes that are asymmetrically expressed by the cardiac primordia suggest that such interaction could occur in the bilateral heart-forming fields. The TGF-β–related genes, *BMP-2* and *-4*, are expressed within the paired heart fields (Chen et al, 1997; Schultheiss et al, 1997; Andree et al, 1998) and have been shown to play an instructive role in recruiting specified cardiac anterior LPM cells into the primary heart tube (Schultheiss et al, 1997; Andree et al, 1998). The ability of BMPs to regulate contribution of cardiogenic mesoderm to the heart is mediated, in part, through induction of *GATA-4* and *Nkx2.5* (Schultheiss et al, 1997; Andree et al, 1998); the latter is required for induction of *HAND* gene expression (Biben and Harvey, 1997). *BMP-4* expression moreover is implicated in generating cardiac left-right asymmetry. Expression of *BMP-4* in the zebrafish heart is initially bilaterally symmetric in the cardiac fields; however, just prior to the onset of looping of the nascent heart tube, *BMP-4* expression becomes greatly enhanced on the left side, with stronger expression observed on the left side of the heart tube than the right (Chen et al, 1997). Inhibition of BMP signaling by expression of a dominant-negative *BMP-2/-4* receptor disrupts looping morphogenesis, and ectopically induced overexpression of *BMP-4* leads to reversed heart looping (Chen et al, 1997). Consistent with these findings, targeted disruption of *ALK5* expression (another member of the TGF-β transmembrane receptor family) in cardiac cells of the mouse also inhibits cardiac looping (Charng et al, 1998). Other genes that are aymmetrically expressed within the cardiac fields encode extracellular matrix proteins, including fibrillin-2 (Smith et al, 1997; Rongish et al, 1998) on the right, and *hLAMP1* (Smith et al, 1997) and flectin (Tsuda et al, 1998) on the left. Expression of each of these genes persists in the fusing heart tube, and *hLAMP-1* (Sinning et al, 1992, 1995) and fibrillin-2 (Wunsch et al, 1994) proteins are thought to be involved in the later induction of endocardial cushion tissues that are formed in a restricted anteroposterior pattern in the looped, tubular heart. Whether *hLAMP-1* and fibrillin-2 also are functionally involved in execution of left-right patterning remains to be determined; however, asymmetric *hLAMP-1* and fibrillin-2 expression by the opposite heart field in chick embryos can be elicited by right-sided application of retinoic acid, a treatment that also results in left-right heart reversals (Smith et al, 1997).

Specification of Cardiac Left-Right Asymmetry and the Mechanics of Heart Looping

The identification of asymmetric gene expression in the cardiac primordia provides molecular confirmation of much earlier studies that demonstrated that inherent differences exist between cells derived from the left and right cardiac fields. Cardiomyocyte differentiation, as assessed by beating (Patten and Kramer, 1933) and striated myofibril formation (Lindner, 1960), appears to occur slightly earlier on the right, and the right cardiogenic fold also appears somewhat more advanced than the left when examined at early stages in chick (Stalsberg and DeHaan, 1969). When cardia bifida is elicited by preventing fusion of the cardiac fields, two hearts form that show left-right differences: the anterior portion of the heart tube is larger in the right heart structure and the posterior portion is larger in the left heart structure (Van Praagh and DeHaan, 1967). How the morphologic and molecular asymmetry of the heart fields is related to the biomechanics of looping is still poorly understood (see Taber et al, 1995 for review). Systematic examination of mitotic indices in anterior and posterior regions of the chick heart has failed to support a role for left-right asymmetric growth of the heart tube in promoting looping (Stahlsberg, 1969b). The ability of explanted precardiac mesoderm to undergo looping (Stahlsberg, 1970; Danos and Yost, 1996) is inconsistent with the idea that looping is driven by physical constraints imposed by surrounding tissues. Differential cell shape changes in the inner and outer curvatures might be envisioned to play a role, and inhibition of actin polymerization in chick embryos results in arrested cardiac looping (Itasaki et al, 1991). As discussed below, asymmetric contributions of cells to the primitive segments is another possibility.

When explants of chick (Stalsberg, 1970) or *Xenopus* (Yost, 1990; Danos and Yost, 1996) precardiac mesoderm and its associated ectoderm are cultured in isolation, these tissues undergo formation of a tube-shaped structure that somewhat resembles the primary heart tube and that undergoes looping morphogenesis. Depending on the developmental stage of the embryo from which the tissues are harvested, the direction of looping is either stochastic or normal. In *Xenopus*, the ability of explants to consistently loop in a normal, rightward direction increases as the age of the donor embryo reaches late neurulation stages (Danos and Yost, 1996). These results suggest not only that heart looping proceeds when early precardiac tissue is separated from axial information provided by the embryo, but also that left-right patterning information from the body axis becomes transmitted to the cardiac progenitors by the end of neurulation. It should be emphasized that at the stages when left-right specification of cardiac cells is occurring in vivo, the primary heart tube has not yet initiated its formation, nor are extracardiac mesenchyme cells that normally populate the heart present at these stages. Thus, as cells in the cardiac fields give rise to the primitive segments and become incorporated into the fusing heart tube, it appears that these cells are already specified for left-right identities.

Examination of the relative contributions of the cardiac fields to the chick heart suggests that the anterior and posterior primitive segments do vary in their composition of cells that were derived from each field (Stahlsberg, 1969a). Specifically, a greater proportion of cells derived from the right heart field is found to incorporate into the heart tube at stages when primitive ventricular segments are being added, and a lesser proportion is found to join in the heart tube at stages encompassing incorporation of the AV canal and sinoatrial segments (Stalsberg, 1969a; Markwald et al, 1998). In this context, it is notable that progenitors to the posterior-most segments of the primary heart tube (i.e., AV canal, sinoatrium) are probably the same subset of cardiac cells that previously expressed *nodal*, *lefty-2*, and *cSnR* as residents of the anterior LPM expression domains. Because the AV canal and sinoatrial segments differ from the primitive ventricular segments in that they simultaneously contribute to both left- and right-sided structures in the mature heart (i.e., left and right inlets), the early asymmetric gene expression by cells that eventually contribute to the posterior primitive segments may be quite significant with respect to their later establishment of distinct left-right identities, in addition to the overall process of directed heart looping.

EARLY LEFT-RIGHT PATTERNING EVENTS

Origin of the Left-Right Axis

The One-Cell Stage

By comparison to other aspects of the left-right developmental pathway, much less is known about the earliest mechanisms involved in establishing initial left-right asymmetry and whether these processes are evolutionarily conserved among vertebrates. With respect to formation of the dorsoventral axis and the overall vertebrate body plan, much information has been elucidated using *Xenopus* as a model

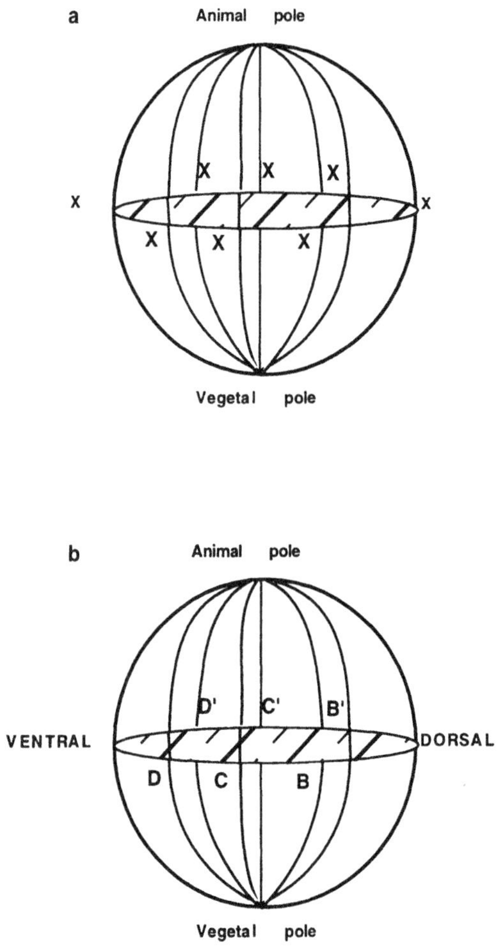

FIGURE 3.4. Establishing bilateral symmetry and left-right asymmetry in *Xenopus*. (a) Before fertilization, each line between the animal (anterior) pole and the vegetal (posterior) pole of the cylindrically symmetric *Xenopus* egg has equivalent developmental potential (indicated by "X's"). (b) After fertilization the dorsoventral midline is determined by cortical rotation of the cytoplasm. The dorsoventral midline divides the embryo into left and right halves that are both symmetric (C and C′ are equidistant from the midline) and asymmetric (C is not equal to C′). How the process of cortical rotation concomitantly generates symmetry and left-right asymmetry is not known. It is possible that a preexisting asymmetry is brought into alignment with the midline as it is established. (Modified from Yost, 1995.)

system. In *Xenopus*, the unfertilized egg is cylindrically symmetric, and the position of the dorsal midline results from a microtubule-dependent cortical rotation within the first cell cycle (Gerhart et al, 1989; Moon and Kimelman, 1998). It is important to note that prior to fertilization and cortical rotation, each point along the presumptive dorsoventral midline has equivalent developmental potential (reviewed by Yost, 1995) (Figure 3.4a). Thus, the establishment of the dorsoventral midline would be expected to result in an embryo with bilaterally symmetric left and right halves (Figure 3.4b). However, experimental perturbations of the one-cell *Xenopus* embryo suggest otherwise.

Cortical rotation can be inhibited in *Xenopus* by ultraviolet (UV) irradiation, which in turn results in decreased dorsoanterior development (Gerhart et al, 1989; Moon and Kimelman, 1998). Following this treatment, left-right heart reversals also are observed, and the frequency of reversed heart looping directly correlates (increases) with the severity of dorsoventral patterning defects (Danos and Yost, 1995). Cortical rotation, and hence normal dorsoventral axis formation, can be restored to UV-irradiated embryos by manually tilting embryos prior to the first cell cycle (Gerhart et al, 1989). However, a significant percentage of such dorsoventral "rescued" embryos typically exhibit subsequent inversion of left-right development (situs inversus), suggesting that the microtubule-dependent mechanism of cortical rotation also may be involved in establishing the left-right axis (Yost, 1991). How the process of cortical rotation might function in breaking molecular symmetry in the one-cell embryo is not known; with respect to the dorsoventral midline, it has been suggested that cortical rotation results in an uneven distribution of cytoplasmic determinants that are required for initiation of the dorsoventral axis (Moon and Kimelman, 1998). The observation that ectopic *Xwnt-8* expression both disrupts dorsoanterior development and elicits left-right heart reversals in *Xenopus* might provide entry into a molecular pathway putatively linking these two events (Danos and Yost, 1995).

Blastula and Gastrula Stages

Both the generation of left-right asymmetry and the concordant alignment of asymmetry relative to the other body axes are susceptible to experimental perturbations during blastula and gastrula stages of development. A role for gap junctional communication in the latter process is implicated by studies in *Xenopus.* In the 16- to 64-cell-stage blastula, gap junctional communication is more active in dorsal cells than in ventral cells (Guthrie, 1984; Guthrie and Gilula, 1989). Introduction of ectopic wild-type connexins into ventral cells of the animal (anterior) pole results in altered *nodal* expression by the LPM and in stochastic heart looping (Levin and Mercola, 1998c). Inhibition of gap junction activity in dorsal animal pole cells prior to neural fold stages results in similar effects on heart looping (Levin and Mercola, 1998c). Interestingly, modification of gap junctional communication in either dorsal or ventral cells is effective in eliciting heart reversals only when cells to both sides of the midline are targeted, suggesting that it is the continuity of gap junctional activity in dorsal cells that is required for normal left-right development. Moreover, when synthetic RNA encoding a serine-to-proline mutation at residue 364 of the putative regulatory domain of connexin43 is used to target either dorsal or ventral animal cells of the *Xenopus* blastula, reversed heart looping is observed in either case, suggesting that this mutation can result in both antimorphic and hypomorphic activities (Levin and Mercola, 1998c). The identity of the small molecules that pass through gap junction channels to contribute to left-right patterning is presently unknown, as is the exact timing of their activity, which appears to be sometime between late blastula stages and gastrulation (Levin and Mercola, 1998c).

Dorsoventral differences in cells of the *Xenopus* blastula also are observed in the context of the other mechanism required for initiation of the left-right axis, i.e., the establishment of early molecular asymmetry. A role for a member of the TGF-β family of secreted growth factors, Vg1, in this process is suggested by a series of ectopic expression experiments. Injection of RNA encoding the mature form of Vg1 results in stochastic heart reversals when targeted to dorsal right vegetal cells of the 16- to 32-cell-stage blastula (Hyatt et al, 1996; Hyatt and Yost, 1998). However, if the same RNA is targeted to ventral right vegetal cell lineages, then complete inversion of the left-right body axis (situs inversus) and right-sided *nodal* misexpression is observed (Hyatt and Yost, 1998). Thus far, this is the only experimental perturbation to be reported that can predictably elicit either randomization or full inversion of the left-right axis. A role for Vg1 in early steps of left-right axis formation in vivo is additionally supported by the finding that left-sided expression of a dominant-negative TGF-β receptor that is capable of blocking Vg1 activity also results in left-right heart reversals (Hyatt et al, 1996; Hyatt and Yost, 1998). The left-right differences in responsiveness of the early blastula to ectopic Vg1 or the dominant-negative receptor moreover suggest that Vg1 signaling is normally active on the left, but not right, side of the early embryo. Because Vg1 precursor protein is equivalently expressed by left and right vegetal regions of the *Xenopus* blastula, it is presently unclear how Vg1 signaling activity becomes spatially restricted to left cell lineages. By analogy to regulation of TGF-β activity in other developmental systems, it is possible that asymmetric protein processing (Hyatt et al, 1996) or antagonistic, right-sided signals (Hyatt and Yost, 1998) might be involved in promoting this asymmetry.

The Left-Right Coordinator Model

The results obtained in the ectopic expression studies with mature Vg1 have been interpreted to mean that left lateral vegetal cells compose a "left-right coordinator" that functions to establish a global left-right axis in the early embryo (Hyatt and Yost, 1998). Thus, the left-right coordinator is thought to confer a left identity to cells in which Vg1 signaling is activated, and cells of the right vegetal region, which normally are not exposed to Vg1, adopt a right identity. In turn, cell lineages specified for left-right identities are envisioned to transmit left-right positional information to their progeny as well as other tissues, including the Spemann organizer. In agreement with this model, experimental evidence in chick indicates that the exchange of left-right patterning information between the node (functionally analogous to the organizer) and tissues extraneous to the node plays an essential role in establishing normal left-right asymmetry (Pagan-Westphal and Tabin, 1998). Surgical rotation of the node in the stage 4 chick results in embryos with normal expression of *shh* to the left of the node; however, rotating the node at late stage 5 or stage 6 results in embryos with misexpression of *shh* to the right of the node. These results indicate that in chick, the node becomes specified for left-right positional identity between stages 5 and 6. Heterochronic transplants of rotated older staged nodes into younger staged hosts (and vice versa) have demonstrated that left-right identity is not inherent to the node, but rather results from specific interactions with surrounding tissues. Once specified, the left-right positional information carrried by the node is dominant to that of surrounding tissues, and the node signals back to adjacent tissues to regulate asymmetric gene expression such as *nodal* by the LPM (Pagan-Westphal and Tabin, 1998). Together, these results and those of the Vg1 experiments described above suggest that left-right axis formation may be complete by gastrulation, and that the organizer/node plays a pivotal role in continuing transmission of global left-right patterning cues initially set forth by the left-right coordinator.

ROLE OF THE MIDLINE

Organizer/Node

Cloning of the iv *and* inv *Mutations*

Another gap in our understanding of the left-right developmental pathway concerns the mechanism(s) by which the organizer/node interprets left-right specification signals from its surrounding tissues, and the recent cloning of the mouse mutations *iv* and *inv* is anticipated to provide insight into this issue. The *iv* gene is allelic to *legless* (McNeish et al, 1990) and encodes a new member of the axonemal dynein family, termed *left-right dynein* (*lrd*) (Supp et al, 1997). By Northern analysis, *lrd* is detected as early as embryonic day 3.5 and localized expression, determined by in situ hybridization, is reported to occur first at embryonic day 7.5, specifically in ventral cells of the node (Supp et al, 1997). Thereafter, *lrd* expression persists into adult stages and is detected in several tissues, including ciliated epithelia. The *inv* gene encodes a novel protein, termed inversin (Mochizuki et al, 1998; Morgan et al, 1998). Sequence analysis reveals multiple ankyrin-like repeats and potential nuclear localization signals, but no evidence of transmembrane or signal peptide signals, suggesting that inversin, like lrd, is an intracellular protein. Limited in situ hybridization studies suggest a broad domain of inversin expression in the embryonic day 7 mouse, and Northern analysis similarly indicates that inversin is expressed by a wide variety of tissues and organs, including those that do and do not exhibit left-right asymmetries (Mochizuki et al, 1998). Evidence supporting the generation of three possible isoforms by alternative splicing also has been presented (Morgan et al, 1998).

Initial morphologic characterization of *iv/iv* homozygotes suggested an equal mix of situs inversus and situs solitus phenotypes (Hummel and Chapman, 1959); however, subsequent and more detailed analyses additionally have detected two other types of laterality defects, cardiopulmonary isomerism and isolated dextrocardia, in homozygous populations (Layton and Manasek, 1980; Seo et al, 1992). Given the complexity of the *iv/iv* phenotypes, the *iv* mutation theoretically could represent either a failure of transmission of LR patterning cues by tissues extraneous to the node, or alternatively, a failure of the node to respond to this induction. Similarly, the situs inversus phenotype observed in nearly all *inv/inv* homozygotes (Yokoyama et al, 1993) could result from aberrations in either process, such as transmission of reversed laterality cues to the node by its surrounding tissues, or an incorrect (inverted) response by the organizer/node to normal, left-right asymmetric signals.

One of the striking (and perhaps unexpected) findings revealed by the cloning of *lrd* and inversin is that, similar to Vg1 expression in the early blastula, none of these genes appears to show an obvious left-right expression pattern, underscoring the idea that genes critical to the left-right developmental pathway themselves need not be asymmetrically expressed. Prior to the advent of molecular characterization of the left-right pathway, Brown and Wolpert (1990) proposed a model in which orientation of a ubiquitously expressed "handed," or chiral, molecule might result in formation of a left-right directional morphogen gradient. Whether *lrd* and inversin function in this type of mechanism remains to be determined. Dynein proteins are implicated in a variety of cellular processes, including ciliary beating and microtubule-driven vesicular transport (Beckerle and Porter, 1982; Hirokawa, 1998). The association of immotile cilia and lack of ciliary dynein arms in individuals afflicted with a particular type of situs inversus, Kartagener's syndrome, has long predicted a link among ciliary motion, dynein protein function, and laterality (Afzelius, 1976). A recent connection betwen heterotaxia and lack of ciliary function specifically at the node is drawn from observations made in the KIF3B null mouse, in which expression of the microtubule-dependent motor protein, KIF3B, is disrupted (Nonaka et al, 1998). Moreover, null mutation of the winged helix transcription factor hepatocyte nuclear factor/forkhead homolog (HFH-4) results in mice showing stochastic heart reversals, absence of cilia, and absence of *lrd* expression (Chen et al, 1998). For a detailed account of how molecular motors and genes such as *lrd*, inversin, and KIF3B might integrate into the chirality model of left-right development, the reader is directed to several reviews of left-right asymmetry (Levin and Nascone, 1997; Levin and Mercola, 1998a; Supp et al, 1998).

The Sonic Hedgehog/Activin Pathway

Whereas much remains to be elucidated on the subject of left-right specification of the organizer/node, significant progress has been made in identifying some of the signaling molecules that are involved in the communication of left-right patterning information from the specified node to its surrounding tissues. In the chick, a signaling cascade involving activin and sonic hedgehog (shh) has been identified based on asymmetric expression patterns and ectopic expression experiments. This signaling cascade is initiated by activin, which is expressed to the right of the node at stages 3 to 5 of chick development (Levin et al, 1997). Right-sided activin expression is postulated to induce the right-sided expression of *ActRIIa* (a TGF-β receptor family member), while repressing right-, but not left-sided, expression of *shh*. The left-sided expression of *shh*, in turn, results in induction of *nodal* expression just to the left of the node (Levin et al, 1995). Shortly thereafter, *nodal* expression expands into the left lateral plate mesoderm. Ectopic misactivation of the activin or shh signaling pathways in the early-stage chick results in stochastic heart looping and altered *nodal* expression by lateral plate mesoderm (Levin et al, 1995, 1997), as does inhibition of endogenous activin or shh signaling by follistatin (Levin et al, 1997), or antibodies directed against shh or patched (a shh receptor) (Pagan-Westphal and Tabin, 1998). It should be noted that unidentified, secondary signal(s) also are thought to be required in this signaling cascade, because ectopic shh protein induces *nodal* expression in explanted lateral plate mesoderm only if paraxial tissue is included in the culture (Pagan-Westphal and Tabin, 1998). Other genes that are asymmetrically expressed in or around the node in chick include *FGF8* (Boettger et al, 1999) and *Wnt8* (Levin et al, 1995). It has not been tested whether *Wnt8* functions in the left-right pathway; however, *FGF8* expression can be induced by activin, suggesting that *FGF8* might participate in the right-sided signaling pathway that regulates expression of *cSnR* during early stages in chick. Consistent with this proposed role, ectopic expression of *FGF8* to the left of the node in chick results in left-right heart reversals, inhibition of *nodal* expression, and left-sided misexpression of *cSnR* (Boettger et al, 1999).

While it is clear that the activin/shh pathway is required for normal left-sided expression of *nodal* in chick, it is less certain that this pathway is operative in other vertebrates. Whereas null mutations in *ActRIIb* (Oh and Li, 1997) or double heterozygous mutations of *nodal* and *Smad2* (a downstream signal transducer of activin signaling) result in complex, cardiac laterality defects in mouse (Nomura and Li, 1998), null mutations in activin (Matzuk et al, 1995b) or *ActRIIa* (Matzuk et al, 1995a) do not result in altered left-right development, nor does inhibition of activin signaling in *Xenopus* affect laterality (Hyatt et al, 1996; Lohr et al, 1998). Asymmetric expression of *shh* is not detected in mouse (Collignon et al, 1996), *Xenopus* (Lohr et al, 1998), or zebrafish (Schauerte et al, 1998), and null mutation of *shh* does not perturb heart looping in the mouse (Chiang et al, 1996) or zebrafish (Schauerte et al, 1998). From these findings, it is concluded that factors other than, but related to, activin and *shh* might be involved in transmission of patterning cues by the organizer/node, or alternatively, that in the context of left-right development, these upstream signaling pathways simply are not functionally conserved among diverse vertebrate species.

Other Midline Influences

Morphologic and Genetic Studies

The importance of the organizer/node in directing left-right asymmetries is recapitulated at later stages of development when cells of the organizer/node contribute to formation of midline structures such as the notochord and floorplate. Mutant zebrafish lines originally classified based on defects of the notochord, floorplate, or tail, show associated left-right heart reversals and altered *BMP-4* expression patterns in many, but not all, lines examined (Danos and Yost, 1996; Chen et al, 1997). Other studies pointing to a role for the midline in left-right development include the mouse mutations *no turning* and *brachyury*. Mice homozygous for the insertional mutation *no turning* exhibit abnormal embryonic rotation, stochastic heart reversals, and bilateral expression of *nodal* and *lefty-2* in the LPM (Melloy et al, 1998). Null mutations of *brachyury* in mouse (King et al, 1998) and zebrafish (Schulte-Merker et al, 1994; Danos and Yost, 1996) also result in stochastic heart looping, although *nodal* and *lefty-2* expression is not detected in the LPM of homozygous mice (King et al, 1998). Differences in LPM gene expression resulting from these two mutations is likely due to differences in notochord defects; *no turning* mutants show some morphologic and molecular evidence of notochord formation prior to its degeneration, whereas the near complete lack of notochord in *brachyury* mutants is thought to be preceded by, and due to, abnormal node development. Consistent with genetic evidence supporting a role for the midline in left-right development, midline extirpations in *Xenopus* result in cardiac reversals and bilateral *nodal* misexpression if performed prior to closure of the neural tube (Danos and Yost, 1996), in agreement with the earlier described developmental period of cardiac left-right specification.

The Midline Barrier

The altered *nodal* and *lefty-2* expression that accompanies midline perturbations described above suggests that one important function of the midline is to maintain asymmetric gene expression by the LPM. Exactly how the midline functions to maintain LPM gene expression is not clear, and several models of midline regulation are possible. Common to the various models of midline regulation is that the midline is envisioned to serve as a barrier, allowing inductive signaling interactions to occur on one side of the embryo while preventing such activity from occurring on the contralateral side. Whether this results from active signaling by the midline itself or from preventing diffusion of inductive signaling molecules expressed by tissues other than the midline is not known. With respect to the first model, signals emitted from the midline to regulate asymmetric gene expression could themselves be expressed in either a symmetric or asymmetric pattern, and implicit in the second model is that inductive signals emitted by non-midline tissues are asymmetrically expressed relative to the midline. The current limited knowledge of genes expressed by the midline and its adjacent tissues that directly function in left-right development does not exclude, nor unequivocally support, any of these possibilities.

Maintenance of Asymmetric Nodal *and* Lefty-2 *Expression*

Analyses of the role of the notochord in modulating *nodal* expression by the LPM demonstrates the difficulty of experimentally distinguishing between models of midline regulation. When LPM explants derived from chick (Levin and Mercola, 1998b) or *Xenopus* (Levin and Mercola, 1998b; Lohr et al, 1998) embryos are cultured in isolation, *nodal* expression is detected, regardless of whether the explants are harvested from the left or the right side of the embryo. Moreover, cultured chick LPM explants contain regenerated node tissue that expresses *shh*, a signal previously shown to positively regulate *nodal* expression in the chick embryo (Levin et al, 1995; Levin and Mercola, 1998b). Importantly, the expression pattern of *shh* is symmetric in these explants, by contrast to its normal, left-sided expression pattern observed in the chick. One interpretation of these findings is that the midline plays an inductive role for left-sided LPM *nodal* expression, and that in the intact embryo this inductive signal is ordinarily restricted to the left of the midline (Levin and Mercola, 1998b). Thus, given the failure of regenerated node cells to form a normally left-right patterned node structure (with asymmetric *shh* expression), the *nodal* expression detected in right-sided LPM chick explants could be due to *shh*-mediated induction of *nodal*. Evidence of regenerated node cells present in LPM explants derived from *Xenopus* embryos also has been reported (Levin and Mercola, 1998b).

The model of positive-acting, inductive signals emitted by the midline, however, does not appear to be supported by findings made in *Xenopus*, which instead suggest that the notochord serves a repressive role in maintenance of *nodal* asymmetry. When LPM explants derived from either side of the *Xenopus* embryo are cocultured with the notochord, a loss of *nodal* expression in the LPM explants is observed (Lohr et al, 1998). Thus, an alternative interpretation is that the midline might normally prohibit *nodal* expression in adjacent LPM tissues, and that the normal, left-sided expression of *nodal* by the LPM results from an overriding inductive signal that is emitted from tissue(s) other than the midline. The identity of this inductive signal in *Xenopus* is not known, and it is possible that either expression of the inductive signal itself or the repressive activity of the midline exhibits left-right asymmetry. Injection of activin protein into the right LPM of the intact *Xenopus* embryo results in stochastic heart looping as well as right-sided *nodal* expression, suggesting that an activin-like signal is involved in induction of *nodal* (Lohr et al, 1998). However, the inductive signal in *Xenopus* does not appear to be either activin or shh because injection of follistatin or shh protein into the LPM regions of intact embryos does not elicit heart reversals (Lohr et al, 1998). Thus, reconciling the two differing models presented for chick and *Xenopus* is not possible given the current data, and resolution likely will result from answering the earlier described question of whether the upstream activin/shh pathway is involved in left-right development of vertebrates other than chick.

Other evidence demonstrating that the midline is involved in regulating LPM gene expression is derived from studies of yet another TGF-β family member, *lefty-1*, which is most related to *lefty-2* and is expressed by the left half of the ventral neural tube, and to a much lesser extent by the left LPM in mouse (Meno et al, 1996, 1997). *Xenopus* and zebrafish *leftys* also have been recently cloned, and *lefty* expression is detected in midline structures, including the prechordal plate, hypochord, and floor plate (Bisgrove et al, 1999; J.J. Essner and H.J. Yost, unpublished). The most consistent phenotype observed in *lefty-1* null mice is left cardiopulmonary isomerism, which is observed in a majority of animals despite the occurrence of normal, rightward looping of the heart (Meno et al, 1998). This phenotype is accompanied by bilateral misexpression of *lefty-2*, *nodal*, and *Pitx2* by the LPM, suggesting that the normal role of *lefty-1* is to inhibit right-sided expression of these genes, possibly via supression of the yet to be identified paraxial signal that lies upstream of *nodal* in *Xenopus* (Lohr et al, 1998) and that is required in combination with *shh* in chick (Pagan-Westphal and Tabin, 1998). Consistent with this model, ectopic expression of *lefty-1* to the left side of the node in chick inhibits *nodal* and *Pitx2* expression, whereas in similar experiments it induces left-sided misexpression of *cSnR* (Meno et al, 1998). Interestingly, the *lefty-1* null phenotype is strikingly opposite that of *ActRIIb* null mice, which predominantly exhibit right thoracic isomerism (Oh and Li, 1997), raising the possibility that this TGF-β receptor might function in transmitting either *lefty-2* or *nodal* signals.

Both the nonoverlapping expression patterns of *lefty* genes at earlier stages of development (Bisgrove et al, 1999) and the recent identification of the upstream regulatory elements of mouse *lefty-1* and *lefty-2* (Saijoh et al, 1999) indicate that the left-sided, asymmetric expression of these genes results from distinct mechanisms of transcriptional control. The *lefty-1* promoter contains a silencing element that is activated on the right side of the embryo, and conversely, the *lefty-2* promoter contains a positive-acting regulatory element that is activated on the left side. However, as cautioned by Saijoh et al (1999), these results should not necessarily be interpreted as evidence for asymmetrically expressed upstream signals that directly induce *lefty-1* and *-2* on the left and inhibit their expression on the right, because several other scenarios of upstream regulation could be envisioned that would be equally compatible with their results.

Clinical Observations

The ultimate utility of conclusions drawn from animal models of development is the provision of a framework in which to explore underlying causes of human congenital disease. The discovery of the midline influence in regulating left-right development in species ranging from mouse to fish has prompted investigators to test for clinical correlations between midline defects and laterality disturbances. In a study performed with a database of 90 individuals identified for left-right disorders at Massachusetts General Hospital, 28 (31%) also were found to exhibit some type of congenital midline defect, including spina bifida, severe scoliosis, and vertebral fusion (Goldstein et al, 1998). Laterality disorders noted in such individuals included isomerism, heterotaxia, and *situs inversus*, but no obvious correlation was observed between specific types of left-right defects and midline defects, a finding that is not surprising given the large and growing number of genes that are implicated in regulating midline formation and left-right development.

A second study relating the midline to human laterality disease is the identification of frameshift, missense, and nonsense mutations in *ZIC3* in individuals afflicted by X-linked heterotaxia (Gebbia et al, 1997). *ZIC3* is a zinc-finger transcription factor that is expressed in the primitive streak and in neural tissues including the spinal cord in mouse embryos (Nagai et al, 1997). Female individuals heterozygous for *ZIC3* mutations were found to exhibit either normal *situs* or *situs inversus*, and sons of females heterozygous for *ZIC3* mutations exhibited heterotaxia. As noted by the authors of this study, while the incidence of *situs inversus* itself poses no obvious adverse effects to afflicted individuals, it does appear, however, to increase the risk of occurrence of heterotaxia in male progeny of female heterozygous for *ZIC3* mutations, a finding that may be relevant to the earlier described question of why invariant left-right *situs* is evolutionarily conserved. Interestingly, in the same study, a daughter of parents showing normal *situs* and *ZIC3* expression was afflicted with heterotaxia, and upon further examination this individual was found to carry a point mutation in the prodomain of the human homolog of *nodal.*

Finally, it should be mentioned that other clinical studies suggest a possible role for mutations of connexin43 in human laterality disease. Initial studies indicated an association of heterotaxia with a point mutation in the putative regulatory domain in connexin43 (Britz-Cunningham et al, 1995), the same mutation that was tested in *Xenopus* left-right development described above (Levin and Mercola, 1998c). However, it is unclear to what extent mutations in connexin43 play a role in human laterality disorders because follow-up studies based on much larger populations of individuals afflicted with left-right disorders have not revealed this correlation (Splitt et al, 1995; Gebbia et al, 1996).

LEFT-RIGHT DEVELOPMENT IN CONJOINED TWINS

For those who study left-right development, conjoined twins present a fascinating and challenging opportunity in which to test some of the predictions set forth by models of left-right signaling that are described in the previous sections. The experimental ease of generating conjoined twins in *Xenopus*, coupled with the relatively high incidence of sporadic twinning in chick, has led to recent use of these two vertebrates in addressing issues of laterality that are unique to twins. Conjoined twins can be induced in *Xenopus* blastulae by ventral grafts of dorsal organizer tissue or by ventral ectopic expression of molecules implicated in normal dorsoventral axis formation, such as *Xwnt8* (Hyatt et al, 1996; Nascone and Mercola, 1997). These types of manipulations result in formation of a secondary axis containing head and thoracic structures (including a heart), and due to the shared abdominal and tail regions, the secondary axis is often morphologically

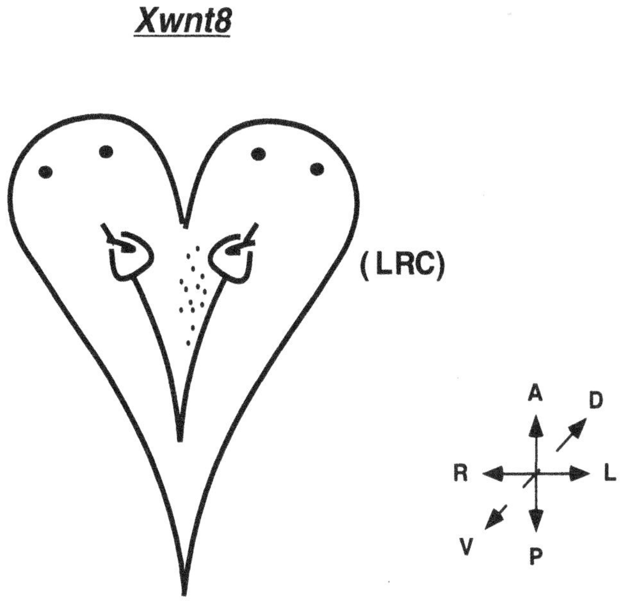

FIGURE 3.5. Schematic illustration (ventral view) of laterality phenotypes of conjoined twins. In spontaneous or experimentally induced (e.g., *Xwnt8*, *disheveled*) conjoined twins, the heart of the left twin is usually normal for left-right *situs*, in contrast to the right twin, which typically shows stochastic heart reversals. The reversed phenotype of the right twin might result from the separation of the right twin from left-right coordinator (LRC) activity and/or inductive signals emitted from the midline of the adjacent twin (stippled marks).

indistinguishable from the primary axis. With few exceptions (described below), experimentally induced or spontaneously occurring conjoined twins follow a general rule of "left twin normal, right twin randomized (stochastic)" with respect to the direction of heart looping.

Two hypotheses have been invoked to explain why the heart undergoes stochastic looping in the right twin. One possibility is that signals emitted by the midline of the left twin, which are required to reinforce global patterning information set forth by the left-right coordinator, diffuse to target cells of the right twin (Levin et al, 1996), thereby inappropriately presenting the left side of the right twin with right-sided laterality cues (Figure 3.5). Consistent with this interpretation, it has been shown in chick that *nodal* and *shh* expression is normal for the left twin (i.e., left-sided), but expression of neither gene typically is detected adjacent to the midline of the right twin (Levin et al, 1996). The other possibility underlying heart reversals of the right twin is that due to the intervening midline of the left twin, the right twin becomes separated from the left-right coordinator, which is positioned to the left side of the left twin and is required to form a normal left-right axis (Hyatt and Yost, 1998) (Figure 3.5). Neither possibility is mutually exclusive, and it would appear based on experiments described below, that both mechanisms are involved in generating heart reversals of the right twin.

To test the possibility that separation of the right twin from left-right coordinator activity causes stochastic heart reversals in this sibling, a left-right coordinator was placed to the left side of the right twin by ectopic expression of mature Vg1, which mimics left-right coordinator activity (Hyatt and Yost, 1998). The right twin

left-right asymmetry must be considered: namely, what is the nature of the patterning process that drives global left-right asymmetry and how do the cardiac primordia receive and respond to this positional information? Less than 5 years ago, virtually nothing was known about the cellular and molecular mechanisms involved in generating vertebrate left-right asymmetry, and the field has since rapidly compensated for this deficit. Genes critical to the left-right developmental pathway, including cell–cell signaling molecules, cell surface receptors, transcription factors, and extracellular matrix proteins, have been identified in a diverse range of vertebrate species, and with few exceptions, these genes appear to be functionally conserved in the pathway. Strikingly, the majority of genes that are implicated in the left-right pathway are expressed prior to organogenesis, demonstrating that developmental processes that at first glance might seem unrelated to cardiac morphogenesis are often quite relevant to the formation of the looped, primary heart tube. Finally, as our understanding of left-right patterning continues to grow, this information is anticipated to have exciting impact on the clinical study of human congenital cardiac and laterality defects, a relationship that already has started to flourish, despite the numerous, outstanding gaps and unanswered questions present in our current models of vertebrate left-right development.

REFERENCES

Afzelius, B.A. (1976). A human syndrome caused by immotile cilia. *Science* 193:317–319.

Andree, B., Duprez, D., Vorbusch, B., Arnold, H.-H., and Brand, T. (1998). BMP-2 induces ectopic expression of cardiac lineage markers and interferes with somite formation in chicken embryos. *Mech Dev* 70:119–131.

Beckerle, M.C., and Porter, K.R. (1982). Inhibitors of dynein activity block intracellular transport in erythrophores. *Nature* 295:701–703.

Biben, C., and Harvey, R.P. (1997). Homeodomain factor Nkx2-5 controls left/right asymmetric expression of bHLH gene eHand during murine heart development. *Genes Dev* 11:1357–1369.

Bisgrove, B., Essner, J., and Yost, H.J. (1999). Regulation of midline development by antagonism of *lefty* and *nodal* signaling. *Development* (*in press*).

Boettger, T., Wittler, L., and Kessel, M. (1999). FGF8 functions in the specification of the right body side of the chick. *Curr Biol* 9:277–280.

Bowers, P.N., Brueckner, M., and Yost, H.J. (1996). Laterality disturbances. *Prog Pediatr Cardiol* 6:53–62.

Britz-Cunningham, S.H., Shah, M.M., Zuppan, C.W., and Fletcher, W.H. (1995). Mutations of the connexin43 gap-junction gene in patients with heart malformations and defects of laterality. *N Engl J Med* 332:1323–1329.

Brown, N.A., and Wolpert, L. (1990). The development of handedness in left/right asymmetry. *Development* 109:1–9.

Burggren, W.W. (1988). Cardiac design in lower vertebrates: what can phylogeny reveal about ontogeny? *Experientia* 44:919–930.

Burn, J. (1991). Disturbance of morphological laterality in humans. In: Bock, G.R., and Marsh, J., eds. *Ciba Foundation Symposium*. Wiley, Chinchester, pp. 282–296.

Campione, M., Steinbeisser, H., Schweickert, A., et al. (1999). The homeobox gene Pitx2: mediator of asymmetric left-right signaling in vertebrate heart and gut looping. *Development* 126:1225–1234.

Charng, M.-J., Frenkel, P.A., Lin, Q., et al. (1998). A constitutive mutation of ALK5 disrupts cardiac looping and morphogenesis in mice. *Dev Biol* 199:72–79.

Chen, J.N., van Eeden, F.J., Warren, K.S., et al. (1997). Left-right pattern of cardiac BMP4 may drive asymmetry of the heart in zebrafish. *Development* 124:4373–4382.
Chiang, C., Litingtung, Y., Lee, E., et al. (1996). Cyclopia and defective axial patterning in mice lacking sonic hedgehog gene function. *Nature* 383:407–413.
Collignon, J., Varlet, I., and Robertson, E.J. (1996). Relationship between asymmetric nodal expression and the direction of embryonic turning. *Nature* 381:155–158.
Danos, M.C., and Yost, H.J. (1995). Linkage of cardiac left-right asymmetry and dorsal-anterior development in *Xenopus. Development* 121:1467–1474.
Danos, M.C., and Yost, H.J. (1996). Role of notochord in specification of cardiac left-right orientation in zebrafish and *Xenopus. Dev Biol* 177:96–103.
Gebbia, M., Ferrero, G.B., Pilia, G., et al. (1997). X-linked situs abnormalities result from mutations in ZIC3. *Nat Genet* 17:305–308.
Gebbia, M., Towbin, J.A., and Casey, B. (1996). Failure to detect connexin43 mutations in 38 cases of sporadic and familial heterotaxy [see comments]. *Circulation* 94:1909–1912.
Gerhart, J., Danilchik, M., Doniach, T., Roberts, S., Rowning, B., and Stewart, R. (1989). Cortical rotation of the *Xenopus* egg: consequences for the anteroposterior pattern of embryonic dorsal development. *Development* 107:37–51.
Goldstein, A.M., Ticho, B.S., and Fishman, M.C. (1998). Patterning the heart's left-right axis: from zebrafish to man. *Dev Genet* 22:278–287.
Guthrie, S. (1984). Patterns of junctional communication in the early amphibian embryo. *Nature* 311:149–151.
Guthrie, S., and Gilula, N. (1989). Gap junctional communication and development. *Trends Neurosci* 12:12–16.
Hirokawa, N. (1998). Kinesin and dynein superfamily proteins and the mechanism of organelle transport. *Science* 279:519–526.
Hummel, K.P., and Chapman, D.B. (1959). Visceral inversion and associated anomalies in the mouse. *J Hered* 50:9–13.
Hyatt, B.A., Lohr, J.L., and Yost, H.J. (1996). Initiation of vertebrate left-right axis formation by maternal Vg1. *Nature* 384:62–65.
Hyatt, B.A., and Yost, H.J. (1998). The left-right coordinator: the role of Vg1 in organizing left-right axis formation. *Cell* 93:37–46.
Isaac, A., Sargent, M.G., and Cooke, J. (1997). Control of vertebrate left-right asymmetry by a snail-related zinc finger gene. *Science* 275:1301–1304.
Itasaki, N., Nakamura, H., Sumida, H., and Yasuda, M. (1991). Actin bundles on the right side in the caudal part of the heart tube play a role in dextro-looping in the embryonic chick heart. *Anat Embryol* 183:29–39.
King, T., Beddington, R.S.P., and Brown, N.A. (1998). The role for the brachyury gene in heart development and left-right specification in the mouse. *Mech Dev* 79:29–37.
Kirby, M.L., and Waldo, K.L. (1995). Neural crest and cardiovascular patterning. *Circ Res* 77:211–215.
Kostetskii, I., Jiang, Y., Kostetskaia, E., Yuan, S., Evans, T., and Zile, M. (1999). Retinoid signaling required for normal heart development regulates GATA-4 in a pathway distinct from cardiomyocyte differentiation. *Dev Biol* 206:206–218.
Layton, W.M., and Manasek, F.J. (1980). Cardiac looping in early iv/iv mouse embryos. In: Van Praagh, R., and Takao, A., eds. *Etiology and Morphogenesis of Congenital Heart Disease*. Futura, Mount Kisco, NY, pp. 109–126.
Levin, M., Johnson, R.L., Stern, C.D., Kuehn, M., and Tabin, C. (1995). A molecular pathway determining left-right asymmetry in chick embryogenesis. *Cell* 82:803–814.
Levin, M., and Mercola, M. (1998a). The compulsion of chirality: toward an understanding of left-right asymmetry. *Genes Dev* 12:763–769.
Levin, M., and Mercola, M. (1998b). Evolutionary conservation of mechanisms upstream of asymmetric nodal expression: reconciling chick and *Xenopus. Dev Genet* 23:185–193.

Levin, M., and Mercola, M. (1998c). Gap junctions are involved in the early generation of left-right asymmetry. *Dev Biol* 203:90–105.

Levin, M., and Nascone, N. (1997). Two molecular models of initial left-right asymmetry generation. *Med Hypotheses* 49:429–435.

Levin, M., Pagan, S., Roberts, D.J., Cooke, J., Kuehn, M.R., and Tabin, C.J. (1997). Left/right patterning signals and the independent regulation of different aspects of situs in the chick embryo. *Dev Biol* 189:57–67.

Levin, M., Roberts, D.J., Holmes, L.B., and Tabin, C. (1996). Laterality defects in conjoined twins. *Nature* 384:321.

Lindner, E. (1960). Myofibrils in the early development of chick embryo hearts as observed with the electron microscope. *Anat Rec* 136:234–235.

Logan, M., Pagan-Westphal, S.M., Smith, D.M., Paganessi, L., and Tabin, C.J. (1998). The transcription factor Pitx2 mediates situs-specific morphogenesis in response to left-right asymmetric signals. *Cell* 94:307–317.

Lohr, J.L., Danos, M.C., Groth, T.W., and Yost, H.J. (1998). Maintenance of asymmetric nodal expression in *Xenopus laevis*. *Dev Genet* 23:194–202.

Lohr, J.L., Danos, M.C., and Yost, H.J. (1997). Left-right asymmetry of a nodal-related gene is regulated by dorsoanterior midline structures during *Xenopus* development. *Development* 124:1465–1472.

Lowe, L.A., Supp, D.M., Sampath, K., et al. (1996). Conserved left-right asymmetry of nodal expression and alterations in murine situs inversus. *Nature* 381:158–161.

Lyons, I., Parsons, L.M., Hartley, L., et al. (1995). Myogenic and morphogenetic defects in the heart tubes of murine embryos lacking the homeo box gene *Nkx2-5*. *Genes Dev* 9:1654–1666.

Markwald, R.R., Trusk, T., Gittenberger-de Groot, A., and Poelmann, R. (1997). Cardiac morphogenesis: formation and septation of the primary heart tube. In: Kavlock, R., and Daston, G., eds. *Drug Toxicity in Embryonic Development*. Springer-Verlag, Heidelberg, pp. 11–33.

Markwald, R.R., Trusk, T., and Moreno-Rodriguez, R. (1998). Formation and septation of the tubular heart: integrating the dynamics of morphology with emerging molecular concepts. In: De la Cruz, M.V., and Markwald, R.R., eds. *Living Morphogenesis of the Heart*. Springer-Verlag, Heidelberg. in press.

Matzuk, M.M., Kumar, T.R., and Bradley, A. (1995a). Different phenotypes for mice deficient in either activins or activin receptor type II. *Nature* 374:356–360.

Matzuk, M.M., Kumar, T.R., Vassalli, A., et al. (1995b). Functional analysis of activins during mammalian development. *Nature* 374:354–356.

McNeish, J.D., Thayer, J., Walling, K., Sulik, K.K., Potter, S., and Scott, W.J. (1990). Phenotypic characterization of the transgenic mouse insertional mutation, *legless*. *J Exp Zool* 253:151–162.

Melloy, P.G., Ewart, J.L., Cohen, M.F., Desmond, M.E., Kuehn, M.R., and Lo, C.W. (1998). No turning, a mouse mutation causing left-right and axial patterning defects. *Dev Biol* 193:77–89.

Meno, C., Ito, Y., Saijoh, Y., et al. (1997). Two closely-related left-right asymmetrically expressed genes, *lefty-1* and *lefty-2*: their distinct expression domains, chromosomal linkage and direct neuralizing activity in *Xenopus* embryos. *Genes Cells* 2:513–524.

Meno, C., Saijoh, Y., Fujii, H., et al. (1996). Left-right asymmetric expression of the TGF-beta-family member lefty in mouse embryos. *Nature* 381:151–155.

Meno, C., Shimono, A., Saijoh, Y., et al. (1998). *lefty-1* is required for left-right determination as a regulator of *lefty-2* and *nodal*. *Cell* 94:287–297.

Mjaatvedt, C.H., Yamamura, H., Wessels, A., Ramsdell, A., Turner, D., and Markwald, R.R. (1999). Mechanisms of segmentation, septation, and remodeling of the tubular heart: endocardial cushion fate and cardiac looping. In: Harvey, R.P., and Rosenthal, N., eds. *Heart Development*. Academic Press, San Diego.

Mochizuki, T., Saijoh, Y., Tsuchiya, K., et al. (1998). Cloning of *inv*, a gene that controls left/right asymmetry and kidney development. *Nature* 395:177–181.
Moon, R.T., and Kimelman, D. (1998). From cortical rotation to organizer gene expression: toward a molecular explanation of axis specification in *Xenopus*. *Bioessays* 20:536–545.
Morgan, D., Turnpenny, L., Goodship, J., et al. (1998). Inversin, a novel gene in the vertebrate left-right axis pathway, is partially deleted in the inv mouse [published erratum appears in *Nat Genet* 1998;20(3):312]. *Nat Genet* 20:149–156.
Nagai, T., Aruga, J., Takada, S., et al. (1997). The expression of the mouse *Zic1*, *Zic2*, and *Zic3* gene suggests an essential role for *Zic* genes in body pattern formation. *Dev Biol* 182:299–313.
Nascone, N., and Mercola, M. (1997). Organizer induction determines left-right asymmetry in *Xenopus*. *Dev Biol* 189:68–78.
Nomura, M., and Li, E. (1998). Smad2 role in mesoderm formation, left-right patterning and craniofacial development [see comments]. *Nature* 393:786–790.
Nonaka, S., Tanaka, Y., Okada, Y., et al. (1998). Randomization of left-right asymmetry due to loss of nodal cilia generating leftward flow of extraembryonic fluid in mice lacking KIF3B motor protein. *Cell* 95:829–837.
Oh, S.P., and Li, E. (1997). The signaling pathway mediated by the type IIB activin receptor controls axial patterning and lateral asymmetry in the mouse. *Genes Dev* 11:1812–1826.
Pagan-Westphal, S.M., and Tabin, C.J. (1998). The transfer of left-right positional information during chick embryogenesis. *Cell* 93:25–35.
Patten, B.M., and Kramer, T.C. (1933). The initiation of contraction in the embryonic chick heart. *Am J Anat* 53:349–375.
Piedra, M.E., Icardo, J.M., Albajar, M., Rodriguez-Rey, J.C., and Ros, M.A. (1998). Pitx2 participates in the late phase of the pathway controlling left-right asymmetry. *Cell* 94:319–324.
Ramsdell, A.F., and Yost, H.J. (1998). Molecular mechanisms of vertebrate left-right development. *Trends Genet* 14:459–465.
Rebagliati, M.R., Toyama, R., Fricke, C., Haffter, P., and Dawid, I.B. (1998). Zebrafish nodal-related genes are implicated in axial patterning and establishing left-right asymmetry. *Dev Biol* 199:261–272.
Rongish, B.J., Drake, C.J., Argraves, W.S., and Little, C.D. (1998). Identification of the developmental marker, JB3-antigen, as fibrillin-2 and its de novo organization into embryonic microfibrous arrays. *Dev Dyn* 212:461–471.
Ryan, A.K., Blumberg, B., Rodriguez-Esteban, C., et al. (1998). Pitx2 determines left-right asymmetry of internal organs in vertebrates. *Nature* 394:545–551.
Saijoh, Y., Adachi, H., Mochida, K., Ohishi, S., Hirao, A., and Hamada, H. (1999). Distinct transcriptional regulatory mechanisms underlie left-right asymmetric expression of *lefty-1* and *lefty-2*. *Genes Dev* 13:259–269.
Sampath, K., Cheng, A.M., Frisch, A., and Wright, C.V. (1997). Functional differences among *Xenopus* nodal-related genes in left-right axis determination. *Development* 124:3293–3302.
Sampath, K., Rubinstein, A.L., Cheng, A.M., et al. (1998). Induction of the zebrafish ventral brain and floorplate requires cyclops/nodal signalling. *Nature* 395:185–189.
Schauerte, H.E., van Eeden, J.M., Fricke, C., Odenthal, J., Strahle, U., and Haffter, P. (1998). Sonic hedgehog is not required for the induction of medial floor plate cells in the zebrafish. *Development* 125:2983–2993.
Schulte-Merker, S., van Eeden, F.J.M., Halpern, M.E., Kimmel, C.B., and Nusslein-Volhard, C. (1994). *No tail* (*ntl*) is the zebrafish homologue of the mouse *T* (*Brachyury*) gene. *Development* 120:1009–1015.
Schultheiss, T.M., Burch, J.B., and Lassar, A.B. (1997). A role for bone morphogenetic proteins in the induction of cardiac myogenesis. *Genes Dev* 11:451–462.

Seo, J.W., Brown, N.A., Ho, S.Y., and Anderson, R.H. (1992). Abnormal laterality and congenital cardiac anomalies. Relations of visceral and cardiac morphologies in the *iv/iv* mouse. *Circulation* 86:642–650.
Sinning, A.R., Hewitt, C.H., and Markwald, R.R. (1995). A subset of SBA lectin binding proteins isolated from myocardial conditioned medium transforms cardiac endothelium. *Acta Anat* 154:111–119.
Sinning, A.R., Krug, E.L., and Markwald, R.R. (1992). Multiple glycoproteins localize to a particulate form of extracellular matrix in regions of the embryonic heart where endothelial cells transform into mesenchyme. *Anat Rec* 232:285–292.
Smith, S.M., and Dickman, E.D. (1997). New insights into retinoid signaling in cardiac development and physiology. *Trends Cardiovasc Med* 7:324–329.
Smith, S.M., Dickman, E.D., Thompson, R.P., Sinning, A.R., Wunsch, A.M., and Markwald, R.R. (1997). Retinoic acid directs cardiac laterality and the expression of early markers of precardiac asymmetry. *Dev Biol* 182:162–171.
Splitt, M.P., Burn, J., and Goodship, J. (1995). Connexin43 mutations in sporadic and familial defects of laterality [letter; comment]. *N Engl J Med* 333:941; discussion 941–942.
Srivastava, D., Cserjesi, P., and Olson, E.N. (1995). A subclass of bHLH proteins required for cardiac morphogenesis. *Science* 270:1995–1999.
Srivastava, D., Thomas, T., Lin, Q., Kirby, M.L., Brown, D., and Olson, E.N. (1997). Regulation of cardiac mesodermal and neural crest development by the bHLH transcription factor, dHAND. *Nat Genet* 16:154–160.
St. Amand, T.R., Ra, J., Zhang, Y., et al. (1998). Cloning and expression pattern of chicken Pitx2: a new component in the SHH signaling pathway controlling embryonic heart looping. *Biochem Biophys Res Commun* 247:100–105.
Stahlsberg, H. (1969a). The origin of heart asymmetry: right and left contributions to the early chick embryo heart. *Dev Biol* 19:109–127.
Stahlsberg, H. (1969b). Regional mitotic activity in the precardiac mesoderm and differentiating heart tube in the chick embryo. *Dev Biol* 20:18–45.
Stahlsberg, H. (1970). Mechanism of dextral looping of the embryonic heart. *Am J Cardiol* 25:265–271.
Stahlsberg, H., and DeHaan, R.L. (1969). The precardiac areas and formation of the tubular heart in the chick embryo. *Dev Biol* 19:128–159.
Supp, D.M., Brueckner, M., and Potter, S.S. (1998). Handed asymmetry in the mouse: understanding how things go right (or left) by studying how they go wrong. *Semin Cell Dev Biol* 9:77–87.
Supp, D.M., Witte, D.P., Potter, S.S., and Brueckner, M. (1997). Mutation of an axonemal dynein affects left-right asymmetry in inversus viscerum mice. *Nature* 389:963–966.
Taber, L.A., Lin, I.E., and Clark, E.B. (1995). Mechanics of cardiac looping. *Dev Dyn* 203:42–50.
Thomas, T., Yamagishi, H., Overbeek, P.A., Olson, E.N., and Srivastava, D. (1998). The bHLH factors, dHAND and eHAND, specify pulmonary and systemic cardiac ventricles independent of left-right sidedness. *Dev Biol* 196:228–236.
Tsuda, T., Majumder, K., and Linask, K.K. (1998). Differential expression of flectin in the extracellular matrix and left-right asymmetry in mouse embryonic heart during looping stages. *Dev Genet* 23:203–214.
Van Praagh, R., and DeHaan, R.L. (1967). Morphogenesis of the heart: mechanism of curvature. In: *Annual Report of the Director, Dept. of Embryology*. Carnegie Institute, Washington, DC, pp. 536–537.
Winer-Muram, H. (1995). Adult presentation of heterotaxic syndromes and related complexes. *J Thorac Imaging* 10:43–57.
Wood, W.B. (1997). Left-right asymmetry in animal development. *Annu Rev Cell Dev Biol* 13:53–82.

Wood, W.B., and Kershaw, D. (1991). Handed asymmetry, handedness reversal and mechanisms of cell fate determination in nematode embryos. *Ciba Found Symp* 162:143–159; discussion 159–164.
Wunsch, A.M., Little, C.D., and Markwald, R.R. (1994). Cardiac endothelial heterogeneity defines valvular development as demonstrated by the diverse expression of JB3, an antigen of the endocardial cushion tissue. *Dev Biol* 165:585–601.
Yasui, H., Morishima, M., Nakazawa, M., and Aikawa, E. (1998). Anomalous looping, atrioventricular cushion dysplasia, and unilateral ventricular hypoplasia in mouse embryos with right isomerism induced by retinoic acid. *Anat Rec* 250:210–219.
Yokoyama, T., Copeland, N.G., Jenkins, N.A., Montgomery, C.A., Elder, F.F., and Overbeek, P.A. (1993). Reversal of left-right asymmetry: a situs inversus mutation. *Science* 260:679–682.
Yoshioka, H., Meno, C., Koshiba, K., et al. (1998). Pitx2, a bicoid-type homeobox gene, is involved in a lefty-signaling pathway in determination of left-right asymmetry. *Cell* 94:299–305.
Yost, H.J. (1990). Inhibition of proteoglycan synthesis eliminates left-right asymmetry in *Xenopus laevis* cardiac looping. *Development* 110:865–874.
Yost, H.J. (1991). Development of the left-right axis in amphibians. *Ciba Found Symp* 162:165–176; discussion 176–181.
Yost, H.J. (1992). Regulation of vertebrate left-right asymmetries by extracellular matrix. *Nature* 357:158–161.
Yost, H.J. (1995). Vertebrate left-right development. *Cell* 82:689–692.

CHAPTER 4

Differentially Expressed Genes and Cardiac Morphogenesis

Jim J.-C. Lin, Da-Zhi Wang, Rebecca S. Reiter, Qin Wang, Jenny L.-C. Lin, and Haley S. Williams

Cardiac morphogenesis is a dynamic, progressive, and intricate process. Defects in this process likely lead to congenital heart malformations. Congenital heart malformation is the most common of all birth defects, with an incidence at about 8 in 1,000 liveborn infants (Hoffman, 1995). Therefore, understanding the molecular mechanisms underlying cardiac morphogenesis is very important and clinically useful.

CARDIAC MORPHOGENESIS IN CHICK AND MOUSE

Several investigators have described early cardiac morphogenesis in the chick and mouse (DeHaan, 1965; DeRuiter et al, 1992; Garcia-Martinez and Schoenwolf, 1993). Originating from the rostral half of the primitive streak, cardiac progenitor cells constitute a pair of heart-forming fields within the lateral plate (splanchnic) mesoderm (DeHaan, 1965; Kaufman and Navaratnam, 1981; DeRuiter et al, 1992; Garcia-Martinez and Schoenwolf, 1993). The splanchnic mesoderm differentiates into a layer of promyocardial cells and a subjacent layer of proendocardial cells. The left and right heart-forming fields migrate ventrally and anteriorly to form a linear heart tube. The newly formed heart tube begins to contract at Hamburger-Hamilton (HH) stage 10 in chick and at day 8.5 postcoitus (p.c.) in mouse and subsequently undergoes rightward looping at HH stage 11 (DeHaan, 1965) in chick and at day 8.5 p.c. in mouse (Kaufman and Navaratnam, 1981; DeRuiter et al, 1992). Morphologically, a linear heart tube can be distinguished as five distinct segments, each composed of endocardium and myocardium (De la Cruz et al, 1989; (Viragh et al, 1989; DeRuiter et al, 1992; Biben and Harvey, 1997; Fishman and Chien, 1997). As development progressively proceeds, the right ventricle (RV), left ventricle (LV), atrioventricular (AV) canal, sinoatrial segment, and conal segment are formed in this order starting in chick from HH stage 9^{-} for the RV to stage 17 for the conal segment (De la Cruz et al, 1989), and in mouse from stage LS-I (day 8.0 p.c.) for the RV and LV to stage LS-III (day 9.5 p.c.) for the sinoatrial and conal segments (Biben and Harvey, 1997). Whether there are spatially and temporally expressed genes controlling their morphogenesis remains to be determined. However, the findings of chamber-specific genes such as atrial myosin heavy chain 1 (*aMHC1*) (Yutzey et al, 1994), ventricular myosin light chain 2

(*MLC2v*) (Ross et al, 1996), and myosin light chain 3f (*MLC3f*) (Franco et al, 1997), and of many differentially expressed genes such as *GATA-4* (Heikinheimo et al, 1994), *Nkx2.5* (Komuro and Izumo, 1993; Lints et al, 1993), *MEF-2c* (Chambers et al, 1994; Edmonson et al, 1994), *eHAND/HAND1* and *dHAND/HAND2* (Cross et al, 1995; Cserjesi et al, 1995; Hollenberg et al, 1995; Srivastava et al, 1995), *Msx-1* and *-2* (Chan-Thomas et al, 1993), *Prx-1* and *-2* (Leussink et al, 1995), bone morphogenetic proteins (*BMP-2* and *-4*) (Lyons et al, 1990; Jones et al, 1991), *NF-ATc* (De la Pompa et al, 1998; Ranger et al, 1998), and *Mf1* (Kume et al, 1998) argue for the existence of such spatially and temporally regulatory genes. This hypothesis is further supported by the recent findings that a single insertional mutation in a gene locus (*hdf*) on mouse chromosome 13 results in a tubular heart failing to develop the most anterior segment (future conus and RV) (Yamamura et al, 1997) and that with a null mutation of *MEF-2c* or *dHAND* no right ventricle forms (Lin et al, 1997; Srivastava et al, 1997).

As soon as the RV and LV have become clearly separated from the heart-forming fields, rightward looping of the heart tube begins at HH stages 10 to 11 in the chick embryo and at day 8.5 p.c. in the mouse embryo. Molecular mechanisms controlling the process and direction of looping remain unclear; however, it may be partly linked to the expression of laterality genes (Isaac et al, 1997; Robertson, 1997; Harvey, 1998). Looping will continue until the two limbs (conus and sinoatrium) are brought together along their hinge point, the inner (or lesser) curvature. As a result, the conus and its outlet septum will align with the AV septum, interventricular septum, and interatrial septum. Also, it will create a common wall, the mitroaortic continuity, between the anterior leaflet of the mitral valve and the outlet of the left ventricle. Therefore, looping can be considered a central event of cardiac morphogenesis, possibly affecting septation and valve formation. Looping has been described as having two phases in chick heart development: the first begins at stage 10^+ and ends at stage 17 with the formation of the S-shaped heart; the second phase begins with the onset of septation at stage 17 and ends at stages 35 to 39 with a fully defined four-chambered heart (Bouman et al, 1995). A comparable period for this to occur in mouse is from day 8.5 p.c. through day 15 p.c. (Webb et al, 1998). During this period, many molecular events occur in the conotruncal and AV canal segments. These include the inductive interaction between myocardium and endocardium at these regions, the epithelial-mesenchymal cell transformation, the proliferation and differentiation of mesenchymal cells, the remodeling of cushion tissues, and myocardialization in the inner curvature (Runyan and Markwald, 1983; Rosenquist et al, 1990; Runyan et al, 1992; Eisenberg and Markwald, 1995; Lamers et al, 1995; Markwald et al, 1996, 1997; Ramsdell and Markwald, 1997). Therefore, the isolation and functional characterization of temporally and spatially expressed genes during this critical period are essential and important steps toward understanding these molecular events.

We previously used the differential messenger RNA (mRNA) display method (Liang and Pardee, 1992; Liang et al, 1994) to compare the differences in mRNA species from the AV regions of developing chick hearts at stages 15 and 21. Using this approach, we cloned several differentially expressed genes (Wang et al, 1996) including a novel gene called *Xin* (former name 21C) (Wang et al, 1999). The chick *Xin* (*cXin*) is specifically expressed in the cardiac progenital cells of embryos as early as stage 8, prior to heart tube formation and looping. Treatment of stages 6 to 8 chick embryos with *cXin* antisense oligonucleotides results in an alteration of cardiac looping. These results suggest a pivotal role for *Xin* in cardiac morphogenesis. Further results from the time course study of *Xin* expression in the explant of anterior medial mesoendoderm induced by BMP-2 and from cotransactivation experiments suggest that *Xin* may be downstream of Nkx2.5 and MEF-2c transcription factors (Wang et al, 1999). This chapter focuses on the characterization of this *Xin* gene in relation to the *BMP–Nkx2.5–MEF-2c* pathway to regulate cardiac cell differentiation and cardiac morphogenesis.

DEVELOPMENTAL EXPRESSION PATTERN OF A NOVEL DIFFERENTIALLY EXPRESSED GENE, *XIN*

We previously reported that *cXin* messages are first detected at stage 8 but not stage 7 in the paired lateral plate mesoderm that forms the primordia of the heart (Wang et al, 1999). As development proceeds, amounts of *cXin* message detected in the heart by ribonuclease (RNase) protection assay increased steadily and reached a maximal peak at stages 21 to 25 (data not shown). A similar expression pattern was observed in developing mouse embryos. At day 8.0 p.c. (five somites) of mouse embryogenesis, the mouse *Xin* (*mXin*) message was distributed uniformly as seen from a ventral view of the embryo (Figure 4.1a; see color insert). However, from a side view, a string-like strong hybridization could be seen (Figure 4.1b). At day 8.5 p.c., when the developing heart starts to loop rightward, expression of *mXin* in the heart was observed with stronger hybridization in the future outflow tract and developing sinus venosus (Figure 4.1c). This expression pattern is very similar to that found in the stage 11 chick embryo (Wang et al, 1999). By day 9.0 p.c., expression of *mXin* was detected in most regions of the developing heart (Figure 4.1d). At day 9.5 p.c., a relatively high level of *mXin* expression could be detected specifically in the myocardial layers of the ventricles and atria (Figure 4.2a,b; see color insert). Interestingly, *mXin* expression was also detected in the developing tongue at days 13 p.c. (data not shown) and 15 p.c. (Figure 4.2c,d). The expression pattern of *mXin* in the tongue is very similar to that observed for *Nkx2.5* (Lints et al, 1993; Reecy et al, 1999).

Whole-mount in situ hybridization of a day 8.0 p.c. (six somites) mouse embryo revealed stronger hybridization on the right side of the developing heart tube than on the left (arrowhead in Figure 4.3f; see color insert). Cranial to caudal serial sections were performed on this embryo (Figure 4.3a–e). Examination of serial sections revealed only slightly asymmetric expression of *mXin* in the ventricles (data not shown). At the region where cardiac progenitor cells were undergoing fusion to form the sinus venosus (SV), more *mXin* message was detected than in either the ventricle (V) or atrium (A) (Figure 4.3a–e). It should be noted that the cell layer continuous with the endocardium of the developing ventricle expressed relatively high levels of *mXin*, as compared to the endocardium of the developing ventricle (arrowhead in Figure 4.3b). In addition, cell layers with strong positive hybridization were always in the inner layer of the myocardium (Figure 4.3a–c). These results suggest that both endocardial and myocardial cells are initially derived from the same Xin-positive progenital cells within the heart-forming fields and then differentiate into two separate cell types.

LOCALIZATION OF THE XIN PROTEIN

Using recombinant proteins from two different regions of the mXin protein, we have obtained polyclonal antibodies specifically recognizing the mXin protein

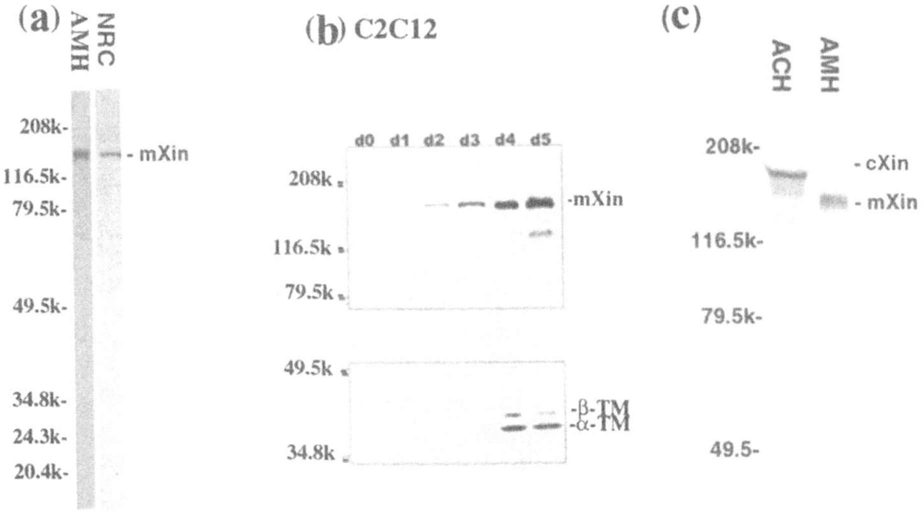

FIGURE 4.4. Western blot analysis with polyclonal antiserum against mXin recombinant protein. (a) Protein extracts were prepared from adult mouse heart (AMH) and cultured neonatal rat cardiomyocytes (NRCs). (b) Protein extracts were prepared from C2C12 cells at various days (d0–d5) after switching into differentiation medium to promote the formation of myotubes. A similar blot was reacted with monoclonal antibody CH1 to muscle α- and β-tropomyosin (TM) (bottom blot). (c) Protein extracts were prepared from adult chick (ACH) and mouse (AMH) hearts.

from adult mouse hearts, cultured neonatal rat cardiomyocytes, and a differentiated skeletal muscle cell line C2C12, as demonstrated by Western blot analysis. A representative Western blot is shown in Figure 4.4. A major band with relative molecular mass (M_r) of a 155 kd was detected with the anti-mXin antibody from total extracts of adult mouse heart (lane AMH in Figure 4.4a,c), neonatal rat cardiomyocytes (lane NRC in Figure 4.4a), and C2C12 differentiated myotubes (Figure 4.4b). It is worth noting that mXin expression precedes the expression of sarcomeric α- and β-tropomyosin (TM) during C2C12 muscle cell differentiation (Figure 4.4b). The anti-mXin antibodies also specifically recognized a protein band with an M_r of 180 kd from total extract of adult chick heart (Figure 4.4c).

Immunofluorescence microscopy on frozen sections of adult mouse hearts with the anti-mXin antibody showed a major localization of the Xin protein at the intercalated disks. Most of mXin proteins are colocalized with a Ca^{2+}-dependent adhesion molecule, N-cadherin, and with a gap junction protein, connexin-43 (Wang et al, 1999). However, a minor fraction of the mXin protein shows diffuse staining that is not colocalized with either N-cadherin or connexin-43. The basis for this staining pattern remains to be determined. The intercalated disk–staining pattern was also observed by immunofluorescence microscopy with the anti-mXin antibody on frozen sections of adult chick hearts (Figure 4.5b). Furthermore, the mXin protein is colocalized with N-cadherin but not connexin-43 in the hearts of day 8.5 and 14.5 p.c. mouse embryos (data not shown), suggesting that Xin may be part of the N-cadherin signaling pathway. In support of this hypothesis, a consensus sequence for an Src homology 3 (SH3)-binding motif (Yu et al, 1994) was found in both cXin and mXin proteins. This SH3-binding motif may mediate protein–protein interactions in a signal transduction pathway. When immunoflu-

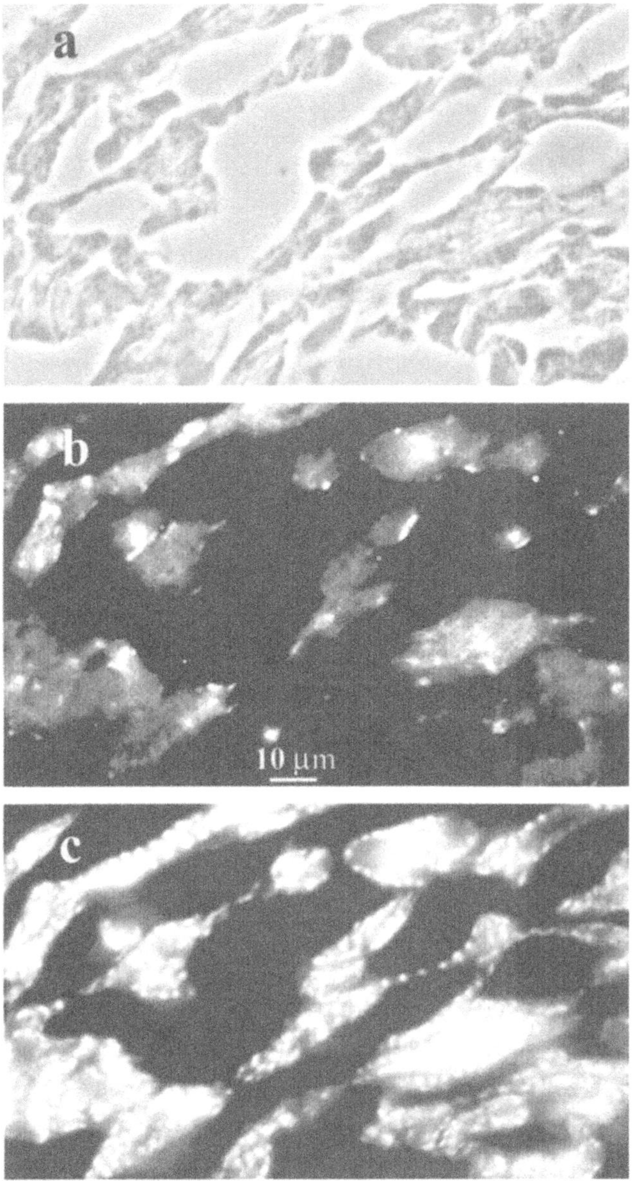

FIGURE 4.5. Double-label immunofluorescence microscopy on frozen section of adult chick heart. Section was first incubated with an antibody mixture of rabbit polyclonal antibody against mXin and mouse monoclonal antibody against cardiac troponin T and subsequently with a mixture of rhodamine-conjugated goat antirabbit immunoglobulin G (IgG) and fluorescein-conjugated goat antimouse IgG. (a) Phase-contrast micrograph and (b,c) fluorescent micrographs. (b) Section viewed selectively for rhodamine fluorescence to allow the localization of cXin to be visualized. (c) Same field seen in b, except it is viewed selectively for fluorescein fluorescence to allow the cardiac troponin T to be visualized.

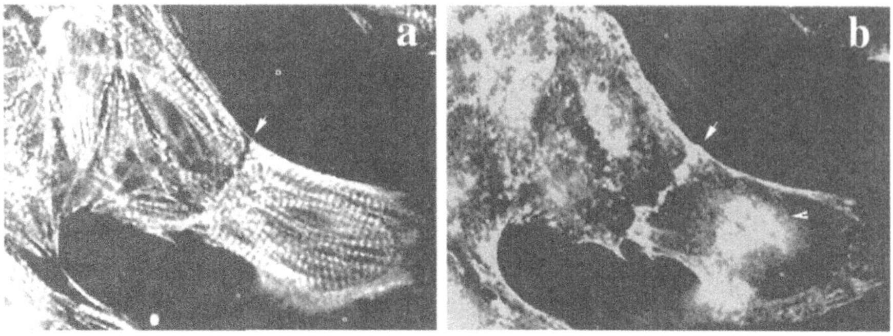

FIGURE 4.6. Double-label immunofluorescence microscopy on cultured neonatal rat cardiomyocytes. Neonatal cardiomyocytes were prepared from 3-day-old rat hearts as described previously (Wang et al, 1994), grown on collagen-coated glass coverslips for 3 days and then processed for double-label indirect immunofluorescence with monoclonal antibody against cardiac troponin T (a) and rabbit polyclonal antibody against mXin (b). The Xin is localized in both intercellular junctions (arrow) and nucleus (arrowhead).

orescence microscopy was performed on cultured neonatal rat cardiomyocytes, both intercellular junction and nuclear staining patterns could be detected (Figure 4.6b). The significance of mXin localization in the nucleus of cultured cardiomyocytes remains to be determined; however, this together with the presence of a DNA-binding motif and a nuclear localization signal in the predicted Xin protein (Wang et al, 1999) may suggest a transcriptional regulatory role for Xin. Therefore, Xin may have dual roles in cardiac myogenesis. Alternatively, the nuclear localization detected by the anti-mXin antibody on cardiomyocytes may represent an in vitro culture artifact. Thus, demonstrations of the in vivo nuclear localization and DNA-binding activity of Xin protein during cardiac development are essential for settling this issue.

XIN IS DOWNSTREAM OF *NKX2.5* AND *MEF-2C* IN CARDIAC MYOGENESIS

Using anterior medial mesoendoderm explant of stage 6 chick embryo, we demonstrated that the expression of *cXin* can be induced by BMP treatment and that the time required for this induction is significantly longer than that for Nkx2.5 and MEF-2c, but shorter than that for the ventricular myosin heavy chain (*vMHC*) (Wang et al, 1999). Transactivation experiments further demonstrate that *MEF-2c* or *Nkx2.5* alone is able to activate the expression of a reporter gene driven by the *mXin* promoter in nonmuscle cells (Wang et al, 1999). These results suggest that *Xin* lies downstream of MEF-2c and Nkx2.5 in a cardiogenic pathway. Consistent with this hypothesis, in a preliminary study with MEF-2c knockout embryos, we have found that the expression of *mXin* was drastically reduced (data not shown). To further show the specificity of transactivation by Nkx2.5, MEF-2c, and others such as GATA-4, experimentally defining the *cis*-regulatory elements for these factors will be needed. An increasing body of evidence suggests that the activation of many cardiac-specific/restricted genes is likely controlled by combinatorial interactions of many transcription factors (Sartorell et al, 1990; Ip et al, 1994; Chen

TABLE 4.1. Transactivation of *mXin-luc* and *cTnT-luc* by GATA-4, MEF2C, Nkx2.5, or combinations

Expression vectors used in cotransfection	Fold increase in promoter activity in fibroblasts	
	mXin-luc	*cTnT-luc*
GATA-4	0.93 ± 0.08	1.46 ± 0.18
MEF2C	9.84 ± 0.45	1.05 ± 0.07
Nkx2.5	3.17 ± 0.21	1.03 ± 0.04
GATA-4 + Nkx2.5	1.02 ± 0.02	5.05 ± 0.27
MEF2C + GATA-4	0.99 ± 0.03	1.28 ± 0.15
MEF2C + Nkx2.5	6.11 ± 0.62	1.10 ± 0.09

pGL3-mXin and pcTNT-497-luc carrying the firefly *luciferase* gene driven by *mXin* promoter (*mXin-luc*) and *cTnT* promoter (*cTnT-luc*), respectively, were used in cotransfections of 10T1/2 cells.

and Schwartz, 1996; Ross et al, 1996; Durocher et al, 1997; Firulli and Olson, 1997; Durocher and Nemer, 1998; Sepulveda et al, 1998). Recently, we have demonstrated that *cardiac troponin T* (*cTnT*) may use an MEF2-independent program to achieve its cardiac-specific expression, although an MEF2-like motif is present in its promoter and appears to be required for its expression in cultured cardiomyocytes and in transgenic mice (Wang et al, 2000). On the other hand, cardiac-specific expression of *mXin* is an MEF2-dependent program. Table 4.1 summarizes differences between these two promoters in response to MEF-2c, Nkx2.5, GATA-4, or a combination of two factors. Figure 4.7 shows schematical diagrams of *mXin* and *cTnT* promoters. There are three MEF2-binding motifs, three GATA sites, and an Nkx2.5-binding site present in the *mXin* promoter, whereas the *cTnT* promoter contains an MEF2-like motif, an Nkx2.5 site and three GATA sites including one located within the MEF2-like motif. The transcription factors that recognize these putative sites may be directly or indirectly involved in the cardiac-restricted expression of *mXin* and *cTnT*. To investigate this possibility further, we have constructed a plasmid, pGL3-mXin, that contains the *mXin* promoter fused to a firefly luciferase reporter (*mXin-luc*) and a plasmid, pcTNT-497-luc, that contains the *cTnT* promoter fused to a firefly luciferase reporter (*cTnT-luc*). In addition to the 1,848–base pair (bp) proximal promoter, pGL3-mXin contains exon 1 (135 bp), intron 1 (3,774 bp) and exon 2 (first 7 bp) of the *mXin* gene. The pcTNT-497-luc contains the 497-bp proximal promoter, 59 bp of exon 1, and 133 bp of intron 1 of the rat *cTnT* gene. The promoter activities of pGL3-mXin and pcTNT-497-luc in cultured neonatal rat cardiomyocytes are 420 and 145 times stronger than that in fibroblasts, respectively. As can be seen in Table 4.1, the muscle enriched transcription factor MEF-2c or Nkx2.5 alone is able to activate the *mXin-luc* gene but not the *cTnT-luc* gene in nonmuscle cells. On the other hand, GATA-4 alone significantly enhances the expression of *cTnT-luc* but not the *mXin-luc* fusion gene. Although Nkx2.5 alone does not have significant effects on the expression of *cTnT-luc*, coexpression of Nkx2.5 and GATA-4 results in synergistic activation of the *cTnT* promoter in fibroblasts. This phenomenon is very similar to that previously reported for the cardiac atrial natriuretic factor (*ANF*), brain natriuretic peptide (*BNP*) (Durocher et al, 1997; Lee et al, 1998) and the cardiac α-*actin* promoters (Sepulveda et al, 1998). On the other hand, a combination of Nkx2.5 and GATA-4 cannot activate the *mXin* promoter, but rather represses the activation stimulated by Nkx2.5 alone (1.02 versus 3.17, $p < .001$). The repression of *mXin* promoter

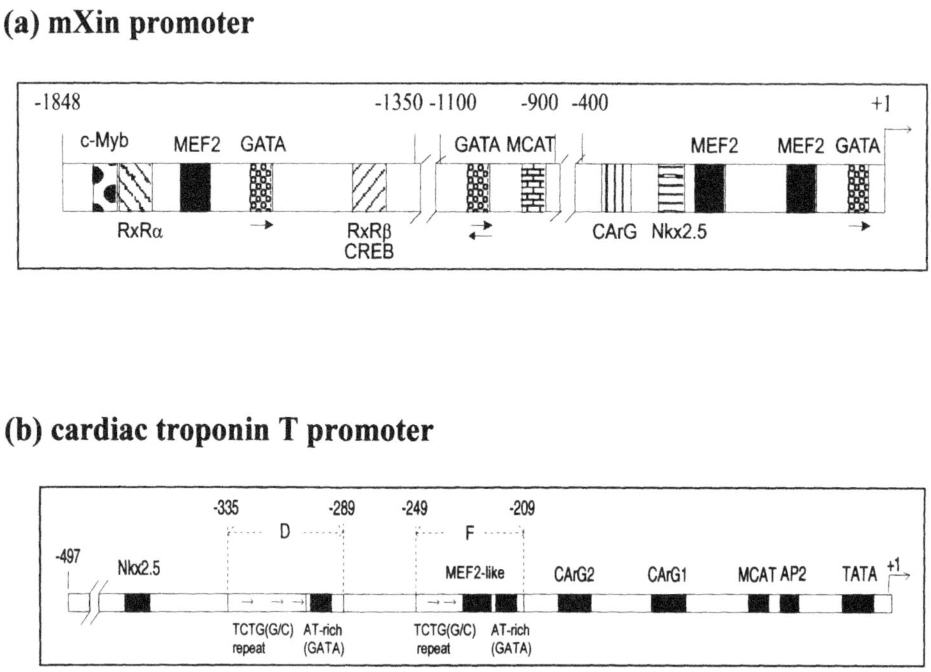

FIGURE 4.7. Diagrams of *mXin* and rat *cTnT* promoters. (a) The *mXin* promoter contains a Nkx2.5 site, three-MEF2 sites and three-GATA-4 motifs. The arrows indicate the position and direction of the GATA-4 consensus sites. (b) Both D and F modules in the *cardiac troponin T* promoter contain at least one A/T-rich/GATA site and TCTG(G/C) direct repeat (indicated by arrows). The MEF2-like motif in the F module contains a consensus MEF2 binding site. However, its A/T-rich core also shows a consensus GATA-binding site.

activity stimulated by MEF-2c alone is also seen with a combination of MEF-2c and GATA-4 (0.99 versus 9.84, $p < .001$). Interestingly, a combination of Nkx2.5 and MEF-2c factors do not increase further but instead decrease the *mXin* promoter activity that is already enhanced by MEF-2c alone (9.84 versus 6.11, $p < .01$). Neither a combination of MEF-2c and GATA-4 nor a combination of MEF-2c and Nkx2.5 has any effect on the *cTnT* promoter. Therefore, cardiac-specific gene expression is regulated by a delicate balance of a set of transcription factors, which themselves are cardiac-restricted and differentially expressed.

SIGNALS THAT INDUCE CARDIAC CELL DIFFERENTIATION

Several groups have previously shown that an interaction between mesoderm and endoderm is required to induce the expression of cardiac-specific proteins and to obtain a beating heart. A diffusible endoderm factor has been further implicated in inducing terminal differentiation in mesoderm (Jacobson, 1960; Jacobson and Sater, 1988; Antin et al, 1994; Sugi and Lough, 1994; Gannon and Bader, 1995; Nascone and Mercola, 1995). To accurately map potential inducing and responding areas of the avian embryo, Schultheiss et al (1995) developed and used a chick/quail coculture system. By employing this system, they have determined that the cardiac inducing activity is present in the endoderm from the anterior lateral plate and is absent from that of the posterior lateral plate. This activity can induce heart muscle differentiation in cells that are normally fated to give rise to blood and extraembryonic tissues. The same group of investigators (Schultheiss et al, 1997) and others (Andree et al, 1998) reported that BMP-2 is able to induce full ectopic cardiac differentiation in explant cultures of chick stage 6 anterior medial mesoendoderm, a tissue that is normally not cardiogenic. Cardiac genes induced by BMP-2 include *GATA-4*, *Nkx2.5*, *MEF-2c*, and *vMHC* (Schultheiss et al, 1997) as well as *cXin* (Wang et al, 1999). Interestingly, not all areas of the embryo are competent to express *Nkx2.5* in response to BMP-2 beads, for example, paraxial mesodermal cells are incompetent. Therefore, in addition to BMP, other signaling molecules may be required for cardiac specification. Alternatively, there may exist an inhibitory mechanism for early cardiac differentiation. BMP-2, in combination with fibroblast growth factor-4 (FGF-4), has been shown to induce cardiac differentiation in cultured stage 6 chick posterior lateral plate tissue, which does not normally form heart (Lough et al, 1996). In *Drosophila*, the *wingless* (*Wg*) signal was found to be a direct signal for heart formation and components of the *Wg* signaling pathway, such as *disheveled* and *armadillo*, are mediating the cardiogenic *Wg* signal (Park et al, 1996). The requirement of this *Wg* gene product for heart development is at a time period distinct from its function in segmentation and neurogenesis. In addition to *Wg*, the *tinman* (*Nkx2.5* homolog) gene is also required for heart formation in the early gastrula *Drosophila* embryo. The maintenance of *tinman* expression in the dorsal mesoderm for heart formation is dependent on the *dpp* signal [a member of the transforming growth factor-β (TGF-β) superfamily] (Staehling-Hampton et al, 1994; Frasch, 1995; Xu et al, 1998). Thus, both *dpp* and *Wg* signals are required for *Drosophila* heart formation. Whether both TGF-β and Wnt signals are also required for vertebrate cardiac morphogenesis remains to be determined.

DIFFERENTIALLY EXPRESSED GENES SUCH AS CARDIAC-ENRICHED TRANSCRIPTION FACTORS CONTROLLING CARDIAC MYOGENESIS: *BMP–Nkx2.5–MEF-2C* PATHWAY

The establishment and maintenance of the cardiac phenotype require the activation of many cardiac-specific/restricted genes in a tightly regulated temporal and spatial manner. This activation process is likely controlled by combinatorial interactions of many transcription factors including cardiac-restricted factors such as Nkx2.5, MEF-2c, GATA-4, etc. and ubiquitous factors such as serum response factor (SRF), transcriptional enhancer factor-1 (TEF-1), etc. Transcriptional regulation of cardiac-specific/restricted genes during development (cardiogenesis) is a very complex process that is believed to be essential for cardiac morphogenesis.

Although cardiac and skeletal muscle are both derived from mesoderm, divergent myogenesis pathways are believed to regulate the expression of a single contractile protein or muscle enzyme gene in a cardiac versus skeletal context. In skeletal muscle, the expression of cell-specific contractile protein genes appears to be controlled by the MyoD family proteins, that are themselves, skeletal muscle-specific. In vertebrates, four members of the *MyoD* family of genes have been identified. Each of them can activate the program for skeletal muscle differentiation when introduced into a variety of nonmuscle cell types (Olson, 1990; Weintraub et al, 1991; Sassoon, 1992; Wright, 1992; Emerson, 1993; Molkentin and Olson, 1996). These MyoD family proteins have not been detected in the heart. Furthermore, studies with heterokaryons between muscle cells and fibroblasts provide clear evidence that the cardiac muscle phenotype is recessive, in contrast to a dominant skeletal muscle phenotype (Evans et al, 1994). These results also suggest that cardiac myogenesis (cardiogenesis) may occur through a combinatorial pathway, as distinct from a master-regulatory mechanism in skeletal myogenesis. Promoter studies of muscle protein genes such as *MLC2v* (Lee et al, 1992; O'Brien et al, 1993; Lee et al, 1994; Ross et al, 1996), *α-MHC* (Adolph et al, 1993; Molkentin and Markham, 1993; Rindt et al, 1995), *muscle creatine kinase* (Amacher et al, 1993; Frasch, 1995; Donoviel et al, 1996; Shield et al, 1996), *cardiac/slow troponin C* (*cTnC*) (Parmacek et al, 1992; Ip et al, 1994), and *cardiac troponin T* (*cTnT*) (Iannello, 1991; Wang et al, 1994), also documented that distinct *cis*-regulatory elements in these promoters are required for muscle specificity in cardiac versus skeletal muscle. Thus, the question remains as to how cardiac-specific gene expression is controlled, and what cardiac-specific *trans*-acting factors are required. Some of the known transcription factor genes important for cardiogenesis are reviewed below.

Nkx2.5

In *Drosophila*, mutations in the homeobox-containing gene *tinman* result in the loss of the dorsal vessel (the insect equivalent of the vertebrate heart) as well as other visceral mesodermal structures (Azpiazu and Frasch, 1993; Bodmer, 1993). During development, *tinman* is first expressed in the visceral mesoderm and then becomes localized to the heart. The mouse homologue of *tinman* was cloned and named *Nkx2.5* or *Csx* (Komuro and Izumo, 1993; Lints et al, 1993). *Nkx2.5* is expressed in the cardiac primordia and in the myocardium throughout development prior to the activation of myofibrillar structural genes. Disruption of the mouse *Nkx2.5* gene results in defects in cardiac morphogenesis (Lyons et al, 1995; Tanaka et al, 1999). However, in mouse *Nkx2.5–/–* embryos, the heart tube forms and most cardiac contractile protein genes are expressed except for the *MLC2v* gene (Lyons et al, 1995). On the other hand, in the *Nkx2.5* knockout background, the left/right asymmetric expression of a *bHLH* gene, *eHAND* (*HAND1*), is altered, suggesting that *Nkx2.5* may control region-specific expression of another transcription factor (Biben and Harvey, 1997). In addition, expression of ANF and BNP, as well as several transcription factors such as c-myc, MEF-2c, HAND1 and Msx2, has been recently shown to be greatly disturbed in the Nkx2.5-deficient heart (Tanaka et al, 1999). It has consistently been shown that a combination of Nkx2.5 and GATA-4 factors can directly and synergistically activate expression of the *ANF* and *BNP* genes in vitro (Durocher et al, 1997; Durocher and Nemer, 1998; Lee et al, 1998). Using *Nkx2.5* null mutant embryos, Tanaka et al. (1999) have further shown that expression of *ANF* and *BNP* in the ventricle but not in the atrium is directly regulated by *Nkx2.5*. This provides additional evidence that the differentially expressed *Nkx2.5* gene can control region-specific expression of its downstream genes. In *Drosophila*, a single copy of the *MEF2* (*D-mef2*) gene has been shown to be a direct target of *tinman* (Gajewski et al, 1997). In *Nkx2.5* null mutant hearts, expression of *MEF-2c* is also downregulated (Tanaka et al,

1999). Recently, it has been reported that human *Nkx2.5* mutations dominantly cause cardiac malformations and atrioventricular conduction abnormalities (Schott et al, 1998). The mechanism for how the human *Nkx2.5* mutation leads to atrial septal defects and AV conduction defects is not clear; however, it may relate to the fact that *Nkx2.5* can control the expression of *Msx2*, whose gene product has been implicated in the formation of the cardiac septum and conduction system (Eisenberg and Markwald, 1995).

MEF-2c

Members of the myocyte enhancer factor-2 (*MEF2*) family of MADS (MCM1, aqamous, deficiens, and serum response factor) box transcription factors have also been shown to play important roles in cardiogenesis. In *Drosophila*, the single *MEF2* (*D-mef2*) gene is expressed in the cardiac, somatic, and visceral muscle lineages (Olson et al, 1995; Olson and Srivastava, 1996). In *D-mef2* mutant embryos, none of these types of muscle cells differentiate, but their precursor cells appear to be normally specified and positioned. These results clearly demonstrate that *MEF2* is required for differentiation of cardiac muscle (Bour et al, 1995; Lilly et al, 1995; Ranganayakula et al, 1995). Further genetic and mutational analyses suggest that *tinman* acts at an early step in cardiogenesis, whereas *D-mef2* acts later to control the expression of cardiac structural genes. This hierarchical relationship between these two genes has been confirmed (Gajewski et al, 1997). Therefore, it appears that *D-mef2* lies downstream of *tinman* in the genetic cascade controlling cardiogenesis.

In vertebrates, there are four *MEF2* genes, termed *MEF-2a–2d*, all of which are expressed in precursors of the cardiac, somatic, and smooth muscle lineages, as well as in certain other cell types (Chambers et al, 1994; Edmonson et al, 1994; Molkentin et al, 1996b). All four MEF2 proteins bind as homo- and heterodimers to the A/T-rich DNA sequence, (C/T) TA $(A/T)_4$ TA (G/A) (Gossett et al, 1989; Cserjesi and Olson, 1991; Yu et al, 1992; Molkentin et al, 1996a). This consensus MEF2 site is found in the control regions of numerous cardiac muscle genes, such as *muscle creatine kinase*, *α-MHC* (Gulick et al, 1991; Thompson et al, 1991; Molkentin and Markham, 1993), *MLC2v* (Navankasattusas et al, 1992), *cTnT* (Iannello, 1991; Wang et al, 1994), *phosphoglycerate mutase* (*PGAM*) (Nakatsuji et al, 1992), and *mXin* (Wang et al, 1999). Deletion and point mutations of the MEF2 binding site of these cardiac muscle genes impair their promoter activity in transient assays in cultured muscle cells, and/or in the apex of the adult heart. However, in a whole animal context (transgenic mice), point mutations in the MEF2-binding site of the *α-MHC* gene cause no effect on the expression of *MHC* in cardiac muscle (Adolph et al, 1993). Similarly, the *muscle creatine kinase* enhancer contains an A/T-rich site, which has been shown in gel shift assays to bind to ubiquitously expressed factors such as Oct-1 (Horlick et al, 1990), Mhox (Cserjesi et al, 1992), or MEF2 (Cserjesi et al, 1994). Although mutational studies in cultured muscle cells suggest that the A/T-rich element is essential for the *muscle creatine kinase* promoter activity, mutation of the A/T-rich site does not seem to significantly affect cardiac muscle expression in transgenic mice (Donoviel et al, 1996). These results raise two issues: (1) differences in the promoter activity of a regulatory element in a cardiac muscle gene between in vitro (cultured muscle cells) and in vivo (transgenic mice) models can exist. Therefore, analysis of cardiac muscle gene regulatory elements in a transgenic setting is a necessary step. (2) An MEF2-binding site or an A/T-rich site found in a cardiac muscle gene may not function through the MEF2 factor for cardiac-specific expression. In this regard, a MEF2 consensus site is found in the *MLC2v* promoter and called HF1b/MEF2. In transgenic mice, mutations in this HF1b or its upstream flanking site, termed HF1a, lead to downregulation of a transgene and relaxation of ventricular-chamber specificity (Lee et al, 1994), suggesting that the MEF2 binding site plays an important role in cardiac-specific *MLC2v* gene expression. Surprisingly, a novel zinc finger protein is found to bind in a sequence-specific manner to the HF-1b site of the rat *MLC2v* promoter (Zhu et al, 1993). Moreover, from a different laboratory, the binding factor (BBF-1), distinct from the MEF2 protein, has been detected to bind the MEF2 site of the chicken cardiac *MLC2v* promoter (Zhou et al, 1993). Taken together, these results suggest that *trans*-acting factors distinct from the MEF2 proteins may bind to MEF2-binding sites and/or its flanking sequences and control the cardiac-specific expression of certain muscle protein genes, such as *α-MHC*, *muscle creatine kinase*, and *MLC2v*. Interestingly, the expression of *MLC2v* and its ventricular chamber specificity are not affected in mouse embryos carrying an *MEF-2c*–null mutation (Lin et al, 1997). Although it is still possible that other members of *MEF2* expressed in the heart may regulate the expression of *MLC2v*, the result from the *MEF-2c* knockout embryos is nevertheless consistent with the idea that a MEF-2c–independent program may exist for the cardiac-specific expression of certain muscle genes. We have also identified a protein family with molecular masses of 40 to 42-kd, which are distinct from the MEF2 and HF1b proteins and are capable of binding to the MEF2-like motif and its flanking sequence in the rat *cTnT* promoter (Wang et al, 2000). It will be of interest to investigate what roles the 40 to 42-kd proteins might play in cardiac myogenesis.

GATA-4

Another differentially expressed transcription factor, GATA, appears to participate in cardiac morphogenesis. Six GATA family members have been identified in vertebrates so far. Within the GATA family, *GATA-4/5/6* are expressed in an overlapping pattern in the extraembryonic endoderm, precardiac mesoderm, embryonic and adult heart, and gut epithelium (Heikinheimo et al, 1994; Morrisey et al, 1996, 1997; Jiang et al, 1998). The GATA-binding sites, WGATAR, have been shown to be important for the cardiac activity of several promoters such as *ANF*, *BNP*, *cTnI*, *cTnC*, *MLC-3*, *α-MHC*, *α-actin* and *cardiac m2 muscarinic acetylcholine* genes (Grepin et al, 1994; Ip et al, 1994; Molkentin et al, 1994; McGrew et al, 1996; Durocher et al, 1997; Murphy et al, 1997; Rosoff and Nathanson, 1998; Sepulveda et al, 1998). Results from the spatial and temporal expression studies and various functional studies suggest an important role in cardiogenesis for *GATA-4* (Durocher and Nemer, 1998). However, recent studies with mice carrying a *GATA-4* null mutation suggest that the *GATA-4* gene is not required for specification or differentiation of myocardial, endocardial, or endodermal lineages, but is necessary for proper morphogenetic movements to form a primitive heart tube (Kuo et al, 1997; Molkentin, 1997). Chimeric analysis also reveals that mutant embryonic stem (ES) cells are able to contribute to a normal heart within a wild-type background (Kuo et al, 1997). Moreover, a cardiogenic role of *GATA-4* may require *Nkx2.5*, since GATA-4/5 are capable of interacting with Nkx2.5 and then regulating the expression of a subset of cardiac-specific genes (Durocher et al, 1997; Durocher and Nemer, 1998; Lee et al, 1998; Sepulveda et al, 1998). Recently, studies of upstream regulatory regions of the *Nkx2.5* gene reveal that an enhancer containing a GATA-4 binding site is essential for early expression of *Nkx2.5* in the developing heart (Lien et al, 1999; Reecy et al, 1999). In addition, potential autoregulatory elements in the *Nkx2.5* promoter, which can be cooperatively transactivated by Nkx2.5 and GATA-4, have been detected. However, coactivation of the *Nkx2.5* promoter by a combination of Nkx2.5 and GATA-4 in vitro is not as robust as that seen for *ANF* and *cardiac* α*-actin* promoters. Furthermore, GATA factors are also expressed in regions, in which *Nkx2.5* is not found during the early development of the heart. Therefore, activation and maintenance of *Nkx2.5* expression most likely involves other factors in addition to Nkx2.5 and GATA-4. In this regard, a recently cloned gene, *FOG-2* (friend of GATA-1), acting as a cardiac and brain-restricted modulator for GATA transcription factors may play important roles in regulating cardiac gene expression and morphogenesis (Lu et al, 1999; Svensson et al, 1999; Tevosian et al, 1999).

HAND

Other cardiac-restricted transcription factors, such as eHAND and dHAND (or HAND1 and HAND2, respectively) have also been implicated in the regulation of cardiac gene expression. In chick, expression of these two basic helix-loop-helix transcription factor genes overlaps completely throughout the heart tube. In chick embryo culture, treatment with antisense oligonucleotides specific to either *eHAND* or *dHAND* has no effect on the development of embryos. However, when both *HAND* genes are inactivated by antisense oligonucleotides, the heart tube forms correctly but is unable to complete looping morphogenesis (Srivastava et al, 1995). In mouse, the expression patterns of these two genes diverge after the heart tube has formed. Transcripts of *dHAND* are found throughout the myocardium of the heart tube and then become restricted to the future right ventricle. *eHAND* expression is complementary to the expression pattern of *dHAND*, thus becoming restricted to the future left ventricle. Both *dHAND* and *eHAND* are also expressed in the forming aortic sac and outflow tract. In addition, *eHAND* is expressed in the extraembryonic mesoderm and posterior lateral plate mesoderm, as well as in tissues in close relationship with the heart, the pericardium, and the epithelium of the oral cavity (Srivastava and Olson, 1997). Studies with knockout embryos and knockout fetuses derived from tetraploid chimeras suggest that *eHAND* is essential for differentiation of both trophoblast and cardiomyocytes (Firulli et al, 1998; Riley et al, 1998), whereas *dHAND* is required for formation of the right ventricle of the heart and the aortic arch arteries (Srivastava et al, 1997).

N-CADHERIN SIGNALING PATHWAY

Using a specific antibody against the mXin protein, we have shown that in adult mouse heart, Xin is colocalized with N-cadherin at the intercalated disks. This, together with the presence of a SH3-binding motif in the mXin protein, suggests that the Xin protein may function as an intracellular signaling molecule in the N-cadherin signaling pathway through interactions with SH3-containing proteins. This signaling transduction role may allow the Xin protein to regulate cell shape and maintain cytoskeletal and adherens junction integrity.

In the chick embryo, N-cadherin is expressed in various regions such as the neural tube, notochord, and mesoderm, and later becomes restricted to the heart and eye (Hatta et al, 1987). In the cardiac lineage, N-cadherin expression is first detected at the time the cardiogenic mesoderm forms an epithelioid myocardium (stage 5 chick embryo) (Linask, 1992). N-cadherin expression continues in myocytes during the epithelial-mesenchymal transformation (stages 15–21) and formation of trabeculae (embryonic days 5–7, E5–7). The N-cadherin–null mice fail to form the epithelioid myocardium of the tubular heart and die by day 10 p.c., prior to the initiation of trabeculation (Radice et al, 1997). Using a replication-defective retrovirus vector carrying a dominant-negative mutant of chick N-cadherin, Ong et al. (1998) have demonstrated that N-cadherin appears to play a role in homotypic interactions between migrating myocytes during trabecular formation of embryonic hearts. The N-cadherin level reaches a plateau during the second week of chick development (E10–E18) at the time myocytes initiate intercalated disk formation (Ong et al, 1998). Therefore, N-cadherin and its signaling pathway play important roles in myocyte epithelialization, trabecular formation, and intercalated disk formation. It is possible that Xin may also have functions in these processes, in addition to its function in cardiac looping (Wang et al, 1999).

Cadherins are known to associate with α- and β-catenins to form a cadherin-catenin complex that is essential for cadherins to function as adhesion molecules (Ranscht, 1994; Takeichi, 1995; Huber et al, 1996; Barth et al, 1997). α-Catenin is homologous to an adhesion plaque molecule, vinculin, and is a candidate for linking the cadherin-catenin complex to the actin cytoskeleton. β-Catenin exhibits homology to plakoglobin, a component of desmosomal plaques, and to the product of the *Drosophila* segment polarity gene, *Armadillo*. Like *Armadillo*, β-catenin is thought to be involved in signal transduction and developmental patterning (Heasman et al, 1994; Haegel et al, 1995; Takeichi, 1995; Huber et al, 1996). In the developing chicken heart, N-cadherin has been shown to colocalize with α- and β-catenins and disruption of the N-cadherin function with neutralizing antibodies results in abnormal heart development (Linask et al, 1997), suggesting a direct involvement of β-catenin in cardiac cell differentiation and heart development. The molecular components and mechanisms for these functions remain to be determined.

Our newly cloned gene, *Xin*, may provide an opportunity to uncover these molecular mechanisms, since the cardiac-specific Xin protein is colocalized with N-cadherin in adult and embryonic hearts, and the expression of the *Xin* gene is induced by BMP-2 (a member of the *TGF-β* superfamily). It is known that the *Armadillo*/β-catenin function in determining cell fates (nuclear accumulation) is modulated by the *Wnt/wingless* (*Wg*) signaling pathway (Behrens et al, 1996; Molenaar et al, 1996; Orsulic and Peifer, 1996; Larabell et al, 1997). Furthermore, in *Drosophila*, the development of heart cells has been shown to require at least both *Wg* and *dpp* (like *BMP*, *dpp* is a member of the *TGF-β* superfamily genes) signaling pathways (Park et al, 1996; Bodmer and Venkatesh, 1998; Xu et al, 1998). Therefore, it is possible that the involvement of the *Xin* gene in both *Wg/N-cadherin* and *BMP-2* signaling pathways may modulate cardiac differentiation and development.

SUMMARY

At day 8.0 p.c., mouse *Xin* is strongly expressed in the outflow tract and sinus venosus of the developing heart tube. The mXin-positive cell layer in the developing sinus venosus is continuous with the mXin-negative endocardial layer of ventricle. At day 15 p.c., mXin messages are found in the myocardium but neither in endocardium nor in endocardial cushion. The Xin protein detected from adult rat or mouse heart or mouse skeletal muscle line C2C12 has molecular mass of 155 kd, whereas the size of the Xin protein from chicken heart is significantly larger. The expression of Xin protein in the differentiating C2C12 cells precedes the expression of skeletal muscle-specific tropomyosins. The Xin protein appears to be colocalized with N-cadherin and β-catenin. In myogenesis, the *Xin* gene is downstream of *Nkx2.5* and *MEF-2c* and upstream of myofibrillar protein genes, such as myosin heavy chain and tropomyosins.

ACKNOWLEDGMENTS

We would like to thank Drs. Eric Olson, Sylvia Evans, and David Wilson for providing full-length cDNA clones expressing *MEF-2c*, *Nkx2.5*, and *GATA-4*, respectively. We also thank Vinh Nguyen for his help in frozen section of chick hearts and immunofluorescence staining. This work was supported, in part, by a Specialized Center of Research (SCOR) in congenital heart disease grant HL62178, and grant from the American Heart Association.

REFERENCES

Adolph, E.A., Subramaniam, A., Cserjesi, P., Olson, E.N., and Robbins, J. (1993). Role of myocyte-specific enhancer-binding factor (MEF2) in transcriptional regulation of the α-cardiac myosin heavy chain gene. *J Biol Chem* 268:5349–5352.

Amacher, S.L., Buskin, J.N., and Hauschka, S.D. (1993). Multiple regulatory elements contribute differentially to muscle creatine kinase enhancer activity in skeletal and cardiac muscle. *Mol Cell Biol* 13:2753–2764.

Andree, B., Duprez, D., Vorbusch, B., Arnold, H.-H., and Brand, T. (1998). BMP-2 induces ectopic expression of cardiac lineage markers and interferes with somite formation in chicken embryos. *Mech Dev* 70:119–131.

Antin, P.B., Taylor, R.G., and Yatskievych, T. (1994). Precardiac mesoderm is specified during gastrulation in quail. *Dev Dyn* 200:144–154.

Azpiazu, N., and Frasch, M. (1993). *Tinman* and *bagpipe*: two homeobox genes that determine cell fates in the dorsal mesoderm of *Drosophila*. *Genes Dev* 7:1325–1340.

Barth, A.I., Näthke, I.S., and Nelson, W.J. (1997). Cadherins, catenins and APC protein: interplay between cytoskeletal complexes and signaling pathways. *Curr Opin Cell Biol* 9:683–690.

Behrens, J., von Kries, J.P., Kuhl, M., et al. (1996). Functional interaction of β-catenin with the transcription factor LEF-1. *Nature* 382:638–642.

Biben, C., and Harvey, R.P. (1997). Homeodomain factor Nkx2.5 controls left-right asymmetric expression of bHLH gene eHand during murine heart development. *Genes Dev* 11:1357–1369.

Bodmer, R. (1993). The gene *tinman* is required for specification of the heart and visceral muscles in *Drosophila*. *Development* 118:719–729.

Bodmer, R., and Venkatesh, T.V. (1998). Heart development in *Drosophila* and vertebrates: conservation of molecular mechanisms. *Dev Genet* 22:181–186.

Bouman, H.G.A., Broekhuizen, M.L.A., Baasten, M.J., Gittenberger-deGroot, A.C., and Wenink, A.C.G. (1995). A spectrum of looping disturbances in stage 34 chicken hearts after retinoic acid treatment. *Anat Rec* 243:101–108.

Bour, B.A., O'Brien, M.A., Lockwood, W.L., et al. (1995). *Drosophila* MEF2, a transcription factor that is essential for myogenesis. *Genes Dev* 9:730–741.

Chambers, A.Z., Logan, M., Kotecha, S., Towers, N., Sparrow, D., and Mohun, T.J. (1994). The RSRF/MEF2 protein SL1 regulates cardiac-muscle-specific transcription of a myosin light-chain gene in *Xenopus* embryos. *Genes Dev* 8:1324–1334.

Chan-Thomas, P.S., Thompson, R.P., Robert, B., Yacoub, M.H., and Barton, P.J.R. (1993). Expression of homeobox genes *Msx-1* (*Hox-7*) and *Msx-2* (*Hox-8*) during cardiac development in the chick. *Dev Dyn* 197:203–216.

Chen, C.Y., and Schwartz, R.J. (1996). Recruitment of the *tinman* homolog *Nkx2.5* by serum response factor activates cardiac α-actin gene transcription. *Mol Cell Biol* 16:6372–6384.

Cross, J.C., Flannery, M.L., Blanar, E., et al. (1995). *Hxt* encodes a basic helix-loop-helix transcription factor that regulates trophoblast cell development. *Development* 121:2513–2525.

Cserjesi, P., Brown, D., Lyons, G.E., and Olson, E.N. (1995). Expression of the novel basic helix-loop-helix gene *eHAND* in neural crest derivatives and extraembryonic membranes during mouse development. *Dev Biol* 170:664–678.

Cserjesi, P., Lilly, B., Bryson, L., Wang, Y., Sasson, D.A., and Olson, E.N. (1992). Mhox: a mesodermally restricted homeodomain protein that binds an essential site in the muscle creatine kinase enhancer. *Development* 115:1087–1101.

Cserjesi, P., Lilly, B., Hinkley, C., Perry, M., and Olson, E.N. (1994). Homeodomain protein Mhox and MADS protein myocyte enhancer-binding factor-2 converge on a common elements in the muscle creatine kinase enhancer. *J Biol Chem* 269:16740–16745.

Cserjesi, P., and Olson, E.N. (1991). Myogenin induces the myocyte-specific enhancer binding factor MEF-2 independently of other muscle-specific gene products. *Mol Cell Biol* 11:4854–4862.

De la Cruz, M.V., Sanchez-Gomez, C., and Palomino, M.A. (1989). The primitive cardiac regions in the straight tube heart (stage 9–) and their anatomical expression in the mature heart: an experimental study in the chick heart. *J Anat* 165:121–131.

De la Pompa, J.L., Timmerman, L.A., Takimoto, H., et al. (1998). Role of the NF-ATc transcription factor in morphogenesis of cardiac valves and septum. *Nature* 392:182–186.

DeHaan, R.L. (1965). Morphogenesis of the vertebrate heart. In: DeHaan, R.L., and Ursprung, H., eds. *Organogenesis*. Holt, Rinehart, and Winston, New York, pp. 377–419.

DeRuiter, M.C., Poelmann, R.E., VanderPlas-deVries, I., Mentink, M.M.T., and Gittenberger-deGroot, A.C. (1992). The development of the myocardium and endocardium in mouse embryos. *Anat Embryol* 185:461–473.

Donoviel, D.B., Shield, M.A., Buskin, J.N., Haugen, H.S., Clegg, C.H., and Hauschka, S.D. (1996). Analysis of muscle creatine kinase gene regulatory elements in skeletal and cardiac muscle of transgenic mice. *Mol Cell Biol* 16:1649–1658.

Durocher, D., Charron, F., Warren, R., Schwartz, R.J., and Nemer, M. (1997). The cardiac transcription factor Nkx2.5 and GATA-4 are mutual cofactors. *EMBO J* 16:5687–5696.

Durocher, D., and Nemer, M. (1998). Combinatorial interactions regulating cardiac transcription. *Dev Genet* 22:250–262.

Edmonson, D.G., Lyons, G.E., Martin, J.F., and Olson, E.N. (1994). Mef2 gene expression marks the cardiac and skeletal muscle lineages during mouse embryogenesis. *Development* 120:1251–1263.

Eisenberg, L.M., and Markwald, R.R. (1995). Molecular regulation of atrioventricular valvuloseptal morphogenesis. *Circ Res* 77:1–6.

Emerson, C.P. (1993). Embryology to the fore. *Curr Opin Genet Dev* 3:265–274.

Evans, S.M., Tai, L.T., Tan, V.P., Newton, C.B., and Chien, K.R. (1994). Heterokaryons of cardiac myocytes and fibroblasts reveal that lack of dominance of the cardiac muscle phenotype. *Mol Cell Biol* 14:4269–4279.

Firulli, A.B., McFadden, D.G., Lin, Q., Srivastava, D., and Olson, E.N. (1998). Heart and extra-embryonic mesodermal defects in mouse embryos lacking the bHLH transcription factor Hand 1. *Nature Genet* 18:266–270.

Firulli, A.B., and Olson, E.N. (1997). Modular regulation of muscle gene transcription: a mechanism for muscle cell diversity. *Trends Genet* 13:364–369.

Fishman, M.C., and Chien, K.R. (1997). Fashioning the vertebrate heart: earliest embryonic decisions. *Development* 124:2099–2117.

Franco, D., Kelly, R., Lamers, W.H., Buckingham, M., and Moorman, A.F.M. (1997). Regionalized transcriptional domain of myosin light chain 3f transgenes in the embryonic mouse heart; morphogenetic implications. *Dev Biol* 188:17–33.

Frasch, M. (1995). Induction of visceral and cardiac mesoderm by ectodermal *Dpp* in the early Drosophila embryo. *Nature* 374:464–467.

Gajewski, K., Kim, Y., Lee, Y.M., Olson, E.N., and Schulz, R.A. (1997). *D-mef2* is a target for tinman activation during *Drosophila* heart development. *EMBO J* 16:515–522.

Gannon, M., and Bader, D. (1995). Initiation of cardiac differentiation occurs in the absence of anterior endoderm. *Development* 121:2439–2450.

Garcia-Martinez, V., and Schoenwolf, G.C. (1993). Primitive-streak origin of the cardiovascular system in avian embryos. *Dev Biol* 159:706–719.

Gossett, L., Kelvin, D., Sternberg, E., and Olson, E. (1989). A new myocyte-specific enhancer binding factor that recognizes a conserved element associated with multiple muscle specific genes. *Mol Cell Biol* 9:5022–5033.

Grepin, C., DagNino, L., Robitaille, L., Haberstroh, L., Antakly, T., and Nemer, M. (1994). A hormone-encoding gene identifies a pathway for cardiac but not skeletal muscle gene transcription. *Mol Cell Biol* 14:3115–3129.

Gulick, J., Subramaniam, A., Neumann, J., and Robbins, J. (1991). Isolation and characterization of the mouse cardiac myosin heavy chain genes. *J Biol Chem* 266:9180–9185.

Haegel, H., Larue, L., Ohsugi, M., Fedorov, L., Herrenknecht, K., and Kemler, R. (1995). Lack of α-catenin affects mouse development at gastrulation. *Development* 121:3529–3537.

Harvey, R.P. (1998). Links in the left/right axial pathway. *Cell* 94:273–276.

Hatta, K., Takagi, S., Fujisawa, H., and Takeichi, M. (1987). Spatial and temporal expression pattern of N-cadherin cell adhesion molecules correlated with morphogenetic processes of chicken embryos. *Dev Biol* 120:215–227.

Heasman, J., Crawford, A., Goldstone, K., et al. (1994). Overexpression of cadherins and underexpression of β-catenin inhibit dorsal mesoderm induction in early *Xenopus* embryos. *Cell* 79:791–803.

Heikinheimo, M., Scandrett, J.M., and Wilson, D.B. (1994). Localization of transcription factor GATA-4 to regions of the mouse embryo involved in cardiac development. *Dev Biol* 164:361–373.

Hoffman, J.I.E. (1995). Incidence of congenital heart disease: II. Prenatal incidence. *Pediatr Cardiol* 16:155–165.

Hollenberg, S.M., Sternglanz, R., Cheng, P.F., and Weintraub, H. (1995). Identification of a new family of tissue-specific basic helix-loop-helix proteins with a two-hybrid system. *Mol Cell Biol* 15:3813–3822.

Horlick, R.A., Hobson, G.M., Patterson, J.H., Mitchell, M.T., and Benfield, P.A. (1990). Brain and muscle *creatine kinase* genes contain common TA-rich recognition protein-binding regulatory elements. *Mol Cell Biol* 10:4826–4836.

Huber, O., Bierkamp, C., and Kemler, R. (1996). Cadherins and catenins in development. *Curr Opin Cell Biol* 8:685–691.

Iannello, R., Mar, J.H., and Ordahl, C.P. (1991). Characterization of a promoter element required for transcription in myocardial cells. *J Biol Chem* 266:3309–3316.

Ip, H.S., Wilson, D.B., Heinkinheimo, M., et al. (1994). The GATA-4 transcription factor transactivates the cardiac muscle-specific troponin C promoter-enhancer in nonmuscle cells. *Mol Cell Biol* 14:7517–7526.

Isaac, A., Sargent, M.G., and Cooke, J. (1997). Control of vertebrate left-right asymmetry by a snail-related zinc finger gene. *Science* 275:1301–1304.

Jacobson, A.G. (1960). Influences of ectoderm and endoderm on heart differentiation in the newt. *Dev Biol* 2:138–154.

Jacobson, A.G., and Sater, A.K. (1988). Features of embryonic induction. *Development* 104:341–359.

Jiang, Y., Tarzami, S., Burch, J.B.E., and Evans, T. (1998). Common role for each of the *cGATA-4/5/6* genes in the regulation of cardiac morphogenesis. *Dev Genet* 22:263–277.

Jones, M.C., Lyons, K.M., and Hogan, B.L.M. (1991). Involvement of bone morphogenetic protein-4 (BMP-4) and Vgr-1 in morphogenesis and neurogenesis in the mouse. *Development* 111:531–542.

Kaufman, M.H., and Navaratnam, V. (1981). Early differentiation of the heart in mouse embryos. *J Anat* 133:235–246.

Komuro, I., and Izumo, S. (1993). Csx: a murine homeobox-containing genes specifically expressed in the developing heart. *Proc Natl Acad Sci USA* 90:8145–8149.

Kume, T., Deng, K.-Y., Winfrey, V., Gould, D.B., Walter, M.A., and Hogan, B.L.M. (1998). The forkhead/winged helix gene *Mf1* is disrupted in the pleiotropic mouse mutation *congenital hydrocephalus*. *Cell* 93:985–996.

Kuo, C.T., Morrisey, E.E., Anandappa, R., et al. (1997). GATA4 transcription factor is required for ventral morphogenesis and heart tube formation. *Genes Dev* 11:1048–1060.

Lamers, W.H., Viragh, S.Z., Wessels, A., Moorman, A.F.M., and Anderson, R.H. (1995). Formation of the tricuspid valve in the human heart. *Circ Res* 91:111–121.

Larabell, C.A., Torres, M., Rowning, B.A., et al. (1997). Establishment of the dorso-ventral axis in *Xenopus* embryos is presaged by early asymmetries in β-catenin that are modulated by the Wnt signaling pathway. *J Cell Biol* 136:1123–1136.

Lee, K.J., Hickey, R., Zhu, H., and Chien, K.R. (1994). Positive regulatory elements (HF-1a and HF-1b) and a novel negative regulatory element (HF-3) mediate ventricular muscle-specific expression of myosin light-chain 2-luciferase fusion genes in transgenic mice. *Mol Cell Biol* 14:1220–1229.

Lee, K.J., Ross, R.S., Rockman, H.A., et al. (1992). Myosin light chain-2 luciferase transgenic mice reveal distinct regulatory programs for cardiac and skeletal muscle-specific expression of a single contractile protein gene. *J Biol Chem* 267:15875–15885.

Lee, Y., Shioi, T., Kasahara, H., et al. (1998). The cardiac tissue-restricted homeobox protein Csx/Nkx2.5 physically associates with the zinc finger protein GATA4 and cooperatively activates atrial natriuretic factor gene expression. *Mol Cell Biol* 18:3120–3129.

Leussink, B., Brouwer, A., El Khattabi, M., Poelmann, R.E., Gittenberger-deGroot, A.C., and Meijlink, F. (1995). Expression patterns of the paired-related homeobox genes *Mhox/Prx1* and *S8/Prx2* suggest roles in development of the heart and the forebrain. *Mech Dev* 52:51–64.

Liang, P., and Pardee, A.B. (1992). Differential display of eukaryotic messenger RNA by means of the polymerase chain reaction. *Science* 257:967–971.

Liang, P., Zhu, W., Zhang, X., et al. (1994). Differential display using one-base anchoral oligo-dT primers. *Nucleic Acids Res* 22:5763–5764.

Lien, C.-L., Wu, C., Mercer, B., Webb, R., Richardson, J.A., and Olson, E.N. (1999). Control of early cardiac-specific transcription of *Nkx2–5* by a GATA-dependent enhancer. *Development* 126:75–84.

Lilly, B., Zhao, B., Ranganayakulu, G., Paterson, B.M., Schulz, R.A., and Olson, E.N. (1995). Requirement of MADS domain transcription factor D-MEF2 for muscle formation in *Drosophila*. *Science* 267:688–693.

Lin, Q., Schwarz, J., Bucana, C., and Olson, E.N. (1997). Control of mouse cardiac morphogenesis and myogenesis by transcription factor MEF2C. *Science* 276:1404–1407.

Linask, K.K. (1992). N-cadherin localization in early heart development and polar expression of Na^+, K^+-ATPase and integrin during pericardial coelom formation and epithelialization of the differentiating myocardium. *Dev Biol* 151:213–224.

Linask, K.K., Knudsen, K.A., and Gui, Y.-H. (1997). N-cadherin-catenin interaction: necessary component of cardiac cell compartmentalization during early vertebrate heart development. *Dev Biol* 185:148–164.

Lints, T.J., Parsons, L.M., Hartley, L., Lyons, I., and Harvey, R.P. (1993). Nkx2.5: a novel murine homeobox gene expressed in early heart progenitor cells and their myogenic descendants. *Development* 119:419–431.

Lough, J., Barron, M., Brogley, M., Sugi, Y., Bolender, D.L., and Zhu, X.L. (1996). Combined BMP-2 and FGF-4 but neither factor alone, induces cardiogenesis in non-precardiac embryonic mesoderm. *Dev Biol* 178:198–202.

Lu, J.-R., McKinsey, T.A., Xu, H., Wang, D.-Z., Richardson, J.A., and Olson, E.N. (1999). FOG-2: a heart- and brain-enriched cofactor for GATA transcription factors. *Mol Cell Biol* 19:4495–4502.

Lyons, I., Parsons, L.M., Hartley, L., et al. (1995). Myogenic and morphogenetic defects in the heart tubes of murine embryos lacking the homeobox gene Nkx2.5. *Genes Dev* 9:1654–1666.

Lyons, K.M., Pelton, R.W., and Hogan, B.L.M. (1990). Organogenesis and pattern formation in the mouse: RNA distribution patterns suggest a role for bone morphogenetic protein-2A (BMP-2A). *Development* 109:833–844.

Markwald, R.R., Eisenberg, C., Eisenberg, L., Trusk, T., and Sugi, Y. (1996). Epithelial-mesenchymal transformations in early avian heart development. *Acta Anat* 156:173–186.

Markwald, R.R., Fitzharris, T.P., and Manasek, F.J. (1997). Structural development of endocardial cushions. *Am J Anat* 148:85–120.

McGrew, M.J., Bogdanova, N., Hasegawa, K., et al. (1996). Distinct gene expression patterns in skeletal and cardiac muscle are dependent on common regulatory sequences in the MLC1/3 locus. *Mol Cell Biol* 16:4524–4538.

Molenaar, M., vande Wetering, M., Oosterwegel, M., et al. (1996). XTcf-3 transcription factor mediates β-catenin-induced axis formation in *Xenopus* embryos. *Cell* 86:391–399.

Molkentin, J.D., Black, B.L., Martin, J.F., and Olson, E.N. (1996a). Mutational analysis of the DNA binding, dimerization, and transcriptional activation domains of MEF2C. *Mol Cell Biol* 16:2627–2636.

Molkentin, J.D., Firulli, A.B., Black, B.L., et al. (1996b). MEF2B is a potent transactivator expressed in early myogenic lineages. *Mol Cell Biol* 16:3814–3824.

Molkentin, J.D., Kalvakolanu, D.V., and Markaham, B.E. (1994). Transcription factor GATA-4 regulates cardiac tissue-specific expression of the α-myosin heavy-chain gene. *Mol Cell Biol* 14:4947–4957.

Molkentin, J.D., Lin, Q., Duncan, S.A., and Olson, E.N. (1997). Requirement of the transcription factor GATA4 for heart tube formation and ventral morphogenesis. *Genes Dev* 11:1061–1072.

Molkentin, J.D., and Markham, B.E. (1993). Myocyte-specific enhancer-binding factor (MEF2) regulates α-cardiac myosin heavy chain gene expression in vitro and in vivo. *J Biol Chem* 268:19512–19520.

Molkentin, J.D., and Olson, E.N. (1996). Defining the regulatory networks for muscle development. *Curr Opin Genet Dev* 6:445–453.

Morrisey, E.E., Ip, H.S., Lu, M.M., and Parmacek, M.S. (1996). GATA-6: a zinc finger transcription factor that is expressed in multiple cell lineages derived from lateral mesoderm. *Dev Biol* 177:309–322.

Morrisey, E.E., Ip, H.S., Tang, Z., Lu, M.M., and Parmacek, M.S. (1997). GATA-5: a transcriptional activator expressed in a novel temporally and spatially-restricted pattern during embryonic development. *Dev Biol* 183:21–36.

Murphy, A.M., Thompson, W.R., Peng, L.F., and Jones, L. (1997). Regulation of the rat *cardiac troponin I* gene by the transcription factor GATA-4. *Biochem J* 322:393–401.

Nakatsuji, Y., Hidaka, K., Tsujino, S., et al. (1992). A single MEF-2 site is a major positive regulatory element required for transcription of the muscle-specific subunit of human *phosphoglycerate mutase* gene in skeletal and cardiac muscle cells. *Mol Cell Biol* 12:4384–4390.

Nascone, N., and Mercola, M. (1995). An inductive role for the endoderm in *Xenopus* cardiogenesis. *Development* 121:515–523.

Navankasattusas, S., Zhu, H., Garcia, A.V., Evans, S.M., and Chien, K.R. (1992). A ubiquitous factor (HF-1a) and a distinct muscle factor (HF-1b/MEF2) form an E-box-independent pathway for cardiac muscle gene expression. *Mol Cell Biol* 12:1469–1479.

O'Brien, T.X., Lee, K.J., and Chien, K.R. (1993). Positional specification of ventricular myosin left chain-2 expression in the primitive murine heart tube. *Proc Natl Acad Sci USA* 90:5157–5161.

Olson, E.N. (1990). The MyoD family, a paradigm for development? *Genes Dev* 4:1451–1461.

Olson, E.N., Perry, M., and Schulz, R.A. (1995). Regulation of muscle differentiation by the MEF2 family of MADS box transcription factors. *Dev Biol* 172:2–14.

Olson, E.N., and Srivastava, D. (1996). Molecular pathways controlling heart development. *Science* 272:671–676.

Ong, L.-L., Kim, N., Mima, T., Cohen-Gould, L., and Milkawa, T. (1998). Trabecular myocytes of the embryonic heart require N-cadherin for migratory unit identity. *Dev Biol* 193:1–9.

Orsulic, S., and Peifer, M. (1996). An in vivo structure-function study of Armadillo, the β-catenin homologue, reveals both separate and overlapping regions of the protein required for cell adhesion and for *wingless* signaling. *J Cell Biol* 134:1283–1300.

Park, M., Wu, X., Golden, K., Axelrod, J.D., and Bodmer, R. (1996). The *wingless* signaling pathway is directly involved in *Drosophila* heart development. *Dev Biol* 177:104–116.

Parmacek, M.S., Vora, A.J., Shen, T., Barr, E., Jung, F., and Leiden, J.M. (1992). Identification and characterization of a cardiac-specific transcriptional regulatory element in the *slow/cardiac troponin C* gene. *Mol Cell Biol* 12:1967–1976.

Radice, G.L., Rayburn, H., Matsunami, H., Knundsen, K.A., Takeichi, M., and Hynes, R.O. (1997). Developmental defects in mouse embryos lacking N-cadherin. *Dev Biol* 181:64–78.

Ramsdell, A., and Markwald, R. (1997). Induction of endocardial cushion tissue in the avian heart is regulated in part, by TGFβ3-mediated autocrine signaling. *Dev Biol* 188:64–74.

Ranganayakula, G., Zhao, B., Dokidis, A., Molkentin, J.D., Olson, E.N., and Schulz, R.A. (1995). A series of mutations in the d-MEF2 transcription factor reveal multiple functions in larval and adult myogenesis in *Drosophila*. *Dev Biol* 171:169–181.

Ranger, A.M., Grusby, M.J., Hodge, M.R., et al. (1998). The transcription factor NF-ATc is essential for cardiac valve formation. *Nature* 392:186–190.

Ranscht, B. (1994). Cadherins and catenins: interactions and functions in embryonic development. *Curr Opin Cell Biol* 6:740–746.

Reecy, J.M., Li, X., Yamada, M., et al. (1999). Identification of upstream regulatory regions in the heart-expressed homeobox gene Nkx2–5. *Development* 126:839–849.

Riley, P., Anson-Cartwright, L., and Cross, J.C. (1998). The Hand 1 bHLH transcription factor is essential for placentation and cardiac morphogenesis. *Nature Genet* 18:271–275.

Rindt, H., Knotts, S., and Robbins, J. (1995). Segregation of cardiac and skeletal muscle-specific regulatory elements of the β-myosin heavy chain gene. *Proc Natl Acad Sci USA* 92:1540–1544.

Robertson, E.J. (1997). Left-right asymmetry. *Science* 275:1280.

Rosenquist, T.H., Fray-Gavalas, C., Waldo, K., and Beall, A.C. (1990). Development of the musculoelastic septation complex in the avian truncus arteriosus. *Am J Anat* 189:339–356.

Rosoff, M.L., and Nathanson, N.M. (1998). GATA factor-dependent regulation of cardiac m2 muscarinic acetylcholine gene transcription. *J Biol Chem* 273:9124–9129.

Ross, R., Navankasattasas, S., Harvey, R., and Chien, K. (1996). An HF-1a/HF-1b/MEF-2 combinatorial element confers cardiac ventricular specificity and establishes an anterior posterior gradient of expression. *Development* 122:1799–1809.

Runyan, R.B., and Markwald, R.R. (1983). Invasion of mesenchyme into three-dimensional gels: a regional and temporal analysis of interaction in embryonic heart tissue. *Dev Biol* 95:108–114.

Runyan, R.B., Potts, J.D., and Weeks, D.L. (1992). TGFβ3 mediated tissue interaction during embryonic heart development. *Mol Reprod Dev* 32:152–159.

Sartorell, V.K., Webster, K.A., and Kedes, L. (1990). Muscle-specific expression for the cardiac a-actin gene requires MyoD, CArG-box binding factor and Sp1. *Genes Dev* 4:1811–1822.

Sassoon, D.A. (1992). Myogenic regulatory factors: dissection of their role and regulation during vertebrate embryogenesis. *Dev Biol* 156:11–23.

Schott, J., Benson, D.W., Basson, C.T., et al. (1998). Congenital heart disease caused by mutations in the transcription factor Nkx2.5. *Science* 281:108–111.

Schultheiss, T.M., Burch, J.B.E., and Lassar, A.B. (1997). A role for bone morphogenetic proteins in the induction of cardiac myogenesis. *Genes Dev* 11:451–462.

Schultheiss, T.M., Xydas, S., and Lassar, A.B. (1995). Induction of avian cardiac myogenesis by anterior endoderm. *Development* 121:4203–4214.

Sepulveda, J.L., Belaguli, N., Nigam, V., Chen, C.-Y., Nemer, M., and Schwartz, R.J. (1998). GATA-4 and Nkx2.5 coactivate Nkx-2 DNA binding targets: role for regulating early cardiac gene expression. *Mol Cell Biol* 18:3405–3415.

Shield, M.A., Haugen, H.S., Clegg, C.H., and Hauschka, S.D. (1996). E-Box sites and a proximal regulatory region of the muscle creatine kinase gene differentially regulate expression in diverse skeletal muscles and cardiac muscle of transgenic mice. *Mol Cell Biol* 16:5058–5068.

Srivastava, D., Cserjesai, P., and Olson, E.N. (1995). A subclass of bHLH proteins required for cardiac morphogenesis. *Science* 270:1995–1999.

Srivastava, D., and Olson, E.N. (1997). The bHLH proteins dHAND and eHAND in cardiac development. In: Olson, E.N., Harvey, R.P., Schulz R.A., and Altman, J.S., eds. *Genetic Control of Heart Development.* Human Frontier Science Program, Strasbourg, pp. 62–68.

Srivastava, D., Thomas, T., Lin, Q., Kirby, M.L., Brown, D., and Olson, E.N. (1997). Regulation of cardiac mesodermal and neural crest development by the bHLH transcription factor, dHAND. *Nature Genet* 16:154–160.

Staehling-Hampton, K., Hoffmann, F.M., Baylies, M.K., Rushton, E., and Bate, M. (1994). *dpp* induces mesodermal gene expression in *Drosophila. Nature* 372:783–786.

Sugi, Y., and Lough, J. (1994). Anterior endoderm is a specific effector of terminal cardiac myocyte differentiation of cells from the embryonic heart forming region. *Dev Dyn* 200:155–162.

Svensson, E.C., Tufts, R.L., Polk, C.E., and Leiden, J.M. (1999). Molecular cloning of FOG-2: a modulator of transcription factor GATA-4 in cardiomyocytes. *Proc Natl Acad Sci USA* 96:956–961.

Takeichi, M. (1995). Morphogenetic roles of classic cadherins. *Curr Opin Cell Biol* 7:619–627.

Tanaka, M., Chen, Z., Bartunkova, S., Yamasaki, N., and Izumo, S. (1999). The cardiac homeobox gene *Csx/Nkx2.5* lies genetically upstream of multiple genes essential for heart development. *Development* 126:1269–1280.

Tevosian, S.G., Deconinck, A.E., Cantor, A.B., et al. (1999). FOG-2: a novel GATA-family cofactor related to multitype zinc-finger proteins friend of GATA-1 and U-shaped. *Proc Natl Acad Sci USA* 96:950–955.

Thompson, W.R., Nadal-Ginard, B., and Maklavi, V. (1991). A MyoD1-independent muscle-specific enhancer controls the expression of the β-myosin heavy chain gene in skeletal and cardiac muscle cells. *J Biol Chem* 266:22678–22688.
Viragh, S.Z., Szabo, E., and Challice, C.E.C. (1989). Formation of the primitive myo- and endocardial tubes in the chicken embryo. *J Mol Cell Cardiol* 21:123–137.
Wang, D.-Z., Hu, X., Lin, J.L.-C., Kitten, G.T., Solursh, M., and Lin, J.J.-C. (1996). Differential display of mRNAs from the atrioventricular region of developing chicken hearts at stages 15 and 21. *Front Biosci* 1:a1–15. (http://www.bioscience.org.)
Wang, D.-Z., Reiter, R.S., Lin, J.L.-C., et al. (1999). Requirement of a novel gene, *Xin*, in cardiac morphogenesis. *Development* 126:1281–1294.
Wang, G., Yeh, H.-I., and Lin, J.J.-C. (1994). Characterization of cis-regulating elements and transactivating factors of the rat *cardiac troponin T* gene. *J Biol Chem* 269:30595–30603.
Wang, Q., Sigmund, C.D., and Lin, J.J.-C. (2000). Identification of cis-elements in the cardiac troponin T gene conferring specific expression in cardiac muscle of transgenic mice. *Circ Res* 86:478–484.
Webb, S., Brown, N.A., and Anderson, R.H. (1998). Formation of the atrioventricular septal strctures in the normal mouse. *Circ Res* 82:645–656.
Weintraub, H., Davis, R., Tapscott, S., et al. (1991). The *MyoD* gene family: nodal point during specification of the muscle cell lineage. *Science* 251:761–766.
Wright, W.E. (1992). Muscle basic helix-loop-helix proteins and the regulation of myogenesis. *Curr Opin Genet Dev* 2:243–248.
Xu, X., Yin, Z., Hudson, J.B., Ferguson, E.L., and Frasch, M. (1998). Smad protein act in combination with synergistic and anatagonistic regulators to target *Dpp* responses to the *Drosophila* mesoderm. *Genes Dev* 12:2354–2370.
Yamamura, H., Zhang, M., Markwald, R.R., and Mjaatvedt, C.H. (1997). A heart segmental defect in the anterior-posterior axis of a transgenic mutant mouse. *Dev Biol* 186:58–72.
Yu, H., Chen, J.K., Feng, S., Dalgargo, D.C., Brauer, A.W., and Schteiber, S.L. (1994). Structural basis for the binding of proline-rich peptides to SH3 domains. *Cell* 76:933–945.
Yu, Y.-T., Breitbart, R.E., Smoot, L.B., Lee, Y., Mahdavi, V., and Nadal-Ginard, B. (1992). Human myocyte-specific enhancer factor 2 comprises a group of tissue-restricted MADS box transcription factors. *Genes Dev* 6:1783–1798.
Yutzey, K.E., Rhee, J.T., and Bader, D. (1994). Expression of the atrial-specific myosin heavy chain AMHC1 and the establishment of anteroposterior pathway in the developing chicken heart. *Development* 120:871–883.
Zhou, M.-D., Goswami, S.K., Martin, M.E., and Siddiqui, M.A.Q. (1993). A new serum responsive, cardiac tissue-specific transcription factor that recognizes the MEF2 site in the myosin light chain-2 promoter. *Mol Cell Biol* 13:1222–1231.
Zhu, H., Nguyen, V.T.B., Brown, A.B., et al. (1993). A novel, tissue-restricted zinc finger protein (HF-1b) binds to the cardiac regulatory element (HF1b/MEF-2) within the rat *myosin light chain-2* gene. *Mol Cell Biol* 13:4432–4444.

CHAPTER 5

Molecular Mechanisms of Cardiac Diversification

Jeffrey D. Croissant, Stacey Carpenter, and David Bader

The heart tube is the first functional organ to form within the developing vertebrate embryo. Proper assembly of the multichambered heart is critical for the viability of the developing organism. The two major components of the mature heart are the atria and ventricles (Figure 5.1; see color insert). These structures are morphogenically distinct, and are generated by many different cell populations. The atria, which store blood coming into the heart during ventricular systole, are composed of a thin myocardial layer of cardiogenic cells. Each ventricle is composed of trabeculated myocardium that is necessary to propel the blood into the circulatory system. The trabeculae are formed by profusion of myocardial bundles of irregular size from the ventricular myocardium. In the human heart, the left ventricle, which is responsible for systemic circulation, is two to three times thicker than the right ventricle, which forces blood into pulmonary circulation. During the development of the heart, the thin-walled atria and thick-walled ventricles are separated by a distinct boundary that will eventually form the valves separating the atrial and ventricular chambers. This chapter focuses on the initial morphogenic and molecular events that establish the diversification of cardiomyocytes into atrial or ventricular cell lineages. Furthermore, this chapter describes (1) the establishment of distinct atrial and ventricular cellular phenotypes during cardiac morphogenesis; (2) molecular markers that are expressed in a chamber-specific manner; (3) experiments using retinoic acid that change the polarity of the cardiogenic field, thereby moving the atrioventricular boundaries; (4) molecular mechanisms involved in chamber-specific gene expression, and (5) gene ablation studies that specifically affect the formation of the functional ventricular chamber.

CARDIAC CHAMBER MORPHOGENESIS

Due to the ease of embryonic manipulation, avian embryos have long been utilized to study the development of cellular structures. Fate-mapping studies in the early chick embryo have been utilized to study the origins of cardiogenic cells in the developing embryo. Current experimental evidence suggests that diversification of cardiomyocytes to an atrial or ventricular cell lineage occurs prior to the formation of the heart tube and appearance of distinct atrial and ventricular morphologies. Cardiac precursor cells, which are initially found in HH stage 3+ (Hamburger and Hamilton, 1951) primitive streak embryos, lie in the epiblast about midway down the length of the streak, and extend laterally from the midline about halfway to the edge of the area pellucida. Initial migration of these cells is medially toward the streak. Cells that are predetermined to a cardiac cell lineage are among the first mesodermal cell types that emerge during gastrulation from the primitive streak (Garcia-Martinez and Schoenwolf, 1993). These cells, as shown by cell labeling experiments in the chick, occupy the rostral half of the primitive streak just caudal to Hensen's node. The sequence of cells migrating from the streak into the mesoderm determines their position within the mature heart, thereby establishing an anteroposterior polarity within the early cardiogenic field. The rostral-most cardiogenic cells that migrate from the streak are destined for the conus arteriosus and ventricle of the mature heart. The caudal-most cardiogenic cells are those destined for the atrium and sinus venosus (Rosenquist, 1985; Garcia-Martinez and Schoenwolf, 1993). As the cells leave the streak, they migrate into the lateral plate mesoderm and spread anterolaterally (Rosenquist and DeHaan, 1966). By the time the precardiac cells leave the streak and complete gastrulation (stage 4), they are specified to differentiate, yet will not do so for several hours (Sater and Jacobson, 1989; Gonzalez-Sanchez and Bader, 1990).

Following the lateral migration, precardiac cells begin to extend anteriorly (stage 5) on both sides of the midline. Recent experimental evidence has defined the cardiogenic field in stage 5 chick embryos as the most lateral portion of the anterior lateral mesoderm. Moreover, this domain correlates with the expression of *Nkx2.5*, a homoebox gene that marks the definitive heart primordia (Ehrman and Yutzey, 1999). The cells of the lateral plate mesoderm split into splanchnic (ventral) and somatic (dorsal) mesoderm (stage 6), separated by the coelom, which forms the amniocardiac vesicles (Patten, 1929). At this stage of development, the first markers for terminal myocardial differentiation are detected in the splanchnic mesoderm (Ruzicka and Schwartz, 1988; Bisaha and Bader, 1991; Han et al, 1992; Logan and Mohun, 1993). Differentiation within the cardiogenic field proceeds in a rostrocaudal direction with anterior progenitors expressing ventricular myosin heavy chain 1 (VMHC1) prior to the induction of atrial myosin heavy chain 1 (AMHC1) in the posterior cardiac progenitor cells (Figure 5.2; see color insert) (Bisaha and Bader, 1991; Yutzey et al, 1994).

The tubular heart forms by the fusion of the two lateral halves of the splanchnic mesoderm that lies on both sides of the midline. The movement of the two lateral halves of the splanchnic mesoderm toward the midline occurs in concert with the endodermal formation of the anterior intestinal portal and the lengthening of the foregut (stages 7 to 8). At these stages, the precardiac mesodermal cells are in contact with the cohesive endodermal layer; hence, the precardiac mesoderm uses this layer as a framework for its own movements. Movement of the cardiogenic mesoderm occurs as a continuous epithelium with no mixing of anterior and posterior cell populations. Subsequently, as the endoderm invaginates to form the anterior intestinal portal (stage 9), the lateral halves of the splanchnic mesoderm migrate ventromedially, meeting at the midline and fusing to form the ventral mesocardium (DeHaan, 1960). Striated myofibrils appear in the myocardium by stage 10 just prior to the onset of muscular contractions (Han et al, 1992). Following fusion of the cardiac primordia, the primitive heart tube begins a rightward looping event, thereby establishing a left/right asymmetry within the heart. The position of distinct ventriculogenic cells within the heart tube is maintained throughout the looping process, with the most anterior ventricular cells being fated to become the right ventricle and the most posterior ventricular cells becoming the left ventricle. Following looping, intraventricular septa are formed that separate the right and left ventricles. Therefore, the position of the cardiogenic cells within the cardiac field of the early embryo establishes both the anteroposterior polarity and left-right asymmetry of the multichambered heart.

DIFFERENTIAL CHAMBER EXPRESSION OF CARDIAC-SPECIFIC GENES

Myosin, the major protein component of the thick filament found in striated muscle, is composed of two heavy chains and four light chains. Numerous heavy and light chain isoforms are expressed in the heart and skeletal muscle of vertebrate organisms during development and throughout adulthood. The first biochemical differences between atrial and ventricular myocytes were detected using monoclonal antibodies raised against various sarcomeric myosin isoforms. Antibodies raised against avian atrial and ventricular specific myosin heavy chains (MHCs) reacted specifically to their respective chambers. Using these antibodies, biochemical differences between the developing atrial and ventricular chambers were detected by stage 15 in the chick (Gonzalez-Sanchez and Bader, 1984; Zhang et al, 1986). Further studies indicated that several distinct MHC isoforms were expressed only in the developing atria, and were not detected in the adult heart (Evans et al, 1988).

Molecular cloning of chamber restricted MHC genes have provided excellent markers for avian atrial and ventricular cell lineages and also have verified that the initiation of chamber diversification occur at the earliest stages of cardiac morphogenesis. To date, the chicken *AMHC1* gene is the only gene isolated that is expressed exclusively in a cardiac chamber specific manner. *AMHC1* expression first occurs in the posterior region of the cardiogenic field prior to heart tube formation in stage 8 chicken embryos. The *AMHC1*-expressing cardiomyocytes from this posterior region of the cardiac crescent will eventually give rise the presumptive atrium. Expression is maintained in the atrium throughout development, with no expression detected in ventricular cardiomyocytes or skeletal muscle (Yutzey et al, 1994; Oana et al, 1998). The slow *MHC3* gene has been isolated from quail and has been shown to be highly homologous to *AMHC1*. However, unlike *AMHC1*, slow *MHC3* is expressed in both chambers of the developing quail heart until day 7, when expression is restricted to the atrium via a repression in the ventricles (Nikovits et al, 1996; Wang et al, 1996). In addition to its expression in the atria, slow *MHC3* gene expression has also been detected in slow skeletal muscle fiber types. A similar molecular MHC isoform switch is also detected in the chick. Expression of ventricular MHC (*VMHC1*) marks the onset of cardiac differentiation in the cardiac crescent at stage 7. Unlike *AMHC1*, *VMHC1* expression is detected in both anterior and posterior population of the cardiogenic field. Both atrial and ventricular cardiomyocytes express *VMHC1* until embryonic day 4 of development, when its expression is restricted to the ventricular chamber (Bisaha and Bader, 1991).

In the rat heart, three distinct MHC isoforms were identified based on differential electrophoretic mobilities (Hoh et al, 1978). These three isoforms were composed of homo- and heterodimers between the *α*- and *β-MHC* genes. Similar to the MHC expression patterns observed in the chick, *α-MHC* and *β-MHC* in the mouse embryo exhibit a dynamic pattern of expression in the atria and ventricles during cardiac development. At 8 days postcoitum (d.p.c.) during the initial stages of cardiac differentiation, transcription of the *α-MHC* and *β-MHC* genes are detected in both atrial and ventricular myocardial cell populations. By 9.5 d.p.c., *β-MHC* expression is dramatically downregulated in the developing atria, while expression of *α-MHC* is maintained at high levels. At birth, only the *α-MHC* gene is detected in the atrium. In the ventricles, *α-MHC* is highly expressed from 8 through 12.5 d.p.c., when transcription is downregulated in the developing ventricular myocytes. This pattern of expression coincides with the formation of the endocardial cushion and the interventricular septum. *β-MHC* expression is detected in the ventricles until birth, when transcript levels are decreasing and *α-MHC* is induced in the ventricles. Thereafter, *α-MHC* is maintained as the dominant MHC expressed in both cardiac chambers (Sinha et al, 1982; Lompre et al, 1984; Mahdavi et al, 1987; Lyons et al, 1990; Ng et al, 1991).

The functional significance of these MHC isoform switches in the developing cardiac chambers is not completely understood. However, clues can be derived from the different mechanical characteristics of the *α-MHC* and *β-MHC* isoforms. The *α-MHC* isoform exhibits higher Ca^{2+} adenosine triphosphatase (ATPase) activity and shortening velocities within the cardiac muscle (Barany, 1967). Cardiac contractility is therefore directly dependent on myosin composition (Schwartz et al, 1981). The consequence of the lower contractile velocities of the *β-MHC* isoforms in the developing ventricles is unknown.

The dynamic expression patterns of *MHC* gene expression in developing mammalian and avian species have provided immunogenic and molecular markers for identifying atrial and ventricular cardiomyocytes. Another chamber-specific molecule has been identified and provides an excellent ventricular marker for murine heart development. The myosin light chain 2V gene (*MLC-2V*) encodes a protein that is part of the contractile apparatus of cardiac and slow skeletal muscle. In cardiac muscle, *MLC-2V* transcripts are detected in the early stages of cardiac morphogenesis, 8.0 d.p.c., exclusively in the region of the cardiac crescent containing ventricular myocytes prior to heart tube formation. By 9.5 d.p.c., *MLC-2V* expression is restricted to the developing ventricular chambers (O'Brien et al, 1993). This restricted expression of *MLC-2V* marks the earliest known expression of a chamber restricted gene in the developing mouse embryo. In the adult mouse heart, another myosin light chain isoform, *MLC-2a*, is expressed solely in the atria. However, unlike *MLC-2V*, *MLC-2a* is expressed throughout the developing heart tube until 9.0 d.p.c., when expression in the ventricle is downregulated, resulting in atrial-specific expression. The repression of *MLC-2a* in the ventricle appears to correspond to the timing of ventricular septation. Molecular cloning of these two genes has provided valuable insight into the molecular mechanisms that may govern chamber diversification. These studies are discussed later in this chapter.

RETINOIC ACID EFFECTS ON CARDIAC DIVERSIFICATION

Retinoic acid confers teratogenic effects on heart morphogenesis and cardiac cell lineage diversification in a variety of developmental systems (Figure 5.3; see color insert). Retinoic acid and related vitamin A derivatives have been known to be required for proper cardiovascular development (Smith and Dickman, 1997). Therefore, exogenous retinoic acid treatments have served as a useful tool in studying the timing of atrioventricular diversification events. Vitamin A deficiencies in quail embryos creates anteriorization of the heart (Heine et al, 1985; Dersch and Zile, 1993; Twal et al, 1995). The importance of retinoid signaling for normal cardiac development in mammals was shown in vitamin A–deficient rats that exhibited ventricular chamber hyperplasia and ventricular septal defects (Wilson and Warkany, 1949). In addition, treatment of stage 7 chick embryos in culture with retinoic acid resulted in a truncation of anterior heart structures and an enlargement of the posterior atrial structures (Osmond et al, 1991). This apparent shift in the anteroposterior axis can be detected at the molecular level by an anterior shift in the expression pattern of *AMHC1*. The expansion of the atrial field occurred at the expense of the presumptive ventricle since there were no overall changes in the size of the myogenic compartment (Yutzey et al, 1994).

The effects of retinoic acid on the posteriorization of the heart field is time and dose dependent, suggesting that the diversification of cardiomyocytes occurs at an early embryonic event occurring prior to cardiomyocyte differentiation. Cardiogenic mesoderm explants removed from stage 4 chicken embryos exhibited diversified phenotypes without the presence of anterior endoderm. Induction of *AMHC1* expression in anterior progenitors was observed in stage 4 explants when treated with retinoic acid. However, when explants were removed from stage 8 embryos, retinoic acid had no effects on cardiac diversification (Yutzey et al, 1995). Similar findings were observed when zebrafish embryos where treated with retinoic acid. These embryos exhibited a posteriorization of the developing heart in a time- and dose-dependent manner (Stainier and Fishman, 1992). Grafting experiments in the mouse embryo have corroborated the time-dependent influence of retinoic acid on cardiac diversification. Grafting of ventricular cells into the atrium did not induce the expression of atrial markers in the grafted cells, suggesting chamber specification is irreversible after differentiation (Gruber et al, 1998).

Recent work in the mouse embryo has established a correlation between endogenous retinoic acid signaling and the generation of the anteroposterior axis within the developing heart field. Using a retinoic acid response element (RARE) linked to a *LacZ* reporter transgene, retinoic acid response was detected only in the posterior structures of the developing heart tube of 8.25 d.p.c. mouse embryos. RALDH2, a critical early enzyme in the retinoic acid synthesis pathway (Zhao et al, 1996), colocalized with *RARE-LacZ* transgene expression in early atrial structures (Moss et al, 1998). How the retinoic acid signaling pathway may lead to diversification changes within the cardiac cell lineage is not known. Changes in the expression of homoebox genes have been shown to affect the anteroposterior axis of a variety of structures in response to retinoic acid in the developing embryo (Conlon, 1995). Therefore, the action of retinoic acid may be to induce the activation and/or the expression of homeobox genes within the cardiogenic mesoderm, thereby inducing the expression of chamber-specific genes.

MOLECULAR REGULATION OF CHAMBER-SPECIFIC GENE EXPRESSION

The atrial-specific expression of *AMHC1* in the chick prior to heart tube formation as well as the effects of retinoic acid on the polarity of the cardiogenic field suggest the diversification decision of cardiomyocytes occurs at the earliest stages of cardiac morphogenesis. Understanding the molecular interactions that control chamber-specific gene expression has been hindered due to the inability to isolate transcription factors that are expressed exclusively in the atria or the ventricles. Recently, one such chamber-specific transcription factor has been isolated and identified in the chick embryo. *Irx4*, a chicken *iroquois*-related homeobox gene, was isolated from an embryonic retinal complement DNA (cDNA) library and shown to be expressed specifically in the ventricular chamber of the heart from stage 10 through day 9 of development. Functional studies showed chamber misexpression of *Irx4* resulted in the induction of *VMHC1* in the atrium, suggesting *Irx4* may play a key role in the establishment of the ventricular phenotype (Bao et al, 1999). Whether *Irx4* is directly involved in the induction of *VMHC1* and/or other ventricular specific genes in developing cardiomyocytes is not clear.

The molecular mechanisms that regulate the earliest steps of cardiac diversification are not fully understood. Characterization of the expression of genes expressed in an atrial- or ventricular-specific manner may provide insight and identify molecules that are involved in cardiac diversification and chamber-specific gene expression. The quail slow myosin heavy chain 3 (*SMyHC3*) gene is initially detected in all chambers of the developing quail heart until day 6, when expression in the ventricle declines, resulting in late-stage atrial-specific expression. Recent genomic cloning of the *SMyHC3* in the quail has identified regulatory sequences that may be important in chamber specific gene expression. A region between –840 and –680 of the 5′ flanking sequence of the *SMyHC3* gene has been termed ARD1 (atrial regulatory domain). Reporter gene constructs containing the ARD1 sequence were expressed at significantly high levels in cultured atrial but not ventricular cardiomyocytes. Several binding sites have been identified within the ARD1 fragment. These sites are similar to HF-1A, M-CAT, E-box, retinoic acid receptor-like (RARE), and GATA binding sites (Wang et al, 1996). Detailed mutational analysis of these binding sites suggested the sequences involved in restricting *SMyHC3* gene expression to atrial cardiomyocytes lie within the RARE. Mutations in the RARE site resulted in an upregulation of the slow *MHC3* reporter gene expression in ventricular but not atrial cardiomyocytes. Immediately downstream to this RARE site lies a positive-acting GATA-binding site that was required for reporter gene expression in both atrial and ventricular cardiomyocytes (Wang et al, 1998). These two sites appear to function independently of one another in conferring atrial expression. *In vivo* experiments utilizing quail embryos infected with replication-competent retroviruses showed the ARD1 cassette was able to recapitulate the chamber-restricted expression of *SMyHC3* in the atrium of day 17 embryos (Wang et al, 1996). Recent cross-species transgenesis experiments indicated that the proximal 840 base pairs (bp) of the quail *SMyHC3* directed reporter gene expression in the developing mouse specifically to the developing atrial chamber. Unlike the endogenous quail gene expression pattern, this reporter gene expression was detected in the posterior population of cardiac cells fated to the atrial cell lineage prior to heart tube formation (7.5 d.p.c.). Similar to studies in the developing chick, treatment of transgenic mice with retinoic acid expanded the posterior boundary of *SMyHC3* reporter gene expression (Xavier-Neto et al, 1999). Taken together, these results establish a model whereby atrial specific expression of *SMyHC3* transcripts occurs through negative transcriptional regulation in the ventricle, acting via the RARE sites located within the 5′ flanking region of the *SMyHC3* gene.

Because of its early restricted expression to ventricular myocytes, the mouse myosin light chain-2V (*MLC-2V*) gene has been effectively utilized to study the molecular mechanisms involved in establishing the ventricular cell lineage. As stated previously, *MLC-2V* is expressed bilaterally in the cardiac crescent, being limited to ventricular cell populations throughout mouse development. Analysis of the *MLC-2V* promoter in transgenic mice has identified a 250-bp element that confers ventricular specificity throughout mouse cardiogenesis (Lee et al, 1992). Within this promoter region lies a 28-bp regulatory element containing binding sites for HF-1a and MEF2 transcription factors (Lee et al, 1994). While the endogenous *MLC-2V* gene is detected in both the left and right ventricular chambers, the HF-1a/MEF2 regulatory region confers expression solely to the right ventricles beginning at 12.5 d.p.c. in the mouse (Ross et al, 1996). Therefore, additional *cis*-acting regulatory elements in the *MLC-2V* gene must be necessary for expression in the left ventricle. Since neither of the factors that bind the HF-1a or MEF2 sites is chamber or heart restricted, combinatorial mechanisms between these and/or other factors may be involved in the chamber-restricted expression of *MLC-2v* and the establishment of ventricular cell lineages. These studies also underlie not only the inherent differences between the atrial and ventricular cell lineages, but also the cellular variations between the left and right chambers of the heart. Identification and analysis of these and additional chamber and cardiac-restricted genes will be required to precisely understand these regulatory mechanisms.

Cloning of the chicken GATA-6 genomic locus has led to the identification of portions of the 5′ flanking region that direct transcription to a subset of cardiogenic cells. The GATA-6 transcription factor is expressed in cardiogenic cells and has been shown to be involved in the activation of numerous cardiac-specific genes (He and Burch, 1997). Transgenic experiments utilizing 10 kilobase (kb) (−9.2/+0.8) of upstream flanking region of the chick *GATA-6* gene directed heart-specific LacZ expression to both atrial and ventricular chambers. In addition, the −1.5/+0.8 distal region of the *GATA-6* locus was able to direct expression solely to myocardial cells located in the atrioventricular canal. These results suggest the presence of several distinct distal enhancers necessary for chamber-specific gene expression, as well as expression within distinct subsets of myocardial cells. Recently, several genes have been isolated that contain *cis*-regulatory regions that can drive gene expression in specific segments of the heart. Both the desmin and *SM-22* genes are expressed throughout the heart, but distinct enhancer regions have been identified that regulate transgene expression to the right ventricle. Likewise, regions of the MLC3F promoter and 3′ enhancer are able to induce expression to the left ventricle and right atrium of the heart.

Recently, two members of the basic helix-loop-helix (bHLH) family of transcription factors, *dHAND* and *eHAND*, have been identified that are expressed in either the left or right ventricle and may be important for cardiac development. In the mouse, *dHAND* expression is restricted to the right ventricle following cardiac looping, whereas *eHAND* is expressed predominantly in the left ventricle. Whether these genes directly activate genes in the respective ventricular chambers has not been determined. However, the requirement for proper *dHAND* expression during cardiogenesis has been shown in *dHAND* null mouse embryos (Srivastava et al, 1997). Mutational analysis has revealed that loss of the *dHAND* allele leads to embryonic lethality at 11 d.p.c. due to the failure of the right ventricle to form. The cardiac looping process appears to occur in mutant mice since the atrial chamber moves dorsally and to the left as in wild-type embryos. Since the linear heart tube appears normal in 8.0 d.p.c. mutant embryos, *dHAND* may be required for the expansion of the right ventricular chamber after the onset of cardiac looping.

GENE ABLATION AND CARDIAC DIVERSIFICATION

Proper formation of the atrial and ventricular chambers of the multichambered heart is essential for the survival of the developing vertebrate embryo. Following the diversification of the atrial and ventricular cell lineages, initiation of ventricular trabeculation, development of septal walls, and expansion of the ventricular compact zone are all key events required for cardiac morphogenesis. Genetic ablation of many widely expressed genes result in early embryonic lethality due to early heart malformations. The RXRα receptor belongs to a large retinoid receptor family that includes three retinoic acid receptors (RARs) and three distinct retinoid X receptors (RXRs). Members of the RXR receptor family have been shown to heterodimerize with many nonsteroid nuclear receptors, including RARs, thyroid hormone receptors, and vitamin D receptors, suggesting RXRs play key regulatory roles in a wide variety of signaling pathways (Yu et al, 1991). To this end, RXRα receptor expression is abundantly expressed in the adult mouse in the heart, intestine, skeletal muscle, skin, kidney, and liver (Mangelsdorf et al, 1992). Gene targeted deletion of the RXRα receptor led to embryonic lethality between E13.5 to E16.5 due to a wide variety of abnormalities in the developing ventricles. RXRα –/– mice exhibit defects in ventricular hypoplastic development, resulting in a loss of ventricular trabeculation and gross septal malformations (Sucov et al, 1994). Further analysis of these mice indicate the thinned-walled ventricles exhibit an atrial-like phenotype both physiologically and biochemically. Expression of *MLC-2a*, which is normally repressed in the ventricles of 10 d.p.c. embryos, was maintained in the thin-walled ventricles of 13.5 d.p.c. RXRα –/– mice. However, *MLC-2V* expression was still detected in the ventricles of RXRα –/– mice, suggesting only a partial misspecification of cardiac cell fate has occurred (Dyson et al, 1995). Results of these studies suggest a link between retinoid signaling and atrioventricular diversification. The role of RXRα receptors in establishing anteroposterior patterning of the early cardiogenic field has not been determined.

Cardiac phenotypes similar to the RXRα mutant embryos have been observed following the ablation of the ErbB2, ErbB4, and neuregulin alleles. ErbB2 and ErbB4 belong to a family of cell-surface receptors that are required for signaling by neuregulin, a member of the epidermal growth factor (EGF) family of growth factors (Carraway and Burden, 1995). In the developing heart, neuregulin is expressed in the endocardial cell layer, whereas ErbB2 and ErbB4 are expressed in the myocardium of the atria and ventricles. Deletion of any one of these three genes resulted in remarkably similar cardiac abnormalities. ErbB2 –/–, ErbB4 –/–, and neuregulin –/– embryos were embryonic lethal at 10.5 d.p.c. due to gross defects in the formation of trabeculae within the developing ventricle. Because of the poorly developed trabeculae, the ventricles and atrium of mutant embryos showed a similar thin-walled morphology (Gassmann et al, 1995; Lee et al, 1995; Meyer and Birchmeier, 1995). The results of these experiments suggest a neuregulin-mediated paracrine signaling event between the endocardial and myocardial layers of the heart is necessary for trabecular formation in the ventricles.

SUMMARY

The development of the functional vertebrate heart has served as an excellent model system to study the molecular, biochemical, and physiologic regulation of cellular diversification. The atria and ventricles of the mature vertebrate heart are composed of unique subsets of cardiomyocytes that are required for variations in chamber function. While this chapter has explored experimental evidence that has led to a better understanding of the initial critical steps involved in atrial and ventricular diversification, many questions are still unresolved. The atrial-ventricular fate of cardiomyocytes appears to be determined by cellular position within the cardiogenic field and affected by retinoic acid signaling. How retinoic acid and other extracelluar signaling molecules regulate these cellular decisions is not clear. Many chamber-specific molecular and biochemical markers have been isolated. However, additional chamber-specific genes need to be identified to ascertain functional variations between atrial and ventricular myocytes.

REFERENCES

Bao, Z.Z., Bruneau, B.G., Seidman, J.G., Seidman, C.E., and Cepko, C.L. (1999). Regulation of chamber-specific gene expression in the developing heart by Irx4. *Science* 283:1161–1164.

Barany, M. (1967). ATPase activity of myosin correlated with speed of muscle shortening. *J Gen Physiol* 50(suppl):197–218.

Bisaha, J.G., and Bader, D. (1991). Identification and characterization of a ventricular-specific avian myosin heavy chain, VMHC1: expression in differentiating cardiac and skeletal muscle. *Dev Biol* 148:355–364.

Carraway, K.L. 3rd, and Burden, S.J. (1995). Neuregulins and their receptors. *Curr Opin Neurobiol* 5:606–612.

Conlon, R.A. (1995). Retinoic acid and pattern formation in vertebrates. *Trends Genet* 11:314–329.

DeHaan, R.L. (1960). Morphogenesis of the vertebrate heart. In: DeHaan, R.L., and Upspring, H., eds. *Organogenesis.* Holt, Rinhart, and Winston, New York, pp. 377–419.

Dyson, E., Sucov, H.M., Kubalak, S.W., Schmid-Schonbein, G.W., DeLano, F.A., Evans, R.M., Ross, J. Jr, and Chien, K.R. (1995). Atrial-like phenotype is associated with embryonic ventricular failure in retinoid X receptor alpha –/– mice. *Proc Natl Acad Sci USA* 92:7386–7390.

Dersch, H., and Zile, M.H. (1993). Induction of normal cardiovascular development in the vitamin A–deprived quail embryo by natural retinoids. *Dev Biol* 160:424–433.

Ehrman, L.A., and Yutzey, K.E. (1999). Lack of regulation in the heart forming region of avian embryos. *Dev Biol* 207:163–175.

Evans, D., Miller, J.B., and Stockdale, F.E. (1988). Developmental patterns of expression and coexpression of myosin heavy chains in atria and ventricles of the avian heart. *Dev Biol* 127:376–383.

Garcia-Martinez, V., and Schoenwolf, G.C. (1993). Primitive-streak origin of the cardiovascular system in avian embryos. *Dev Biol* 159:706–719.

Gassmann, M., Casagranda, F., Orioli, D., Simon H., Lai, C., Klein, R., and Lemke, G. (1995). Aberrant neural and cardiac development in mice lacking the ErbB4 neuregulin receptor. *Nature* 378:390–394.

Gonzalez-Sanchez, A., and Bader, D. (1984). Immunochemical analysis of myosin heavy chains in the developing chicken heart. *Dev Biol* 103:151–158.

Gonzalez-Sanchez, A., and Bader, D. (1990). In vitro analysis of cardiac progenitor cell differentiation. *Dev Biol* 139:197–209.

Gruber, P.J., Kubalak, S.W., and Chien, K.R. (1998). Downregulation of atrial markers during cardiac chamber morphogenesis is irreversible in murine embryos. *Development* 125:4427–4438.

Hamburger, V., and Hamilton, H. (1951). A series of normal stages in the development of the chick embryo. *J Morphol* 88:49–92.

Han, Y., Dennis, J.E., Cohen-Gould, L., Bader, D.M., and Fischman, D.A. (1992). Expression of sarcomeric myosin in presumptive myocardium of chicken embryos occurs within six hours of myocyte commitment. *Dev Dyn* 193:257–265.

He, C.Z., and Burch, J.B. (1997). The chicken GATA-6 locus contains multiple control regions that confer distinct patterns of heart region-specific expression in transgenic mouse embryos. *J Biol Chem* 272:28550–28556.

Heine, U.I., Roberts, A.B., Munoz, E.F., Roche, N.S., and Sporn, M.B. (1985). Effects of retinoid deficiency on the development of the heart and vascular system of the quail embryo. *Virchows Arch [B]* 50:135–152.

Hoh, J.F., McGrath, P.A., and Hale, P.T. (1978). Electrophoretic analysis of multiple forms of rat cardiac myosin: effects of hypophysectomy and thryoxine replacement. *J Mol Cell Cardiol* 10:1053–1076.

Lee, K.F., Simon, H., Chen, H., Bates, B., Hung, M.C., and Hauser, C. (1995). Requirement for neuregulin receptor erbB2 in neural and cardiac development. *Nature* 378:394–398.

Lee, K.J., Hickey, R., Zhu, H., and Chien, K.R. (1994). Positive regulatory elements (HF-1a and HF-1b) and a novel negative regulatory element (HF-3) mediate ventricular muscle-specific expression of myosin light chain 2-lucifierase fusion genes in transgenic mice. *Mol Cell Biol* 14:1220–1229.

Lee, K.J., Ross, R.S., Rockman, H.A., et al. (1992). Myosin light chain-2 luciferase transgenic mice reveal distinct regulatory programs for cardiac and skeletal muscle-specific expression of a single contractile protein gene. *J Biol Chem* 267:15875–15885.

Logan, M., and Mohun, T. (1993). Induction of cardiac muscle differentiation in isolated animal pole explants of *Xenopus laevis* embryos. *Development* 118:865–875.

Lompre, A.M., Nadal-Ginard, B., and Mahdavi, V. (1984). Expression of the cardiac ventricular alpha- and beta-myosin heavy chain genes is developmentally and hormonally regulated. *J Biol Chem* 259:6437–6446.

Lyons, G.E., Schiaffino, S., Sassoon, D., Barton, P., and Buckingham, M. (1990). Developmental regulation of myosin gene expression in mouse cardiac muscle. *J Cell Biol* 111:2427–2436.

Mahdavi, V., Izumo, S., and Nadal-Ginard, B. (1987). Develpmental and hormonal regulation of sarcomeric myosin heavy chain gene family. *Circ Res* 60:804–814.

Mangelsdorf, D.J., Borgmeyer, U., Heyman, R.A., Zhou, J.Y., Ong, E.S., Oro, A.E., Kakizuka, A., and Evans, R.M. (1992). Characterization of three RXR genes that mediate the action of 9-cis retinoic acid. *Genes Dev* 6:329–344.

Meyer, D., and Birchmeier, C. (1995). Multiple essential functions of neuregulin in development. *Nature* 378:386–390.

Moss, J.B., Xavier-Neto, J., Shapiro, M.D., et al. (1998). Dynamic patterns of retinoic acid synthesis and response in the developing mammalian heart. *Dev Biol* 199:55–71.

Ng, W.A., Grupp, I.L., Subramaniam, A., and Robbins, J. (1991). Cardiac myosin heavy chain mRNA expression and myocardial function in the mouse heart. *Circ Res* 68:1742–1750.

Nikovits, W., Wang, G.F., Feldman, J.L., et al. (1996). Isolation and characterization of an avian slow myosin heavy chain gene expressed during embryonic skeletal muscle fiber formation. *J Biol Chem* 271:17047–17056.

Oana, S., Machida, S., Hiratsuka, E., Furutani, Y., Momma, K., Takao, A., and Matsuoka, R. (1998). The complete sequence and expression patterns of the atrial myosin heavy chain in the developing chick. *Biol Cell* 90(9):605–613.

O'Brien, T.X., Lee, K.J., and Chien, K.R. (1993). Positional information of ventricular myosin light chain 2 expression in the primitive murine heart tube. *Proc Natl Acad Sci USA* 90:5157–5161.

Osmond, M.K., Butler, A.J., Voon, F.C., and Bellairs, R. (1991). The effects of retinoic acid on heart formation in the early chick embryo. *Development* 113:1405–1417.

Patten, B.M. (1929). *The Early Embryology of the Chick*. Blakiston, Philadelphia.

Rosenquist, G.C. (1985). Migration of precardiac cells from their origin in the epiblast until they form the definitive heart in the chick embryo. In: Ferrans, V.J., Rosenquist, G., and Weinstein, C., eds. *Cardiac Morphogenesis*. Elsevier, New York, pp. 44–53.

Rosenquist, G.C., and DeHaan, R.L. (1966). Migration of precardiac cells in the chick embryo: a radiographic study. *Carnegie Inst Wash Contrib Embryol* 38:111–121.

Ross, R.S., Navankasattusas, S., Harvey, R.P., and Chien, K.R. (1996). An HF-1a/HF-1b/MEF-2 combinatorial element confers cardiac ventricular specificity and established an anterior-posterior gradient of expression. *Development* 122:1799–1809.

Ruzicka, D.L., and Schwartz, R.J. (1988). Sequential activation of alpha-actin genes during avian cardiogenesis: vascular smooth muscle alpha-actin gene transcripts mark the onset of cardiomyocyte differentiation. *J Cell Biol* 107:2575–2586.

Sater, A.K., and Jacobson, A.G. (1989). The specification of heart mesoderm occurs during gastrulation in *Xenopus laevis*. *Development* 105:821–830.

Schwartz, K., Lecarpentier, Y., Martin, J.L., Lompre, A.M., Mercadier, J.J., and Swynghedauw, B. (1981). Myosin isoenzymic distribution correlates with speed of myocardial contraction. *J Mol Cell Cardiol* 13:1071–1075.

Sinha, A.M., Umeda, P.K., Kavinsky, C.J., et al. (1982). Molecular cloning of mRNA sequences for cardiac alpha- and beta-form myosin heavy chains: expression in ventricles of normal, hypothryoid, and thyrotoxic rabbits. *Proc Natl Acad Sci USA* 79:5847–5851.

Smith, S., and Dickman, E.D. (1997). New insights into retinoid signalin in cardiac development and physiology. *Trends Cardiovasc Med* 7:324–329.

Srivastava, D., Thomas, T., Lin, Q., Kirby, M.L., Brown, D., and Olson, E.N. (1997). Regulation of cardiac mesodermal and neural crest development by the bHLH transcription factor, dHAND. *Nat Genet* 16:154–160.

Stainier, D.Y., and Fishman, M.C. (1992). Patterning the zebrafish heart tube: aquisition of anteroposterior polarity. *Dev Biol* 153:91–101.

Sucov, H.M., Dyson, E., Gumeringer, C.L., Price, J., Chien, K.R., and Evans, R.M. (1994). RXR alpha mutant mice establish a genetic basis for vitamin A signaling in heart morphogenesis. *Genes Dev* 8:1007–1018.

Twal, W., Roze, L., and Zile, M.H. (1995). Anti-retinoic acid monoclonal antibody localizes all-trans-retinoic acid in target cells and blocks normal development in early quail embryo. *Dev Biol* 168:225–234.

Yu, V.C., Delsert, C., Andersen, B., Holloway, J.M., Devary, O.V., Naar, A.M., Kim, S.Y., Boutin, J.M., Glass, C.K., and Rosenfeld, M.G. (1991). RXR beta: A coregulator that enhances binding of retinoic acid, thyroid hormone, and vitamin D receptors to their cognate response elements. *Cell* 67:1251–1266.

Wang, G.F., Nikovits, W., Schleinitz, M., and Stockdale, F.E. (1996). Atrial chamber-specific expression of the slow myosin heavy chain 3 gene in the embryonic heart. *J Biol Chem* 271:19836–19845.

Wang, G.F., Nikovits, W., Schleinitz, M., and Stockdale, F.E. (1998). A positive GATA element and a negative vitamin D receptor-like element control atrial chamber-specific expression of a slow myosin heavy-chain gene during cardiac morphogenesis. *Mol Cell Biol* 18:6023–6034.

Wilson, J.G., and Warkany, J. (1949). Aortic arch and cardiac anomalies in the offspring of vitamin A deficient rats. *Am J Anat* 85:113–155.

Xavier-Neto, J., Neville, C.M., Shapiro, M.D., et al. (1999). A retinoic acid-inducible transgenic marker of sino-atrial development in the mouse heart. *Development* 126:2677–2687.

Yutzey, K., Gannon, M., and Bader, D. (1995). Diversification of cardiomyogenic cell lineages in vitro. *Dev Biol* 170:531–541.

Yutzey, K.E., Rhee, J.T., and Bader, D. (1994). Expression of the atrial-specific myosin heavy chain AMHC1 and the establishment of anteroposterior polarity in the developing chicken heart. *Development* 120:871–883.

Zhang, Y., Shafiq, S.A., and Bader, D. (1986). Detection of a ventricular-specific myosin heavy chain in adult and developing chicken heart. *J Cell Biol* 102:1480–1484.

Zhao, D., McCaffery, P., Ivins, K.J., et al. (1996). Molecular identification of a major retinoic-acid-synthesizing enzyme, a retinaldehyde-specific dehydrogenase. *Eur Journ Biochem* 240:15–22.

CHAPTER 6

Neural Crest and Heart Development

Margaret L. Kirby

Ablation of cardiac neural crest results in a unique set of morphologic and function changes in cardiovascular development. The most dramatic morphologic changes are seen after the embryonic period is completed and include defective septation of the cardiac outflow tract and mispatterning (or interruption) of the great arteries, which is associated with hypoplastic development of the pharyngeal glands. These phenotypic changes are accompanied by myocardial functional alterations that include defective excitation-contraction coupling from the earliest time of myocardial function, resulting in heart failure. Abnormal ventricular function begins prior to the time when neural crest cells normally reach the heart in an intact embryo. Because the cardiac neural crest cells migrate initially into the pharyngeal region, where they support normal development of the aortic arch arteries, it was initially thought that abnormal development of the aortic (pharyngeal) arch arteries was likely to impact on ventricular development and would thus explain poor myocardial function. The pharyngeal arteries carry all of the cardiac output during a significant portion of early heart development. Loss of neural crest cells that support the endothelial walls of these vascular channels could change the properties of their walls. However, several attempts to document hemodynamic abnormalities in the aortic arch arteries following cardiac neural crest ablation have failed. Coculture of myocardium with endoderm leads to myocardial functional abnormalities similar to those seen in neural crest–ablated embryos. This has led to the hypothesis that an interaction of neural crest cells with pharyngeal endoderm may be required for inhibition or sequestration of a fibroblast growth factor (FGF)-like factor that is deleterious to myocardial development. This recent finding has produced a dramatic revision in classical thinking about the function of cardiac neural crest cells and the interactions that influence normal heart development.

CARDIAC NEURAL CREST

Specification

Neural crest cells are specified in the zone between the lateral margins of the neural plate and the dorsal ectoderm. Dorsoventral axis formation is critical for determination of the neural crest population. Thus, many genes that play a role in neural crest specification have been identified because they are part of the dorsoventral axis coordinate system of genes. Juxtaposition of neural and nonneural ectoderm induces expression of the dorsal markers Wnt1, Wnt3a, and Slug (Dickinson et al, 1995). One of the earliest dorsal determinants identified was the Pax3 transcription factor, which is characterized by both paired box and homeobox binding domains (Goulding et al, 1991; Chalepakis et al, 1993). The induced mutation of Pax3 in the Splotch mutant mouse is a good phenotypic illustration of an animal lacking normal Pax3 expression (Franz and Kothary, 1993; Conway et al, 1997). While some neural crest cells form, the peripheral manifestation of these cells is quite variable, with some completely missing and others diminished. In addition to dorsoventral axis formation, inductive signals may originate from the dorsolateral mesoderm (Marchant et al, 1998). Cells appear to originate from both neural plate and ectoderm (Mancilla and Mayor, 1996), and it has recently been shown that the helix-loop-helix repressor, Id2, is important for sustaining epidermal traits at the margin between the ectoderm and neural crest (Martinsen and Bronner-Fraser, 1998). It has been suggested that modulation of bone morphogenetic protein (BMP) activity by noggin, an inhibitor of BMP, causes induction of the neural crest from competent ectoderm (Marchant et al, 1998).

Craniocaudal axis specification is also important in neural crest determination. The axial origin of the neural crest determines whether it can produce mesenchymal cells. Traditionally, the neural crest has been designated as trunk or cranial, with the separation designated at somite 5. Cranial neural crest generates much of the mesenchyme for the face and pharyngeal arches, in addition to the peripheral nervous system of the head, neck, and viscera (Horstadius, 1950). A vagal region that generates the enteric and cardiac plexuses is recognized as extending from somite 1 to somite 7 (Le Douarin, 1982). In addition, the region between the midotic placode and somite 3, corresponding to rhombomeres 6, 7, and 8, was designated the cardiac region, not because it contributes exclusively to development of cardiovascular structures, but because its contribution to these structures is essential for viability (Kirby et al, 1985).

Robust regeneration of the preotic neural crest in the preotic caudal midbrain/rostral hindbrain was shown by Bronner-Fraser and colleagues (Scherson et al, 1993). Removal of the cardiac neural crest, which is postotic, results in cardiovascular abnormalities and no reconstitution of the neural crest cells has been shown, although cells from the nodose placodes migrate along similar pathways and settle at the same terminal sites as cardiac neural crest cells in intact embryos (Kirby, 1988a,b). Direct experimental evidence was obtained using cardiac neural crest ablation combined with DiI labeling. No regeneration occurred from either the cardiac crest region or the regions cranial or caudal to it after removal of the neural crest by either incision or laser ablation (Suzuki and Kirby, 1997). This supports the idea that the cardiac region of the neural tube and crest (rhombomeres 6 to 8) is a transition region between trunk and cranial tube/crest, and shares characteristics of both (Kirby, 1993).

Migration

Migration involves a classic epithelial-to-mesenchymal conversion of the neural crest cells as they escape the neural folds (Erickson and Reedy, 1998). A number of factors have been identified in the migrating neural crest cells, including snail, slug, and the winged-helix transcription factor Hfh2, which are all expressed in the premigratory and migrating neural crest (Essex et al, 1993; Mayor et al, 1995; Jiang et al, 1998; Labosky and Kaestner, 1998). The function of the proteins encoded by these genes is still unclear, although in some of the species examined they appear to be involved in the epithelial-to-mesenchymal conversion. RhoB, a member of the rho family of guanosine triphosphate (GTP)-binding proteins, is also expressed in the dorsal neural tube in response to BMP, and persists transiently in migrating neural crest cells (Liu and Jessell, 1998). Inhibition of rhoB by C3 exotoxin prevents the delamination of neural crest cells from the neural tube but does not affect cell lineage decisions or later migration (Liu and Jessell, 1998). The c-*ret* proto-oncogene, which encodes a receptor tyrosine kinase, has been used as a marker for enteric neurons, but it also is present in the vagal neural tube prior to migration (Robertson and Mason, 1995). The spatial restriction of c-*ret* is modulated by retinoic acid, which also affects development of the neural crest arising from the vagal region (Robertson and Mason, 1995).

Neural crest cells migrate as sheets rather than as single cells. This may be important in normal migration because another active gene in migrating neural crest codes for connexin43 (Cx43), a gap junction protein. While it seems counterintuitive that migrating cells would be connected by gap junctions, the expression of Cx43 is accompanied by cell coupling that is maintained until well after the cells stop migrating (Lo et al, 1997; Huang and Lo, 1998). Neural crest cells that are unable to express Cx43 or to form gap junctions, migrate slower than cells with normal gap junctions, while cells that overexpress Cx43 and are hypercoupled migrate faster than normal (Huang and Lo, 1998). While the use of this coupling is not known presently, Lo and colleagues speculate that it plays a role in signal delivery to the entire population of migrating cells, and thus equalizes the availability of signaling information throughout the population (Lo, personal communication).

Cell Lineage

Cell lineage in the neural crest is determined at a number of sites along the migratory pathway. The neural crest cells have mixed potentials as they leave the neural folds, in that some retain pluripotency while others are already committed to a cell lineage (Sieber-Blum and Zhang, 1997). Neurotrophins and many other growth factors affect survival, proliferation, and differentiation into different lineages. These factors work synergistically rather than as individual factors (Sieber-Blum and Zhang, 1997). Stem cell factor is trophic for pluripotent neural crest cells. Basic fibroblast growth factor (bFGF) is mitogenic both for pluripotent cells and committed melanogenic cells. Transforming growth factor β_1 (TGF-β_1) is an antimitotic signal for all neural crest cells and can override a bFGF/neurotrophin proliferative signal (Sieber-Blum and Zhang, 1997). Vascular smooth muscle cells are derived either from the neural crest (great arteries) or mesoderm (all other vessels). Interestingly, cultured vascular smooth muscle cells respond differentially to TGF-β after pretreatment with platelet-derived growth factor (PDGF), depending on their origin. The neural crest–derived cells proliferate while mesodermally derived cell proliferation is inhibited (Wrenn et al, 1993).

CARDIAC NEURAL CREST AND MYOCARDIAL MATURATION

Cardiac neural crest cells begin to migrate about stage 11 in the chick and 9 days postcoitum (dpc) in the mouse embryo. The cells migrate to the circumpharyngeal region located dorsal to the region where the caudal pharynx will form. Because the migration is precocious with regard to formation of the caudal pharyngeal arches, the crest cells pause in the circumpharyngeal ridge and migrate into the arches as they form (Kuratani and Kirby, 1991). After the cardiac crest cells populate the caudal arches, it is several days before some of the cells begin their migration into the outflow tract to participate in septation (Waldo et al, 1998).

Within 24 hours after cardiac neural crest ablation, myocardial development is abnormal, even though cardiac neural crest cells would not have reached the cardiac outflow tract in an intact embryo. The external features of the looping heart in neural crest–ablated embryos include straight outflow limbs, tighter heart loops, and variable dilations (Leatherbury et al, 1996). These hearts are readily distinguishable from wild-type hearts by their variability, although morphometric measurements have not yielded consistent differences. Beginning at stage 14, myocardial intracellular calcium transients are depressed by 30% to 60%, indicating impaired excitation-contraction coupling (Waldo et al, 1999b). This is accompanied by disorganized myofibrils. The thickness of the cardiac jelly, another indicator of myocardial development, is uneven. The extreme variability in external morphology after neural crest ablation may be due to this uneven distribution of cardiac jelly (Waldo et al, 1999). In contrast to the myocardium, endocardial development appears to be normal, as evidenced by normal expression of fibrillin2, normal seeding of the cardiac jelly by mesenchyme, and formation of trabeculae (Waldo et al, 1999b). Formation of trabeculae depends on a heregulin signal produced by the endocardium (Meyer and Birchmeier, 1995). Therefore, the presence of normal trabeculation supports the observation that endocardial development is unaffected by cardiac neural crest ablation.

Waldo et al (1999) determined that the time when initial myocardial abnormalities can be observed corresponds to the time when the neural crest cells begin to populate the caudal pharyngeal arches. This suggests that the presence of cardiac neural crest cells in these arches is critical to normal myocardial maturation. Many of the early myocardial alterations can be mitigated by back-transplanting the cardiac neural crest (Waldo et al, 1999), showing that the early changes are not due to a surgical effect or some other artifact of embryo manipulation.

Cell tracing studies have shown that the neural crest cells do not interact directly with the myocardium at the stage when these changes are first observed. Furthermore, various attempts to document hemodynamic alterations of the aortic arch arteries that might cause such myocardial changes have been unsuccessful (Leatherbury, Jackson, Stewart, Connuck, unpublished observations). However, we have observed that the pharyngeal endoderm lies in apposition to the myocardium at stages 13 to 15 in cardiac neural crest–ablated embryos (Farrell et al, 2000) until mesenchyme derived from the epipharyngeal placodes and lateral plate mesoderm intervenes (Kirby, 1988b).

Head Organizer

The cranial pharyngeal endoderm, recently recognized as a head organizer (Beddington and Robertson, 1998), releases factors that induce myocardial development from lateral plate mesoderm (Jacobson and Duncan, 1968). The pharyngeal endoderm releases signals for myocardial induction long after they are needed for development of stable sarcomere expression, myofibril organization, and beating (Gannon and Bader, 1995). This signal retains potency in lateral plate mesoderm until formation of the third pharyngeal pouch (Sater and Jacobson, 1990; Farrell and Kirby, unpublished). The evidence for this conclusion is from experiments in which explants of lateral plate mesoderm, when left in the presence of endoderm, show an enhanced rate of myocyte differentiation and shortened delay between expression of myosin heavy chain and onset of beating (Antin et al, 1994). This indicates that the pharyngeal endoderm continues to release factors capable of influencing myocardial development long after the initial inductive period. Although the identity of the particular factors remains unknown, it has been shown that pharyngeal endoderm produces activin-A, FGF-2 and -4, insulin-like growth factor (IGF)-I and -II, insulin, and BMP-2 (Muslin and Williams, 1991; Mangiacapra et al, 1995; Sugi et al, 1995; Antin et al, 1996; Zhu et al, 1996; Schultheiss et al, 1997).

Neural Crest, Endoderm, and Myocardium

The earliest consistent finding after neural crest ablation is altered ventricular function that occurs prior to the time that neural crest cells would normally enter the heart proper. At this time, cardiac neural crest cells populate the caudal pharyngeal arches (Waldo et al, 1996). Myofibrillogenesis and electrical activity begin in the chick myocardium at about stages 7.5 to 8 (Kamino et al, 1981; Tokuyasu and Maher, 1987; Lyons, 1994) with striated myofibrils appearing in scattered cardiomyocytes at about stage 10, when the first weak contractions are seen (Hibbs, 1956; Manasek, 1970; Han et al, 1992; Hiruma and Hirakow, 1985). Anterograde circulation of blood begins at stage 14 (Romanoff, 1960), coincident with the first myocardial abnormalities seen in cardiac neural crest–ablated embryos.

Recent data from our group show that a proliferating population of neural crest cells in the pharynx forms a substantial cellular partition between vascular endothelial and endodermal cells. It is known that neural crest cells can respond to many of the factors released by pharyngeal endoderm (Stocker et al, 1991; Bannerman and Pleasure, 1993; Murphy et al, 1994; Mayor et al, 1997; LaBonne and Bronner-Fraser, 1998). Thus, in the presence of an intact neural crest population, these same factors presumably have a reduced opportunity for influencing either endothelial or myocardial development.

To determine whether pharyngeal endoderm could be responsible for some of the myocardial changes seen in neural crest–ablated embryos, Farrell et al (2000) cocultured stages 12 to 13 myocardium with stage 14 pharyngeal endoderm for 24 hours. They showed that the decrease in the magnitude of the field-stimulated intracellular myocardial calcium transients was similar to that seen in myocardium from stage 14 neural crest–ablated embryos. FGF-2, one of the factors released by the pharyngeal endoderm, also caused a decrease (33%) in the myocardial calcium transient at stage 14, and the effect was abolished by anti–FGF-2 neutralizing polyclonal antibody. Anti–FGF-2 antibody also neutralized the effect of the pharyngeal endoderm on the myocardial calcium transients. In contrast, TGF-β_1 caused an increase in the calcium transients.

These results support the idea that the cardiac neural crest cells interpose between the pharyngeal endoderm and the myocardium and intercept or alter an FGF-like signal that is deleterious to myocardial development. Additional studies indicate that abnormal myocardial development at these early stages is associated with malalignment defects such as dextroposed aorta (Farrell et al, 1999b). This may provide a common mechanism to explain the presence of these defects in a number of experimental paradigms (see below).

NEURAL CREST AND CARDIOVASCULAR PATTERNING

Great Arteries from Pharyngeal Arch Arteries

The third, fourth, and sixth aortic arch arteries develop from endothelial precursors that form incomplete cords abutting the pharyngeal endoderm (de Ruiter et al, 1991). These precursors become connected and lumenized as they are separated from the pharyngeal endoderm by cardiac neural crest cells migrating into the arch and ensheathing the nascent artery (Waldo et al, 1996). In the absence of cardiac neural crest, the arch arteries lumenize but fail to be separated from the pharyngeal endoderm, showing that neural crest is not needed for initial formation of the aortic arch arteries. However, the arteries in crest-ablated embryos quickly become misshapen and sinusoidal with abnormal connections to the dorsal aorta and a marked loss of the bilateral symmetry that characterizes early development of normal aortic arch arteries (Bockman et al, 1987, 1989). This suggests that patterning of the arch arteries is unstable in the absence of neural crest (Waldo et al, 1996). Normally aortic arch arteries 3, 4, and 6 are patent channels that connect the ventrally located aortic sac with the dorsal aorta by stages 19 to 20. By stages 20 to 21, several layers of neural crest cells separate each arch artery and the aortic sac from the pharyngeal endoderm. Each nascent artery is surrounded by a sheath of neural crest cells that begin differentiating into the tunica media of the great arteries (Le Lièvre and Le Douarin, 1975).

Since it appeared from ablation studies that cardiac neural crest cells were not essential for initial patterning of the aortic arch arteries but were critical for correct remodeling of the arch arteries to the great arteries, a method was needed to perturb this patterning function directly. The chick embryo is an excellent candidate for such studies because of the ease of manipulating tissues, i.e., removing and back-transplanting neural crest. To determine directly whether neural crest has a role in repatterning of the aortic arch arteries into the great arteries, expression of paralogous groups of *Hox* genes in pharyngeal arches 3, 4, and 6 was targeted for attenuation (Kirby et al, 1997). Treatment and back-transplantation of cardiac neural folds with antisense oligonucleotides targeted to *Hox* paralogous groups 3, 4 and 5 cause altered repatterning of the aortic arch arteries in the pharyngeal arch where that message is expressed most cranially. In this case, targeting of paralogous group 3 *Hox* message causes aortic arch 3 to regress in a manner similar to aortic arch 2, such that the common carotid artery is missing. Antisense targeted to paralogous group 5 *Hox* message caused the appearance of an additional pharyngeal arch containing a novel and completely independent aortic arch artery that formed a double aortic arch on the right side. In both experiments, the final pattern of the great arteries was abnormal, but outflow septation developed normally. This study demonstrates two things: (1) cardiac neural crest plays an active role in repatterning the aortic arch arteries to form the great arteries, and (2) abnormalities in the aortic arches do not necessarily lead to cardiac outflow anomalies.

A possibility for communication in the cardiovascular system that has been neglected in embryonic development is longitudinal conduction via the endothelial cell networks. Dye coupling via gap junctions is observed between endothelial cells and between endothelial and smooth muscle cells (Little et al, 1995). The cell dimensions and orientation of endothelial cells along with high junctional plaque density (eight times that of smooth muscle cells) indicate that endothelial cells form a permissive pathway for conduction of signals longitudinally (up to 1 mm) through blood vessels (Haas and Duling, 1997). Cardiac neural crest cells are the smooth muscle precursors for the aortic arch arteries and are known to express Cx43 (Lo et al, 1999). They attain an intimate relationship with endothelium or the endothelial precursors of the aortic arch arteries in a time frame that could account for alterations in ventricular function and cardiac looping morphology. Furthermore, longitudinal signal transmission can be propagated for 1 mm or more, which is well within the range of endoderm–neuralcrest–endothelium–myocardium geometry from the pharyngeal region to the myocardial cuff. Thus, it is possible that contact of neural crest cells with endothelium in the pharyngeal region triggers a signal that could be propagated into the heart. Alternatively, sequestration of a factor or release of an inhibitory factor by neural crest cells could provoke or prevent longitudinal propagation of a signal by the endothelium. The endothelium is continued into the heart as the endocardium, which is intimately involved in myocardial development, as evidenced by the recent knockout studies of heregulin and its receptors (Meyer and Birchmeier, 1995).

It has been known for some time that signals pass between the endocardial and myocardial layers of the developing heart. Several identified factors pass from myocardium to endocardium in induction of the epithelial-mesenchymal transformation of endocardial cells to form the cardiac cushions (Eisenberg and Markwald, 1995). A specific signaling system has been described recently that mediates signaling from endocardium to myocardium. This system, composed of the heregulin (HRG, neuregulin) growth factors and their receptors, erbB1, erbB3, and erbB4, is essential for normal trabeculation and cushion development (Meyer and Birchmeier, 1995). Heregulin is expressed by endocardial cells in the heart either as a secreted growth factor or as a membrane-bound form. Myocardial cells express the receptors erbB2, erbB3, and erbB4. When expression of erbB2, erbB4, or heregulin is disrupted, the mutant embryos die after failing to form trabeculae. Disruption of the *erbB3* gene results in cardiac cushion abnormalities leading to death at embryonic day 13.5. These experiments show the potential for existence of communications between the endothelium and myocardium.

Cardiac Outflow Tract Septation

It has been possible to study cardiac neural crest in outflow septation in chicks, using quail-chick chimeras (Waldo et al, 1998), and in mice using a transgenic mouse in which a portion of the Cx43 promoter drives expression of lacZ (Lo et al, 1997; Huang et al, 1998; Waldo et al, 1999). Outflow tract septation begins when a condensed shelf of neural crest cells bulges into the dorsal wall of the aortic sac between the origins of aortic arch arteries 4 and 6. The condensed mesenchyme, composed of neural crest cells, straddles the aortic sac. Two prongs of condensed neural crest cells extend into the truncus. As the shelf elongates at the expense of the prongs, the distal lumen of the outflow tract is divided into the aorta and pulmonary trunks. This septation complex ends abruptly just below the level where the semilunar valves form. However, numerous neural crest cells continue into the conal cushions that are being invaded by myocardial cells prior to (chick) or just after (mouse) closure of the proximal outflow tract septum. It has been suggested that the neural crest cells in the conal cushions are not only responsible for orchestrating conal septation but they attract myocardial cells needed for muscularization of the conal cushions, although there is currently no direct evidence that this is the case (Waldo et al, 1998; Poelmann et al, 1998). The conal septum closes zipper-like at the same time and as a single unit with the ventricular septum. The entire seam of closure was underlaid with neural crest cells. Crest cells extend to the boundary of the pulmonary infundibulum with the right ventricular wall (Waldo et al, 1998; Waldo et al, 1999a).

Cardiac Outflow Tract Alignment

One of the phenotypes associated with cardiac neural crest ablation is overriding aorta. In many cases the aorta overrides the ventricular septum to such an extent that more than 50% of the blood comes from the right ventricle. In this case the heart is described as having double outlet right ventricle (Lev et al, 1972; Anderson et al, 1974). This type of defect is not characterized by absence of the outflow septum but rather malalignment of the outflow tract with respect to the ventricles. These defects have been found after a number of diverse experimental manipulations and are a rather common feature in mutant or transgenic mice. Because of the diversity of conditions that can cause these malalignments, it is thought that neural crest ablation is only one of the precipitating factors in their etiology, and that a common pathway must be identified to understand their genesis. Tomita et al (1991) documented that hearts with persistent truncus arising from the right ventricle and with double outlet right ventricle show poor ventricular contractility well before the neural crest arrives in the outflow tract to begin the septation process. The depressed contractility is specifically correlated with cardiac neural crest ablation. In a recent study, antisense treatment of the cardiac neural crest targeted to HIRA with subsequent replacement of the neural folds resulted in a significant elevation of the incidence of persistent truncus arteriosus. Interestingly, the cardiac output and ejection fraction at stage 18 in the HIRA antisense treatment groups did not show the characteristics of neural crest ablation. This provides the first evidence that nonseptation of the outflow tract can be separated from the myocardial dysfunction seen after neural crest ablation (Farrell et al, 1999). It also suggests that the presence of cardiac neural crest in the pharyngeal region is important in preventing myocardial dysfunction. The persisting single vessels in this study arose from the normal position of the aorta and pulmonary trunk rather than originating from the right ventricle, as is seen after neural crest ablation. Thus, the outflow tract was correctly aligned even though the outflow septum failed to form. This suggests that outflow tract alignment may be correlated with early myocardial dysfunction and may provide a common etiology for the double outlet right ventricle seen in a number of experimental models.

SUMMARY AND CONCLUSIONS

Early studies of cardiac neural crest have focused on its role in cardiac outflow tract septation. The steps in this process are now well described in both chick and mouse embryonic development. However, almost nothing is known about the molecular program used by neural crest cells in orchestrating outflow tract septation.

The role of cardiac neural crest cells in the pharynx is quite complex and has taken longer to discern. The crest cells are necessary for appropriate remodeling of the aortic arch arteries into the great vessels, but in addition, are required for normal functional maturation of the myocardium. Patterning instructions from the cardiac neural crest cells may involve expression of the *Hox* genes. The cells may interact with pharyngeal endoderm to intercept an FGF-2–like signal that would otherwise disturb development of the myocardial calcium transient needed for efficient contraction. In addition, the neural crest cells release TGF-β or an activating protease that that may be needed for development and maturation of the sarcoplasmic reticulum. These findings show that heart development is influenced and fine-tuned by signaling from extracardiac tissues. The head organizer, which plays an essential role in cranial development, may be a major factor in these processes, due to its proximity to the cardiogenic mesenchyme during myocardial induction and to the inflow and outflow segments of the cardiac tube later (Beddington and Robertson, 1998). Thus, gene expression and morphogenesis of the cardiogenic mesenchyme and early heart tube are most likely controlled by interactions of the head organizer (Jacobson, 1960) with other noncardiac tissues such as the neural crest.

ACKNOWLEDGMENTS

My sincere thanks to all of my colleagues who have contributed to what we know about cardiac neural crest and heart development and especially to Ms. Karen Waldo and Dr. Michael Farrell, for continuing discussion and assistance with this manuscript, and to Dr. Rhea Beth Markowitz, for editorial assistance. This work was supported by National Institutes of Health grants HL36059 and HD17063.

REFERENCES

Anderson, R.H., Wilkinson, J.L., Arnold, Becker, A.E., and Lubkiewicz, K. (1974). Morphogenesis of bulboventricular malformations. II. Observations on malformed hearts. *Br Heart J* 36:948.

Antin, P.B., Taylor, R.G., and Yatskievych, T. (1994). Precardiac mesoderm is specified during gastrulation in quail. *Dev Dyn* 200:144–154.

Antin, P.B., Yatskievych, T., Dominguez, J.L., and Chieffi, P. (1996). Regulation of avian precardiac mesoderm development by insulin and insulin-like growth factors. *J Cell Physiol* 168:42–50.

Bannerman, P.G., and Pleasure, D. (1993). Protein growth factor requirements of rat neural crest cells. *J Neurosci Res* 36:46–57.

Beddington, R.S., and Robertson, E.J. (1998). Anterior patterning in mouse. *TIG* 14: 277–284.

Bockman, D.E., Redmond, M.E., and Kirby, M.L. (1989). Alteration of early vascular development after ablation of cranial neural crest. *Anat Rec* 225:209–217.
Bockman, D.E., Redmond, M.E., Waldo, K., Davis, H., and Kirby, M.L. (1987). Effect of neural crest ablation on development of the heart and arch arteries in the chick. *Am J Anat* 180:332–341.
Chalepakis, G., Stoykova, A., Wijnholds, J., Tremblay, P., and Gruss, P. (1993). Pax: gene regulators in the developing nervous system. *J Neurobiol* 24:1367–1384.
Conway, S.J., Henderson, D.J., Kirby, M.L., Anderson, R.H., and Copp, A.J. (1997). Development of a lethal congenital heart defect in the splotch (Pax3) mutant mouse. *Cardiovasc Res* 36:163–173.
de Ruiter, M.C., Hogers, B., Poelmann, R.E., Vaniperen, L., and Gittenberger-de Groot, A.C. (1991). The development of the vascular system in quail embryos. A combination of microvascular corrosion casts and immunohistochemical identification. *Scanning Microsc* 5:1081–1090.
Dickinson, M.E., Selleck, M.A.J., McMahon, A.P., and Bronner-Fraser, M. (1995). Dorsalization of the neural tube by the non-neural ectoderm. *Development* 121:2099–2106.
Eisenberg, L.M., and Markwald, R.R. (1995). Molecular regulation of atrioventricular valvuloseptal morphogenesis. *Circ Res* 77:1–6.
Erickson, C.A., and Reedy, M.V. (1998). Neural crest development: the interplay between morphogenesis and cell differentiation. *Curr Top Dev Biol* 40:177–209.
Essex, L.J., Mayor, R., and Sargent, M.G. (1993). Expression of *Xenopus* snail in mesoderm and prospective neural fold ectoderm. *Dev Dyn* 198:108–122.
Farrell, M.J., Stadt, H.A., Wallis, K.T., Scambler, P., Hixon, R., Wolfe, R.R., Leatherbury, L., and Kirby, M.L. (1999). HIRA, a DiGeorge syndrome candidate gene is required for normal outflow tract septation. *Circ Res* 84:127–135.
Farrell, M.J., Burch, J.L., Rowley, L., Kumiski, D., Stadt, H., Godt, R.E., Creazzo, T.L., and Kirby, M.L. (2000). Pharyngeal endoderm produces a factor that suppresses development of myocardial calcium transients (submitted).
Franz, T., and Kothary, R. (1993). Characterization of the neural crest defect in Splotch (Sp^{1H}) mutant mice using a lacZ transgene. *Dev Brain Res* 72:99–105.
Gannon, M., and Bader, D. (1995). Initiation of cardiac differentiation occurs in the absence of anterior endoderm. *Development* 121:2439–2450.
Goulding, M.D., Chalepakis, G., Deutsch, U., Erselius, J.R., and Gruss, P. (1991). Pax-3, a novel murine DNA binding protein expressed during early neurogenesis. *EMBO J* 10:1135–1147.
Han, Y., Dennis, J.E., Cohen-Gould, L., Bader, D.M., and Fischman, D.A. (1992). Expression of sarcomeric myosin in the presumptive myocardium of chicken embryos occurs within six hours of myocyte commitment. *Dev Dyn* 193:257–265.
Haas, T.L., and Duling, B.R. (1997). Morphology favors an endothelial cell pathway for longitudinal conduction within arterioles. *Microvasc Res* 53:113–120.
Hibbs, R.G. (1956). Electron microscopy of developing cardiac muscle in chick embryos. *Am J Anat* 99:17–52.
Hiruma, T., and Hirakow, R. (1985). An ultrastructural topographical study on myofibrillogenesis in the heart of the chick embryo during pulsation onset period. *Anat Embryol* 172:325–329.
Horstadius, S. (1950). *The Neural Crest. Its Properties and Derivatives in the Light of Experimental Research.* Oxford University Press, London.
Huang, G.Y., Cooper, E.S., Waldo, K., Kirby, M.L., Gilula, N.B., and Lo, C.W. (1998). Gap junction-mediated cell-cell communication modulates mouse neural crest migration. *J Cell Biol* 143:1725–1734.
Huang, G.-Y., and Lo, C.W. (1998). Modulation of neural crest migration by Cx43 mediated gap junctional communication.
Jacobson, A.G., and Duncan, J.T. (1968). Heart induction in salamanders. *Exp Zool* 167: 79–103.

Kamino, K., Hirota, A., and Fujii, S. (1981). Localization of pacemaking activity in early embryonic heart monitored using voltage-sensitive dye. *Nature* 290:595–597.

Jiang, R.L., Lan, Y., Norton, C.R., Sundberg, J.P., and Gridley, T. (1998). The slug gene is not essential for mesoderm or neural crest development in mice. *Dev Biol* 198:277–285.

Kirby, M.L. (1988a). Nodose placode contributes autonomic neurons to the heart in the absence of cardiac neural crest. *J Neurosci* 8:1089–1095.

Kirby, M.L. (1988b). Nodose placode provides ectomesenchyme to the developing heart in the absence of cardiac neural crest. *Cell Tissue Res* 252:17–22.

Kirby, M.L. (1993). Cellular and molecular contributions of the cardiac neural crest to cardiovascular development. *Trends Cardiovasc Med* 3:18–23.

Kirby, M.L., Hunt, P., Wallis, K.T., and Thorogood, P. (1997). Normal development of the cardiac outflow tract is not dependent on normal patterning of the aortic arch arteries. *Dev Dyn* 208:34–47.

Kirby, M.L., Turnage, K.L., and Hays, B.M. (1985). Characterization of conotruncal malformations following ablation of "cardiac" neural crest. *Anat Rec* 213:87–93.

Kuratani, S.C., and Kirby, M.L. (1991). Initial migration and distribution of the cardiac neural crest in the avian embryo: an introduction to the concept of the circumpharyngeal crest. *Am J Anat* 191:215–227.

LaBonne, C., and Bronner-Fraser, M. (1998). Neural crest induction in *Xenopus*: evidence for a two-signal model. *Development* 125:2403–2414.

Labosky, P.A., and Kaestner, K.H. (1998). The winged helix transcription factor Hfh2 is expressed in neural crest and spinal cord during mouse development. *Mech Dev* 76:185–190.

Leatherbury, L., Yun, J.S., and Wolfe, R. (1996). Association of abnormal configuration of the heart tube with depressed contractility after cardiac neural crest ablation. *Ped Res* 39:62A.

Le Douarin, N. (1982). *The Neural Crest*. Cambridge University Press, Cambridge.

Le Lièvre, C.S., and Le Douarin, N.M. (1975). Mesenchymal derivatives of the neural crest. Analysis of chimaeric quail and chick embryos. *J Embryol Exp Morphol* 34:125–154.

Lev, M., Bharati, S., Meng, L., Liberthson, R.R., Paul, M.H., and Idriss, F.A. (1972). A concept of double-outlet right ventricle. *J Thorac Cardiovasc Surg* 64:271.

Little, T.L., Beyer, E.C., and Duling, B.R. (1995). Connexin 43 and connexin 40 gap junctional proteins are present in arteriolar smooth muscle and endothelium in vivo. *Am J Physiol* 268:H729–H739.

Liu, J.P., and Jessell, T.M. (1998). A role for rhoB in the delamination of neural crest cells from the dorsal neural tube. *Development* 125:5055–5067.

Lo, C.W., Cohen, M.F., Huang, G.Y., et al. (1997). Cx43 gap junction gene expression and gap junctional communication in mouse neural crest cells. *Dev Genet* 20:119–132.

Lo, C.W., Waldo, K.L., and Kirby, M.L. (1999). Gap junction communication and the modulation of cardiac neural crest cells. *Trends Cardiovasc Med* 9 (3–4):63–69.

Lyons, G.E. (1994). In situ analysis of the cardiac muscle gene program during embryogenesis. *Trends Cardiovasc Med* 4:70–77.

Manasek, F.J. (1970). Histogenesis of the embryonic myocardium. *Am J Cardiol* 25:149–168.

Mancilla, A., and Mayor, R. (1996). Neural crest formation in *Xenopus laevis*: mechanisms of Xslug induction. *Dev Biol* 177:580–590.

Mangiacapra, F.J., Fransen, M.E., and Lemanski, L.F. (1995). Activin A and transforming growth factor-β stimulate heart formation in axolotls but do not rescue cardiac lethal mutants. *Cell Tissue Res* 282:227–236.

Marchant, L., Linker, C., Ruiz, P., Guerrero, N., and Mayor, R. (1998). The inductive properties of mesoderm suggest that the neural crest cells are specified by a BMP gradient. *Dev Biol* 198:319–329.

Martinsen, B.J., and Bronner-Fraser, M. (1998). Neural crest specification regulated by the helix-loop-helix repressor Id2. *Science* 281:988–991.

Mayor, R., Guerrero, N., and Martínez, C. (1997). Role of FGF and noggin in neural crest induction. *Dev Biol* 189:1–12.

Mayor, R., Morgan, R., and Sargent, M.G. (1995). Induction of the prospective neural crest of *Xenopus*. *Development* 121:767–777.
Meyer, D., and Birchmeier, C. (1995). Multiple essential functions of neuregulin in development. *Nature* 378:386–390.
Murphy, M., Reid, K., Furness, J.B., and Bartlett, P.F. (1994). FGF2 regulates proliferation of neural crest cells, with subsequent neuronal differentiation regulated by LIF or related factors. *Development* 120:3519–3528.
Muslin, A.J., and Williams, L.T. (1991). Well-defined growth factors promote cardiac development in axolotl mesodermal explants. *Development* 112:1095–1101.
Poelmann, R.E., Mikawa, T., and Gittenberger-de Groot, A.C. (1998). Neural crest cells in outflow tract septation of the embryonic chicken heart: differentiation and apoptosis. *Dev Dyn* 212:373–384.
Robertson, K., and Mason, I. (1995). Expression of ret in the chicken embryo suggests roles in regionalisation of the vagal neural tube and somites and in development of multiple neural crest and placodal lineages. *Mech Dev* 53:329–344.
Romanoff, A.L. (1960). The Avian Embryo: Structural and Functional Development. New York: The Macmillan Company.
Sater, A.K., and Jacobson, A.G. (1990). The restriction of the heart morphogenetic field in *Xenopus laevis*. *Dev Biol* 140:328–336.
Scherson, T., Serbedzija, G., Fraser, S., and Bronner-Fraser, M. (1993). Regulative capacity of the cranial neural tube to form neural crest. *Development* 118:1049–1062.
Schultheiss, T.M., Burch, J.B.E., and Lassar, A.B. (1997). A role for bone morphogenetic proteins in the induction of cardiac myogenesis. *Genes Dev* 11:451–462.
Sieber-Blum, M., and Zhang, J.M. (1997). Growth factor action in neural crest cell diversification. *J Anat* 191:493–499.
Stocker, K.M., Sherman, L., Rees, S., and Ciment, G. (1991). Basic FGF and TGF-β1 influence commitment to melanogenesis in neural crest-derived cells of avian embryos. *Development* 111:635–645.
Sugi, Y., Sasse, J., Barron, M., and Lough, J. (1995). Developmental expression of fibroblast growth factor receptor-1 (cek-1;flg) during heart development. *Dev Dyn* 202:115–125.
Suzuki, H.R., and Kirby, M.L. (1997). Absence of neural crest cell regeneration from the postotic neural tube. *Dev Biol* 184:222–233.
Tokuyasu, T.K., and Maher, P.A. (1987). Immunocytochemical studies of cardiac myofibrillogenesis in early chick embryos. I. Presence of immunofluorescent titin spots in premyofibril stages. *J Cell Biol* 105:2781–2793.
Tomita, H., Connuck, D.M., Leatherbury, L., and Kirby, M.L. (1991). Relation of early hemodynamic changes to final cardiac phenotype and survival after neural crest ablation in chick embryos. *Circulation* 84:1289–1295.
Waldo, K.L., Kumiski, D., and Kirby, M.L. (1996). Cardiac neural crest is essential for the persistence rather than the formation of an arch artery. *Dev Dyn* 205:281–292.
Waldo, K.L., Lo, C.W., and Kirby, M.L. (1999a). Cx43 expression reflects neural crest patterns during cardiovascular development. *Dev Biol* 208:307–323.
Waldo, K.L., Miyagawa-Tomita, S., Kumiski, D., and Kirby, M.L. (1998). Cardiac neural crest cells provide new insight into septation of the outflow tract: aortic sac to ventricular septal closure. *Dev Biol* 196:2129–2144.
Waldo, K.L., Zdanowicz, M., Burch, J., Kumiski, D.H., Godt, R.E., Creazzo, T.L., and Kirby, M.L. (1999b). A novel role for cardiac neural crest in heart development. *J Clin Invest* 103:1499–1507.
Wrenn, R.W., Raeuber, C.L., Herman, I.E., Walton, W.J., and Rosenquist, T.H. (1993). Transforming growth factor-beta: Signal transduction via protein kinase C in cultured embryonic vascular smooth muscle cells. *In Vitro Cell Dev Biol* 29A:73–78.
Zhu, X., Sasse, J., McAllister, D., and Lough, J. (1996). Evidence that fibroblast growth factors 1 and 4 participate in regulation of cardiogenesis. *Dev Dyn* 207:429–438.

CHAPTER 7

Cardiac Conduction System Development

Takashi Mikawa, Robert G. Gourdie, Jeanette Hyer, and Kimiko Takebayashi-Suzuki

A rhythmic heart beat is coordinated by pacemaking impulses generated and conducted through the excitation-conduction system. The physiologic, anatomic, and cellular characteristics of this specialized cardiac tissue have been studied in great detail in the last decades. The advent of recent molecular approaches has now enabled us to address the mechanism that regulates differentiation and patterning of the conduction system within the developing embryo. Information accumulated in the last few years has proven that the conduction system is established by conversion of beating myocytes into conducting cells. The differentiation and patterning of the conducting cell network appears to be regulated by molecular signals derived from nonmuscle cardiac cell populations.

EXCITATION-CONDUCTION SYSTEM NETWORK

A rhythmic heart beat is initiated and coordinated by the precisely timed sequence of pacemaking impulse propagation through the cardiac exitation-conduction system (Tawara, 1906; Goldenberg and Rothberger, 1936; Bozler, 1942). This specialized cardiac tissue consists of several distinct subcomponents (Figure 7.1), including the sinoatrial (SA) node, the atrioventricular (AV) node and ring, the AV bundle, and Purkinje fibers. The cardiac pacemaking impulse is rhythmically provoked at the SA node (Keith and Flack, 1907; Brooks and Lu, 1972) and is conducted across the atrial chambers inducing their contraction (Wenckebach, 1906; Thörel, 1909; Robb and Petri, 1961; Brooks and Lu, 1972). The pacemaking impulses do not spread randomly to the ventricle, but instead first converge on the AV node (Tawara, 1906) and AV ring, a cellular bundle circumscribing the right and left atrioventricular valves. From the AV node, impulses are rapidly propagated along the AV bundle (His, 1893) and its branched limbs (Tawara, 1906), spreading into the working ventricular muscle via the Purkinje fiber network (Purkinje, 1845; Kölliker, 1902; Tawara, 1906). Dysfunction of this essential cardiac tissue in the adult is a direct cause of arrhythmias, leading to sudden death. Defects in conduction system development result in pediatric arrhythmias and conduction block in infants and children. Future therapeutic approaches to repair pacemaking and impulse-conducting tissue after heart injury or congenital disease will

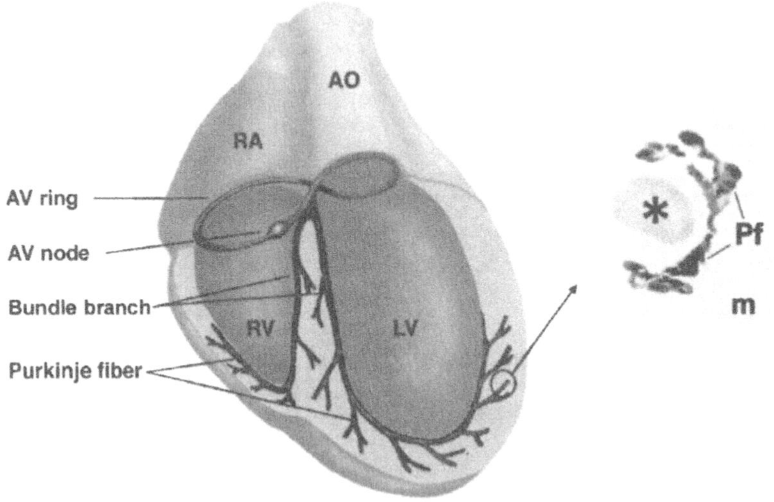

FIGURE 7.1. Diagram of the conduction system of the chicken heart. The central conduction network consists of the atrioventricular (AV) ring, node, and bundle branches. The peripheral conduction system is composed of subendocardial and periarterial Purkinje fiber networks (inset). Asterisk indicates the lumen of intramural coronary artery. AO: aorta, LV: left ventricle, m: myocardium, Pf: Purkinje fiber, RA: right atrium, RV: right ventricle.

benefit from a clear understanding of the mechanisms that regulate differentiation of this essential tissue.

TOPOLOGICAL SHIFT OF IMPULSE CONDUCTION PATHWAY DURING HEART DEVELOPMENT

The primitive heart begins its beating as a double-walled tube—the outer myocardium and the inner endocardium (Manasek, 1968). By this stage, epithelioid myocytes all become electrically active, but action potentials are evoked most frequently by myocytes in the posterior inflow tract (the presumptive sinus venosus and atrium) (Kamino et al, 1981; Yada et al, 1985; Kamino, 1991). These impulses propagate to the anterior end of the heart through gap junctions between the epithelioid myocytes, generating a caudal-to-rostral contractile wave in the tubular heart (Figure 7.2). Thus, at this stage, except for the pacemaker cells, the cardiac conduction system has not yet developed. The primitive tubular heart partitions further into four chambers; atrial and ventricular chambers are first confined by atrioventricular septa and then divided into right and left chambers by interatrial and interventriular septa. Once the four-chambered heart is established, electrical impulses from atria need to be transmitted to the ventricular apex, avoiding direct propagation to the basal part of the ventricle (Fig. 7.2). Recent electrophysiologic studies of developing chick hearts have demonstrated that propagation of pacemaking impulses in the ventricle changes from base-to-apex to apex-to-base when the interventricular septation is completed at embryonic day 7 (E7) (Chuck et al, 1997). This topologic shift of the impulse-transmission pathway in the ventricle depends on the differentiation and patterning of the cardiac conduction system.

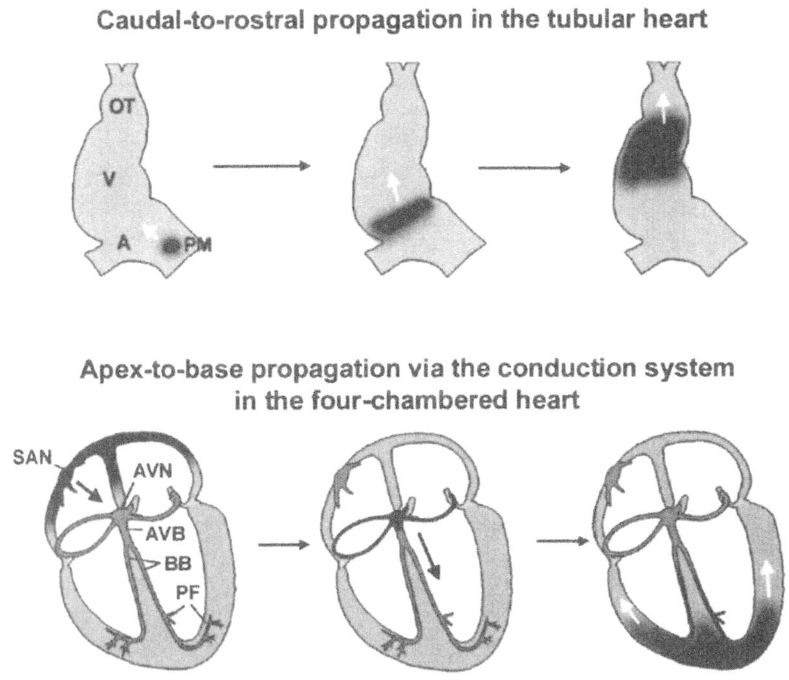

FIGURE 7.2. Topological shift of impulse-conducting pathways between the two-chambered tubular heart and the four-chambered heart (modified from Mikawa, 1998). While (top) and black (bottom) arrows: direction of propagation of action potentials that induce contractile wave. Filled area: action potential-positive myocytes. A: atria, AVB: AV-bundle, AVN: AV-node, BB: bundle branches, OT: outflow tract, PF: Purkinje fibers, SAN: sinoatrial node, V: ventricle.

DEVELOPMENTAL PATHWAY OF THE CARDIAC CONDUCTION SYSTEM

Cells in the conduction system are usually characterized by a diameter considerably greater than cardiac muscle cells. Conduction cells have a unique set of ion channels (Cavalie et al, 1983; Callewaert et al, 1986; Hagiwara et al, 1988) and channel-associated proteins, such as minK (Vetter et al, 1996; McDonald et al, 1997; Lei et al, 1998; Kupershmidt et al, 1999), and unique connexins for gap-junctional electrical couplings between conduction cells (van Kempen et al, 1991; Gourdie et al, 1992, 1993a,b; Bastide et al, 1993; Kanter et al, 1993; Gros et al, 1994; Coppen et al, 1999). In addition, several cardiac muscle-specific genes are not expressed in conduction cells (Schiaffino, 1997; Alyonycheva et al, 1997; Moorman et al, 1998; Gourdie et al, 1998). Instead, genes usually functioning in neural or skeletal muscle tissues, including neurofilament peptides, brain-associated glycoproteins, and slow skeletal myosin heavy chain, are expressed in cells of the conduction system (reviewed in Gorza et al, 1994; Mikawa and Fischman, 1996; Schiaffino, 1997; Moorman et al, 1998; Gourdie et al, 1999a).

By probing for the unique cellular and molecular characteristics, the development of the cardiac conduction system in embryonic hearts (Viragh and Challice, 1982) has been studied in several species including human (Wessels et al, 1992) rat

(Gourdie et al, 1992, 1993b; Coopen et al, 1999) and chick (Davies, 1930; Vassal-Adams, 1982; Chan-Thomas et al, 1993; Gourdie et al, 1993a). A ring-like cluster of cells, often referred as the "primary conduction ring" (Wessels et al, 1992), at the junction of presumptive right and left ventricles, has been proposed as the putative initiation site of conduction system formation. In the chicken, the primary conduction ring can be identified at E2 as cells expressing a homeobox gene, *Msx-2* (Chan-Thomas et al, 1993). *Msx-2* expression extends to the AV node, but does not occur in the peripheral conduction network, including the bundle branches and Purkinje fibers (Chan-Thomas et al, 1993) even when their differentiation becomes detectable by other markers at E10 (Gourdie et al, 1993a). Based on the proximal-distal wave of conduction system development, a model has been proposed wherein a single primary conduction ring contains precursor cells that provide progeny to differentiate into the entire conduction network (Wessels et al, 1992; Lamers et al, 1991; Chan-Thomas et al, 1993; Fishman and Chien, 1997; Moorman et al, 1998). This model, however, has been questioned, since non–DNA-synthesizing cells in the tubular heart of early chick embryos map to primitive conduction tissues (Thompson et al, 1990).

In addition to debate on the pathway underpinning conduction system development, parental cell types giving rise to conduction cells remain uncertain. Because of the coexpression of neural and muscle genes in cells of the conduction system, two possible origins, myogenic (Pattern and Kramer, 1933; Pattern, 1956) and neural crest (Gorza et al, 1988, 1994; Vitadello et al, 1990), have been suggested for this specialized tissue. These long-standing questions regarding the origin and patterning of conduction system network have recently been settled by retroviral cell lineage studies of the Purkinje fiber network in the chicken embryonic heart, as described below (Gourdie et al, 1995, 1999b; Mikawa and Fischman, 1996; Mikawa, 1998a,b).

MYOCYTE ORIGIN OF THE CARDIAC CONDUCTION SYSTEM

In the chick, all cells in the epithelioid myocardium of the tubular heart differentiate into contractile myocytes by E2 (Manasek, 1968; Han et al, 1992; Linask, 1992), while cardiac neural crest cells begin migration from the embryonic hind brain at E2–3 and enter the heart at E4 (Kirby and Stewart, 1983; Kirby, 1993). The differentiating Purkinje fibers can first be detected as a network of connexin42 (Cx42)-positive cells juxtaposed to the developing coronary arteries and the endocardium after E10 but not before that time (Gourdie et al, 1993a, 1999b). By E15, Purkinje fibers are easily identified by their uniqe gene expression (Fig. 7.1), including induction of a slow twitch muscle myosin heavy chain, sMHC (Sartore et al, 1978), upregulation of a skeletal muscle–type myosin binding protein MyBP-H (Anytecheva et al, 1997), and downregulation of a myosin binding protein cMyBP-C (Gourdie et al, 1998) essential for normal heart muscle contractility (Bonne et al, 1995; Watkins et al, 1995).

The definitive cell lineage relationships among three cell types—conduction cells, myocytes, and neural crest—can be tested by single cell tagging and tracing. Replication-defective variants of the spleen necrosis virus, an evolutionarily new form of the avian reticuloendotheliosis virus (Dougherty and Temin, 1986), have been engineered by replacing viral structural genes with a reporter gene, bacterial

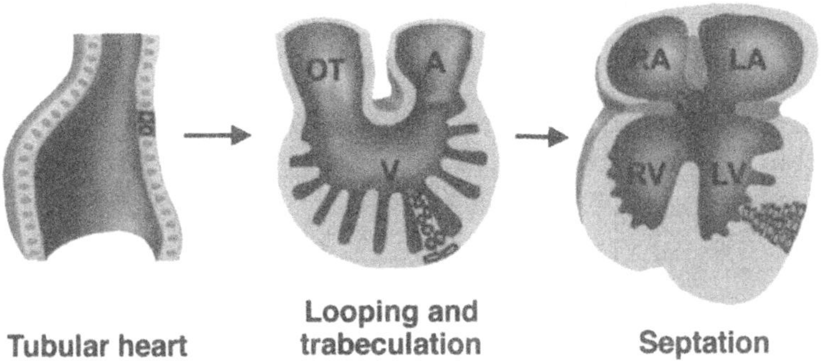

FIGURE 7.3. The main morphogenetic events in the myocyte lineage during formation of the ventricular myocardium (modified from Mikawa, 1995). Tilled cells are a clonal myocyte population. The endoderm is omitted from the diagram.

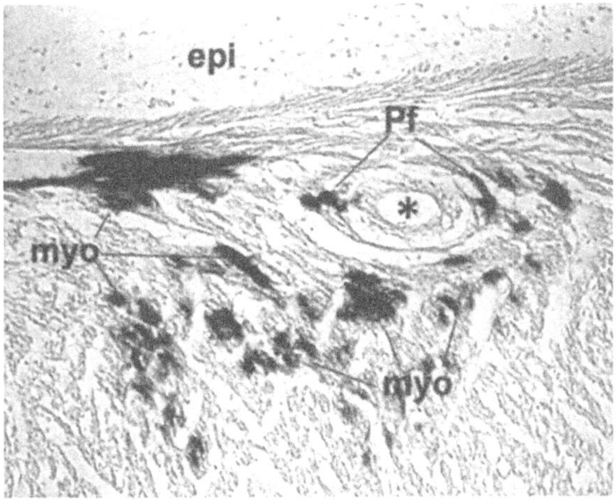

FIGURE 7.4. A clonal population (stained in black) derived from a virally tagged parental myocyte contains ordinary myocytes (myo) and periarterial Purkinje fiber cells (Pf). Asterisk indicates the lumen of coronary artery. Epi: epicardium.

β-galactosidase (β-gal) (Mikawa et al, 1992b, 1996). Using the β-gal viral vector, the fate of cardiac neural crest derivatives and differentiated myocytes has been examined from E2–3 through E18 (Mikawa et al, 1992a,b). Analysis of myocyte clonal expansion during the myocardial formation has revealed that individual myocyte precursor cells give rise to a series of progeny that migrates more vertically than horizontally to form clones that usually span the full thickness of the myocardium, i.e., from epicardial to endocardial surfaces of the muscle wall (Figure 7.3) (Mikawa et al, 1992a; Mikawa, 1995, 1998a,b; Mikawa and Fischman, 1996; reviewed in Mikawa et al, 1996). Because clonally related myocytes develop as a tight cluster, diversification of cell types within each myocyte clone can be examined in detail. β-gal–positive Purkinje fibers are exclusively and frequently found in myocyte clones (Figure 7.4). In contrast, no β-gal–positive conduction cells are produced from cardiac neural crest. In addition, no clone contains both

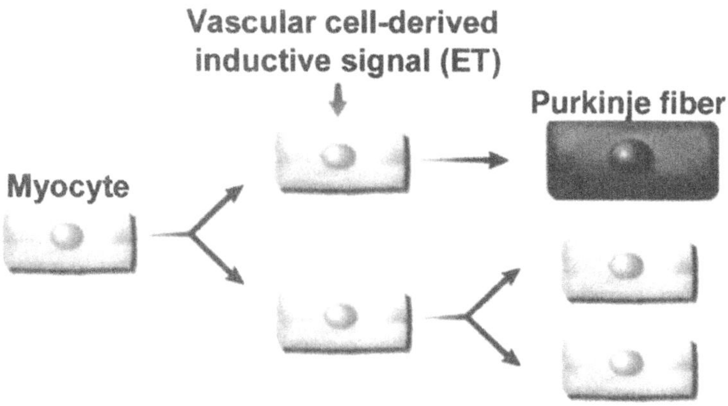

FIGURE 7.5. A model for Purkinje fiber differentiation within the myocyte lineage during chick heart development (modified from Mikawa, 1998). A subpopulation of clonally related contractile myocytes juxtaposed to arteries and subendocardium receive an inductive cue, such as endothelin (ET), to differentiate into Purkinje fiber cells. Contractile myocytes continue to undergo cell division, while Purkinje fiber cells withdraw from the proliferate state (Gourdie et al, 1995).

Purkinje fibers and cells of the proximal conduction system such as the AV node and AV ring (Gourdie et al, 1995, 1999a; Mikawa, 1998a,b).

The results from the retroviral cell lineage studies indicate that (1) Purkinje fibers are transdifferentiated from a subset of contractile myocytes, not from neural crest as previously suggested; (2) the local recruitment of conduction cells from contractile cells occurs at the two restricted sites—periarterial and subendocardial; (3) a unique pattern of gene expression in the conduction cells is mediated by switching the program from contractile to conducting types; (4) Purkinje fibers do not share the same parental lineage with proximal components of the conduction system; and (5) in situ linkage of the peripheral and central components establishes the entire conduction system network (Figure 7.5).

INDUCTIVE ROLES OF ENDOTHELIAL CELL–DERIVED FACTOR FOR PURKINJE FIBER DIFFERENTIATION

The above in vivo evidence that Purkinje fibers are derived from myocyte precursors indicates that Purkinje fiber differentiation can be studied by determining those mechanisms that convert myocytes from a muscle to a conducting phenotype. The recruitment of cells to the conduction system at periarterial and subendocardial sites, within clonally derived populations of differentiated myocytes, has led to the hypothesis that embryonic myocytes may be induced to form Purkinje fibers by receiving paracrine signals originating from endocardial cells and arterial vascular tissues (Gourdie et al, 1995, 1999a; Mikawa and Fischman, 1996; Mikawa, 1998a,b).

Endothelin (ET) is one of such paracrine factors abundant both in the endocardium and arterial system of the heart, and is secreted by endothelial cells in a shear stress–dependent manner (Yanagisawa et al, 1988; Masaki et al, 1991). ET was originally isolated as an endothelial cell–derived vasoconstrictor (Yanagisawa et al, 1988) through its binding to two distinct G-protein–coupled receptors, ET_AR

and ET_BR (Arai et al, 1990; Sakurai et al, 1990). ET is a posttranslationally processed product from its precursor, preproET (Xu et al, 1994). PreproET is first cleaved by furin proteases into bigET, which is further processed with ET-converting enzyme (ECE) into biologically active ET (Xu et al, 1994; Emoto and Yanagisawa, 1995). The expression of ET begins at embryonic stages (Baynash et al, 1994; Kurihara et al, 1995). ET expression at the tubular heart stage is restricted to the outflow tract endocardium and endocardial cushion (Kurihara et al, 1995; Clouthier et al, 1998), while ETR is expressed by all myocytes and absent from the endocardium (Clouthier et al, 1998; Yanagisawa et al, 1998). In contrast, ECE expression is present in a portion of the endocardium and absent from myocytes (Yanagisawa et al, 1998).

In addition to its originally predicted role as a vasoconstrictor (Yanagisawa et al, 1988), ET signaling is now implicated in a diverse set of biologic activities such as cell proliferation, stimulation of hormone release, and the modulation of central nervous activity (Yanagisawa and Masaki, 1989; Yanagisawa, 1994). ET also induces neonatal and adult cardiac myocyte hypertrophy (Ito et al, 1991). Homozygous gene deletions of ET (Baynash et al, 1994; Kurihara et al, 1995), ETR (Hosoda et al, 1994; Clouthier et al, 1998), and ECE (Yanagisawa et al, 1998) exhibit similar cardiac phenotypes, including septal defects and malformations of the great vessels and arch arteries, all of which are seen in cardiac neural crest-ablated chicken embryos (Kirby, 1993).

In these mouse models, conduction system development has not been studied, because of embryonic death prior to detection of definitive Purkinje fiber differentiation with specific molecular markers (Delorme et al, 1995; Gourdie et al, 1999a). However, it has been recently found that exposure of beating myocytes from embryonic chick ventricles to ET in vitro upregulates Purkinje fiber–specific gene products and downregulates a myocyte-specific myofibrillar protein, resembling changes seen during Purkinje fiber differentiation in vivo. By contrast, no detectable Purkinje fiber phenotype is induced following exposure to its precursor big endothelin (big-ET) or by other blood vessel–associated factors, including fibroblast growth factor (FGF) and platelet-derived growth factor (PDGF) (Gourdie et al, 1998). Hence, induction of Purkinje fiber phenotype appears to be specific to ET. These in vitro data demonstrate for the first time that embryonic myocytes are competent to respond to ET, a paracrine factor prominently secreted from cells of the endocardium and the coronary arterial bed, and can be induced to differentiate into Purkinje fiber–like cells.

POTENTIAL MECHANISMS DEFINING THE BRANCHING PATTERN OF PURKINJE FIBERS WITHIN THE MYOCARDIUM

The information described above shows that in the chicken heart, Purkinje fiber recruitment from myocytes occurs exclusively at the subendocardial and periarterial sites and that myocytes can be converted into conducting cells in vitro by a vessel-derived paracrine factor. These data suggest that Purkinje fiber differentiation and its branching pattern may be regulated by instructive cues from vascular endothelial cells. This idea is further supported by the timing and location at which both coronary arterial and Purkinje fiber networks develop in the myocardium.

The tissues giving rise to the coronary vasculature do not arise by outgrowth from the root of the aorta, but migrate into the tubular heart from extracardiac mesenchyme (Mikawa and Fischman, 1992; Poelmann et al, 1993; Mikawa and Gourdie, 1996) along with a proepicardial sheet (Ho and Shimada, 1988; Hiruma and Hirakow, 1989). In the chick, entry of these coronary precursors to the heart begins at E3 (Mikawa and Fischman, 1992; Poelmann et al, 1993). Following inward migration, vasculogenic cells form discontinuous endothelial channels, which subsequently fuse by E6 and connect to the aorta (Bogers et al, 1989; Waldo et al, 1990; Mikawa and Fischman, 1992; Mikawa and Gourdie, 1996). A closed coronary vessel network is established by E14 (Rychter and Ostadal, 1971). Coincident with this early vasculogenic process, recruitment of intramural Purkinje fibers begins exclusively in myocyte subpopulations juxtaposed to developing coronary arteries (Gourdie et al, 1995).

If signals from coronary arteries play a role in the recruitment of Purkinje fibers from contractile myocytes, the vessel network may also define the branching pattern of the intramural conduction network. This hypothesis has been tested by modulating the coronary artery development. Ablation of the cardiac neural crest alters the pattern of the coronary arterial tree (Hood and Rosenquist, 1992); neural crest derivatives are necessary for the survival of branches of the coronary artery system (Waldo et al, 1994). The numbers and profiles of coronary arteries penetrating the ventricular myocardium are dramatically reduced in neural crest–ablated hearts. Associated with suppressed coronary vessel development, differentiation of periarterial Purkinje fibers in the neural crest–ablated embryos is inhibited (Hyer et al, 1999). The data support the idea that an instructive cue from vascular cells regulates Purkinje fiber differentiation and its branching pattern. However, it remains unexplained how intramural Purkinje fiber recruitment is restricted to periarterial sites but not those juxtaposing veins and capillaries. It is also uncertain why only a subpopulation of myocytes juxtaposing the endocardium differentiate into subendocardial Purkinje fibers. Since ET, a shear stress–dependent paracrine factor, has the potential to induce Purkinje fiber differentiation, hemodynamic factors may play a role in diversifying the cardiac endothelial cells into populations that are either active or inactive for secreting an inductive signal, thereby defining Purkinje fiber recruitment site within the developing myocardium (Figure 7.6).

IN SITU LINKAGE OF THE CONDUCTION SYSTEM NETWORK

The retroviral cell lineage data have revealed independent origins for proximal and distal elements of the conduction system (Gourdie et al, 1995). It has also been shown that cells in the primary conduction ring cease their proliferation very early in development (Thompson et al, 1990). Both data indicate that an integrated network of the entire conduction system is established by in situ linkage of the proximal and distal components (Gourdie et al, 1995; reviewed in Mikawa and Fischman, 1996; Mikawa, 1998a,b), and not by outgrowth from the primary conduction ring (Lamers et al, 1991; Wessels et al, 1992; Chan-Thomas et al, 1993; Fishman and Chien, 1997; Moorman et al, 1998). These findings now allow us to study the mechanism that establishes the entire conduction system network by analyzing the process through which locally differentiated conduction elements are connected.

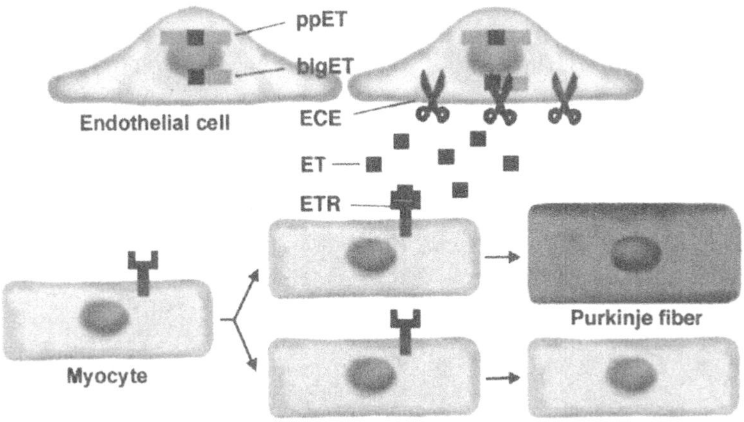

FIGURE 7.6. Model for the ET-dependent, local induction of embryonic myocytes into Purkinje fibers (modified from Takebayashi-Suzuki et al, 2000). In the ventricular myocardium of the embryonic chicken heart, there are two types of endothelial cells, ECE-1 expressing and non-expressing. Only the former can produce mature ET from big-ET. Among ETR positive myocytes, only those adjacent to ET-production can be induced to differentiate into Purkinje fibers.

There are two linkage sites critical for conduction network formation: connection between the AV bundle (a proximal component derived from the primary conduction ring) and the bundle branches (a distal conduction component derived from the interventricular septum), and linkage between subendocardial and intramural Purkinje fibers. In the chicken heart, electrical coupling between the AV bundle and bundle branches occurs at E7 at the same time when fusion of the interventricular septum completes (Chuck et al, 1997). This coupling process has been studied morphologically by using markers specific for the proximal and distal conduction components, respectively (Chuck and Watanabe, 1997). In a preseptated heart, the differentiating bundle branches are detected at the tip and along the sides of the primitive interventricular septum. When the interventricular septum fuses, the anterior tip of the differentiating bundle branches is superiorly covered by cells of the posterior end of the AV bundle. However, it remains unsolved how the two conduction components establish a specific gap junctional connection and how the coupling process is segregated from a potential linkage with adjacent contractile myocytes. By contrast to the complexity seen in coupling between the AV bundle and bundle branches, the subendocardial and periarterial Purkinje fiber networks seem to be linked by a rather simple mechanism. The coupling between the two distal conduction components occurs invariably at the sites where the interventricular arteries and arterioles juxtapose the subendocardial conduction cells. Thus, the branching pattern of developing arteries appear to define both sites, intramural Purkinje fiber recruitment, and its connection with the subendocardial Purkinje fibers, within the ventricular myocardium (Figure 7.7).

CONCLUSION

Except for the pacemaking cells, the lineage relationships of all cell types in the cardiac conduction system have now been identified. Clone-based cell lineage data define the timing and sites at which embryos induce the cardiac conduction system

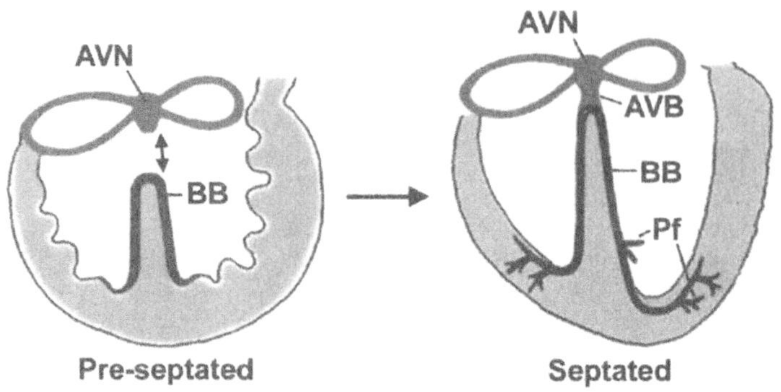

FIGURE 7.7. Schematic illustration of linkage between central and peripheral conduction components during ventricular septation.

lineages and their differentiation and patterning. The new information of the potentiality or plasticity of embryonic myocytes in their lineage commitment and terminal differentiation now allows us to move further to understand the molecular and cell biologic mechanisms involved in these processes. The questions to be addressed include the following: (1) What factors recruit embryonic myocytes to differentiate into the proximal conduction system? (2) What mechanisms govern linkage of the Purkinje fiber network to the proximal conduction system? (3) Is there any common mechanism that regulates both the proximal and distal conduction system differentiation and patterning? (4) What mechanisms diversify the cardiac endothelia into active or inactive in producing an inductive signal and do so in a spatially defined pattern? (5) How are the coronary arterial branching paths that pattern the intramural Purkinje fiber network defined, and what is the role of the cardiac neural crest in this process? (6) Which mechanism defines the sites of electrical coupling between Purkinje fibers and contractile myocytes? (7) What transcriptional factors regulate the expression of a unique set of genes in conduction cells? A clarification of these questions will contribute significantly to our understanding of the development and integrated function of the cardiac conduction system.

ACKNOWLEDGEMENTS

This work is supported by the National Institutes of Health (HL54128, HL56987, HL62175 to T.M.; and HL56728 to R.G.G.), the National Science Foundation (97344046, R.G.G.), the Human Frontier Science Program (K.T.S. and T.M.). The excellent technical assistance of Ms. Sandra Klatt and Lydia Miroff is acknowleged with gratitude. R.G.G. is a Early Career Scholar of the National Science Foundation. T.M. is an Irma T. Hirschl Scholar.

REFERENCES

Alyonycheva, T., Cohen-Gould, L., Siewert, C., Fischman, D.A., and Mikawa, T. (1997). Myosin binding protein-H (MyBP-H) characteristic of skeletal muscle is induced in a subset of cardiac muscle cells differentiated into the conducting Purkinje fibers. *Circ Res* 80:665–672.

Arai, H., Hori, S., Aramori, I., Ohkubo, H., and Nakanishi, S. (1990). Cloning and expression of a cDNA encoding an endothelin receptor. *Nature* 348:730–732.

Bastide, B., Neyses, L., Ganten, D., Paul, M., Willecke, K., and Traub, O. (1993). Gap junction protein connexin40 is preferentially expressed in vascular endothelium and conductive bundles of rat myocardium and is increased under hypertensive conditions. *Circ Res* 73:1138–1149.

Baynash, A.G., Hosoda, K., Giaid, A., et al. (1994). Interaction of endothelin-3 with endothelin-B receptor is essential for development of epidermal melanocytes and enteric neurons. *Cell* 79:1277–1285.

Bogers, A.J.J.C., de Groot, A.C., Poelmann, R.E., and Huysmans, H.A. (1989). Development of the origin of the coronary arteries, a matter of ingrowth or outgrowth? *Anat Embryol* 180:437–441.

Bonne, G., Carrier, L., Bercovici, J., et al. (1995). Cardiac myosin binding protein-C gene splice acceptor site mutation is associated with familial hypertrophic cardiomyopathy. *Nat Genet* 11:438–440.

Bozler, E. (1942). The initiation of impulses in cardiac muscle. *Am J Physiol* 138: 273–282.

Brooks, C. McC., and Lu, H.-H. (1972). *The Sinoatrial Pacemaker of the Heart.* Charles C Thomas, Springfield, IL.

Callewaert, G., Vereecke, J., and Carmeliet, E. (1986). Existence of a calcium-dependent potassium channel in the membrane of cow cardiac Purkinje cells. *Pflugers Arch* 406:424–426.

Cavalie, A., Ochi, R., Pelzer, D., and Trautwein, W. (1983). Elementary currents through Ca^{2+} channels in guinea pig myocytes. *Pflugers Arch* 398:284–297.

Chan-Thomas, P.S., Thompson, R.P., Robert, B., Yacoub, M.H., and Barton, P.J.R. (1993). Expression of homeobox genes *Msx-1* (*Hox-7*) and *Msx-2* (*Hox-8*) during cardiac development in the chick. *Dev Dyn* 197:203–216.

Chuck, E.T., Freeman, D.M., Watanabe, M., and Rosenbaum, D.S. (1997). Changing activation sequence in the embryonic chick heart—implications for the development of the His-Purkinje system. *Circ Res* 81:470–476.

Chuck, E.T., and Watanabe, M. (1997). Differential expression of PSA-NCAM and HNK-1 epitopes in the developing cardiac conduction system of the chick. *Dev Dyn* 209:182–195.

Clouthier, D.E., Hosoda, K., Richardson, J.A., et al. (1998). Cranial and cardiac neural crest defects in endothelin-A receptor-deficient mice. *Development* 125:813–824.

Coppen, S.R., Severs, N.J., and Gourdie, R.G. (1999). Cx45 expression delineates and extended conduction system in the embryonic and adult rodent heart. *Dev Genet* 24:82–90.

Davies, F. (1930). The conducting system of the bird's heart. *J Anat* 64:129–146.

Delorme, B., Dahl, E., Jarry-Guichard, T., et al. (1995). Developmental regulation of connexin 40 gene expression in mouse heart correlates with the differentiation of the conduction system. *Dev Dyn* 204:358–371.

Dougherty, J.P., and Temin, H.M. (1986). High mutation rate of a spleen necrosis virus-based retrovirus vector. *Mol Cell Biol* 7:4387–4395.

Emoto, N., and Yanagisawa, M. (1995). Endothelin-converting enzyme-2 is a membrane-bound, phosphoramidon-sensitive metalloprotease with acidic pH optimum. *J Biol Chem* 270:15262–15268.

Fishman, M.C., and Chien, K.R. (1997). Fashioning the vertebrate heart: earliest embryonic decisions. *Development* 124:2099–2117.

Goldenberg, M., and Rothberger, C.J. (1936). Über des Elektrogramm der spezifischen Herz-muskulature. *Pflugers Arch* 237:295–306.

Gorza, L., Schiaffino, S., and Vitadello, M. (1988). Heart conduction system: a neural crest derivative. *Brain Res* 457:360–366.

Gorza, L., Vettore, S., and Vitadello, M. (1994). Molecular and cellular diversity of heart conduction system myocytes. *Trends Cardiovasc Med* 4:153–159.
Gourdie, R.G., Cheng, G., Thompson, R.P., and Mikawa, T. (1999b). Retroviral cell lineage analysis in the developing heart. In: Tuan, R.S., and Lo, C.W., eds. *Methods in Molecular Biology, vol 1, Developmental Biology Protocols.* (Walker, J., series ed.) Humana Press, in press.
Gourdie, R.G., Green, C.R., Severs, N.J., Anderson, R.H., and Thompson, R.P. (1993a). Evidence for a distinct gap-junctional phenotype in ventricular conduction tissues of the developing and mature avian heart. *Circ Res* 72:278–289.
Gourdie, R.G., Green, C.R., Severs, N.J., and Thompson, R.P. (1992). Immunolabelling patterns of gap junction connexins in the developing and mature rat heart. *Anat Embryol* 185:363–378.
Gourdie, R.G., Kubalak, S., and Mikawa, T. (1999a). Conducting the embryonic heart: orchestrating development of specialized cardiac tissues. *Trends Cardiovasc Med* 9:18–26.
Gourdie, R.G., Mima, T., Thompson, R.P., and Mikawa, T. (1995). Terminal diversification of the myocyte lineage generates Purkinje fibers of the cardiac conduction system. *Development* 121:1423–1431.
Gourdie, R.G., Severs, N.J., Green, C.R., et al. (1993b). The spatial distribution and relative abundance of gap junctional connexin40 and connexin43 correlate to functional. *J Cell Sci* 105:985–991.
Gourdie, R.G., Wei, Y., Kim, D., Klatt, S.C., and Mikawa, T. (1998). Endothelin-induced conversion of heart muscle cells into impulse-conducting Purkinje fibers. *Proc Natl Acad Sci USA* 95:6815–6818.
Gros, D., Jarry-Guichard, T., Ten-Velde, I., et al. (1994). Restricted distribution of connexin40, a gap junctional protein, in mammalian heart. *Circ Res* 74:839–851.
Hagiwara, N., Irisawa, H., and Kameyama, M. (1988). Contribution of two types of calcium currents to the pacemaker potentials of rabbit sino-atrial node cells. *J Physiol* 395:233–253.
Hamburger, V., and Hamilton, H.L. (1951). A series of normal stages in the development of the chick embryo. *J Morphol* 88:49–92.
Han, Y., Dennis, J.E., Cohen-Gould, L., Bader, D.M., and Fischman, D.A. (1992). Expression of sarcomeric myosin in the presumptive myocardium of chicken embryos occurs within six hours of myocyte commitment. *Dev Dyn* 193:257–265.
Hiruma, T., and Hirakow, R. (1989). Epicardial formation in embryonic chick heart: computer-aided reconstruction, scanning, and transmission electron microscopic studies. *Am J Anat* 184:129–138.
His, Wm. Jr. (1893). Die Tätigkeit des embryonalen Herzens und deren Bedeutung für die Lehre von der Herzbewegung beim Erwachsenen. *Arb Med Klin Leipzig* 14.
Ho, E., and Shimada, Y. (1988). Formation of the epicardium studied with the scanning electron microscope. *Dev Biol* 66:579–585.
Hood, L.A., and Rosenquist, T.H. (1992). Coronary artery development in the chick: origin and development of smooth muscle cells, and effects of neural crest ablation. *Anat Rec* 234:291–300.
Hosoda, K., Hammer, R.E., Richardson, J.A., et al. (1994). Targeted and natural (piebald-lethal) mutations of endothelin-B receptor gene produce megacolon associated with spotted coat color in mice. *Cell* 79:1267–1276.
Hyer, J., Johanssen, M., Prasad, A., et al. (1999). Induction of Purkinje fiber differentiation by coronary arterialization. *Proc Natl Acad Sci USA* 96:13214–13218.
Ito, H., Hirata, Y., Hiroe, M., et al. (1991). Endothelin-1 induces hypertrophy with enhanced expression of muscle-specific genes in cultured neonatal rat cardiomyocytes. *Circ Res* 69:209–215.
Kaczmarek, L.K., and Blumenthal, E.M. (1997). Properties and regulation of the minK potassium channel protein. *Physiol Rev* 77:627–641.

Kamino, K. (1991). Optical approaches to ontogeny of electrical activity and related functional organization during early heart development. *Physiol Rev* 71:53–91.
Kamino, K., Hirota, A., and Fujii, S. (1981). Localization of pacemaking activity in early embryonic heart monitored using voltage-sensitive dye. *Nature* 290:595–597.
Kanter, H.L., Laing, J.G., Beau, S.L., Beyer, E.C., and Saffitz, J.E. (1993). Distinct patterns of connexin expression in canine Purkinje fibers and ventricular muscle. *Circ Res* 72:1124–1131.
Keith, A., and Flack, M. (1907). The form and nature of the muscular connections between the primary divisions of the vertebrate heart. *J Anat Physiol* 41:172–189.
Kirby, M.L. (1993). Cellular and molecular contributions of the cardiac neural crest to cardiovascular development. *Trends Cardiovasc Med* 3:18–23.
Kirby, M.L., and Stewart, D.E. (1983). Neural crest origin of cardiac ganglion cells in the chick embryo: identification and extirpation. *Dev Biol* 97:433–443.
Kölliker, A. (1902). Gewebeslehre. 6 Aufl. Lpz.
Kupershmidt, S., Yang, T., Anderson, M.E., et al. (1999). Replacement by homologous recombination of the minK gene with lacZ reveals restriction of minK expression to the mouse cardiac conduction system. *Circ Res* 84:146–152.
Kurihara, Y., Kurihara, H., Oda, H., et al. (1995). Aortic arch malformations and ventricular septal defect in mice deficient in endothelin-1. *J Clin Invest* 96:293–300.
Lamers, W.H., De Jong, F., De Groot, I.J.M., and Moorman, A.F.M. (1991). The development of the avian conduction system, a review. *Eur J Morphol* 29:233–253.
Lei, M., Kohl, P., and Rigg, L. (1998). Immunohistochemical staining of minK in rabbit sino-atrial node cells. *J Physiol* 506P:39–40.
Linask, K.K. (1992). N-cadherin localization in early heart development and polar expression of Na^+, K^+-ATPase, and integrin during pericardial coelom formation and epithelialization of the differentiating myocardium. *Dev Biol* 151:213–224.
Manasek, F.J. (1968). Embryonic development of the heart: a light and electron microscopic study of myocardial development in the early chick embryo. *J Morphol* 125:329–366.
Masaki, T., Kimura, S., Yanagisawa, M., and Goto, K. (1991). Molecular and cellular mechanism of endothelin regulation. Implications for vascular function. *Circulation* 84:1457–1468.
McDonald, T.V., Yu, Z., Ming, Z., et al. (1997). A minK-HERG complex regulates the cardiac potassium current I(Kr). *Nature* 388:289–292.
Mikawa, T. (1995). Retroviral targeting of FGF and FGFR in cardiomyocytes and coronary vascular cells during heart development. *Ann NY Acad Sci* 752:506–516.
Mikawa, T. (1998a). Cardiac lineages. In: Harvey, R.P., and Rosenthal, N., eds. *Heart Development*. Academic Press, New York, pp. 19–33.
Mikawa, T. (1998b). Determination of heart cell lineages. In: Moody, S.A., ed. *Cell Fate Determination*. Academic Press, New York, pp. 449–460.
Mikawa, T., Borisov, A., Brown, A.M.C., and Fischman, D.A. (1992a). Clonal analysis of cardiac morphogenesis in the chicken embryo using a replication-defective retrovirus: I. Formation of the ventricular myocardium. *Dev Dyn* 193:11–23.
Mikawa, T., Cohen-Gould, L., and Fischman, D.A. (1992b). Clonal analysis of cardiac morphogenesis in the chicken embryo using a replication-defective retrovirus. III: Polyclonal origin of adjacent ventricular myocytes. *Dev Dyn* 195:133–141.
Mikawa, T., and Fischman, D.A. (1992). Retroviral analysis of cardiac morphogenesis: discontinuous formation of coronary vessels. *Proc Natl Acad Sci USA* 89:9504–9508.
Mikawa, T., and Fischman, D.A. (1996). The polyclonal origin of myocyte lineages. *Annu Rev Physiol* 58:509–521.
Mikawa, T., and Gourdie, R.G. (1996). Pericardial mesoderm generates a population of coronary smooth muscle cells migrating into the heart along with ingrowth of the epicardial organ. *Dev Biol* 173:221–232.

CHAPTER 8

The Epicardium and the Formation of the Coronary Vasculature

Robert E. Poelmann, Mark Paul F.M. Vrancken Peeters, and Adriana C. Gittenberger-de Groot

In early development the epicardium as an extracardiac organ envelops the initially bare myocardial surface. In this way, a three-layered organ is formed consisting of an inner endocardium, the muscular myocardium, and the covering epicardium (Manasek, 1969; Ho and Shimada, 1978; Viragh and Challice, 1981; Hiruma and Hirakow, 1988). The epicardium proved to be the source for novel cell populations migrating into the myocardial wall, which play major, but not yet fully comprehended, roles in heart development. It has been shown in a number of studies, using knockout mice, that when the epicardium is missing, the coronary vessels fail to develop properly (Kwee et al, 1995; Yang et al, 1995), usually leading to embryonic death at a very specific point in development. It must be concluded that the formation of the coronary vasculature is one of the features of heart development depending on epicardial differentiation. Therefore, we describe and discuss first the differentiation of the epicardial-derived cells (EPDCs), followed by the development of the endothelium that does not derive from the epicardial organ proper, and the smooth muscle cells and adventitial fibroblasts of the coronary vessel wall. Finally, we discuss the differentiation of the vascular tree, starting with a plexiform system of sinusoidal vessels, developing into arteries and veins, interconnected by a capillary network.

DEVELOPMENT OF EPICARDIUM-DERIVED CELLS

The Chimera Approach

In experimental setups the chicken-quail chimera has proven to provide reliable answers for cell lineage studies. A proper quail graft placed at the proper time of development into a proper site of a chick embryo shows after proper survival times the origin, the target, and the differentiation state of the grafted cells. To this aim the application of quail-specific antibodies is of tremendous help in discerning grafted cells from host cells. Furthermore, other antibodies are very helpful in determining the differentiation state of specific cell populations, such as endothelial, smooth muscle, and fibroblastic cells.

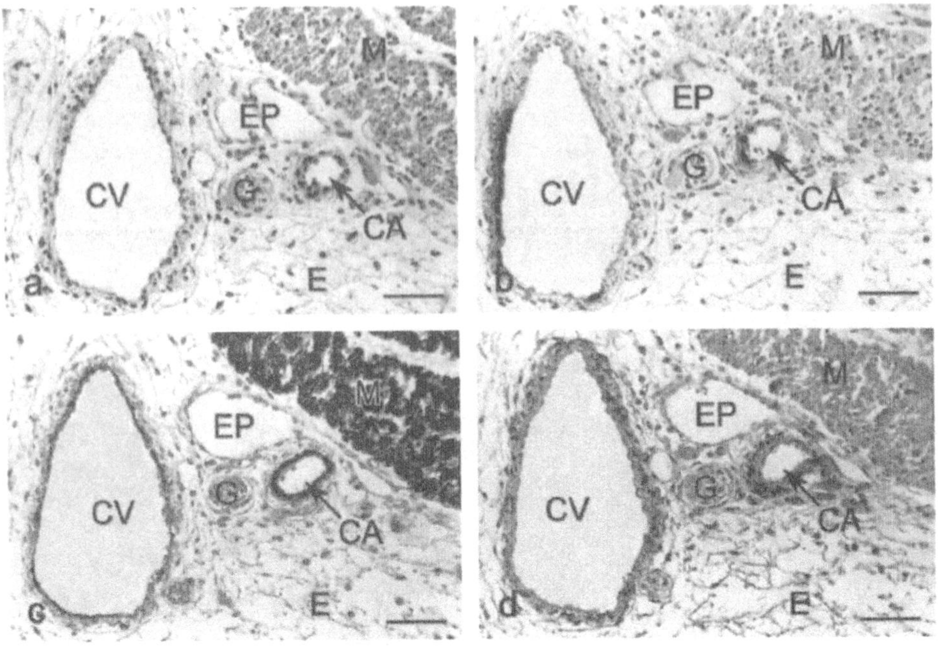

FIGURE 8.1. A stage-17 chicken embryo received a quail proepicardial graft, containing also a piece of sinus venosus and adjoining liver tissue. The chimera was allowed to survive until stage 43. Adjacent sections with antibody staining for antiquail nuclear marker QCPN (a), antiquail endothelium QH1 (b), antiactin HHF35 (Tsukuda et al, 1987) (c), and antiprocollagen M38 (d). Panel a and b show that grafted quail cells have invaded the endothelial, medial, and adventitial layers of a coronary vein (CV) within the subepicardium (E). CA, coronary artery; EP, epicardial pocket; G, cardiac ganglion; M, myocardium. Bar 50 µm.

For studies on the epicardium the eggs have to be windowed to reach the membranes overlying the embryo. The vitelline membrane has to be removed, and the pericardial cavity of the embryo can be reached either through the naturally existing hiatus in the body wall (stages 15 to 16), or by making a slit in the amniotic and pericardial membranes (stages 16 to 18) with watchmakers' forceps. A piece of proepicardial organ including a small adherent segment of liver, harvested from a quail embryo from the same developmental stage, can be inserted into the epicardial cavity. It has been shown that the coronary endothelium derives from the cotransplanted liver sinusoids and inner lining of the sinus venosus (Poelmann et al, 1993) and not from the proepicardial organ per se.

Using this approach we demonstrated in all chimeras grafted quail cells, present not only in the epicardial lining but also in the subepicardium (Figure 8.1). Furthermore, quail cells were found interspersed between cardiomyocytes as deep as the subendocardium, in the atrioventricular (AV) cushions, and in the lining of the developing coronary vessels. We concentrate here on the EPDCs that are related to the coronary vessels, whereas the other EPDCs will be discussed only as background.

Myocardial Invasion

The epicardial lining of chimeras that were allowed to survive until Hamburger-Hamilton (HH) stages 25 to 31 showed a mosaicism of quail- and chick-derived cells. This is also the case for the subepicardial mesenchyme and for the coronary vasculature confined in the subepicardium. Apart from the vasculature, a subset of scattered quail-EPDCs invaded the myocardium. These cells favor a position in the subendocardial region, i.e., adjacent to the endocardial lining. These cells stain positive for M38 (McDonald et al, 1986), which is indicative of procollagen production (Gittenberger-de Groot et al, 1998).

Endocardial Invasion

Chimeras that were allowed to survive until HH stages 32 to 43 also showed the mosaicism in their epicardium and subepicardium. In addition to the quail-derived EPDCs found between the cardiomyocytes, including the trabeculations and the subendocardial myocardium, we observed a large number of EPDCs inside the AV endocardial cushions. Depending on the specific chimera, only the tricuspid or both the mitral and the tricuspid cushions had been invaded. A continuity was usually observed between the subepicardally located graft and the cushion tissue, as the atrioventricular myocardial connection became disrupted by mesenchymal tissue, enabling a route for migration. Two locations in the endocardial cushions were favored—the myocardial border and the subluminal surface. Incorporation of EPDCs in the lining has not been observed. EPDCs in the fibrous annulus and in the subendocardial position were usually M38 positive, demonstrating the formation of procollagen. The outflow tract cushions and semilunar valves did not contain EPDCs (Gittenberger-de Groot et al, 1998).

Coronary Vasculature

To demonstrate quail-derived endothelium in a chimera, the incorporation of a piece of liver tissue was obligatory. We deduced from earlier experiments (Poelmann et al, 1993) that the sinusoidal endothelium of either the liver or the close-by sinus venosus migrates into the blebs of the proepicardial organ and extends into the subepicardium (Figure 8.1B) underneath and in concert with the expanding epicardium. In a double-staining experiment, we proved furthermore that the smooth muscle cell (Figure 8.1C) and fibroblasts (Figure 8.1D) belong to the epicardial lining and the coronary endothelium belongs to a separate lineage (Viragh et al, 1993). Basically, two different mechanisms have been described by which endothelial tube formation is established—angiogenesis (Poole and Coffin, 1988; Noden, 1989) and vasculogenesis (Gonzalez-Crussi, 1971; Hirakow and Hiruma, 1981; Coffin and Poole, 1988; Risau and Flamme, 1995). The main items addressed are whether endothelial cells in an expanding vasculature differentiate through sprouting from existing endothelial stem cells, and/or whether endothelial cells transform in situ from surrounding mesenchymal cells, subsequently organizing into a vascular plexus (Pardanaud et al, 1987). It is unclear which of the described possibilities is responsible for the developing coronary network. Rychter and Ostadal (1971) and Aikawa and Kawano (1982) claimed that the coronary plexus arises from the developing systemic vascular system, whereas Manasek (1971) and Bolender et al (1990) reported that endothelial cells derive independently from the subepicardial mesenchyme by fusion of blood island–like structures. Based on meticulous studies of subsequent phases of development of both the chick and the quail, we have reported that the coronary plexus grows by continuous extension from the sinus venous endocardium, which is an already-existing vessel. Mikawa and Fischman (1992) and Mikawa and Gourdie (1996), however, hold the opinion that endothelial cells could also transform from the subendocardial mesenchyme, as suggested by local marking with a retrovirally transmitted reporter gene. As we have observed scattered QH1-positive cells that were not part of the endothelial lining (Vrancken Peeters et al, 1997a), we conclude that a combination of vasculogenesis and angiogenesis is responsible for the development of the coronary vessels.

It is evident that signaling factors must guide the migration of the coronary endothelial cells first into the subendocardium and only slightly later into the myocardium. Important factors in a growing embryo are probably physical factors, the oxygen level, and the related hypoxia inducible factor (Carmeliet et al, 1998). Interactions with the myocardium are likewise key factors in regulating endothelial proliferation and ingrowth. Among the latter is vascular endothelial growth factor (VEGF), from which a number of splice forms are known. The VEGF-120 mutant mouse shows an impaired myocardial angiogenesis and eventually dies of cardiac failure (Carmeliet et al, 1999). Tomanek et al (1999) demonstrated a correlation between sites of high VEGF expression in the rat myocardium and the ingrowth pattern of endothelial tubes. They also concluded that coronary formation results from the combination of angiogenesis and vasculogenesis.

Formation of the Smooth Muscle Cells and Fibroblasts

The established endothelium-lined network allows for real circulation, before or coincident with formation of the media and the adventitia of the coronary vessels. Vrancken Peeters et al (1997b) demonstrated that in stage 32 the subepicardial vessels are connected both to the right atrium and to the aortic sinuses of Valsalva. Because in earlier stages the microvascular labyrinth was always in contact with the lumen of sinus venosus and liver sinusoids (Vrancken Peeters et al, 1997a), we have to conclude that the coronary system is filled with blood as soon as it is formed in about stage 20. The blood pressure is probably not constant, but varies with the pressure in the sinus venosus and atrial segment.

The first signs of formation of the media and adventitia can be observed at the moment the connection with the aorta is made. We assume that the pressure inside the system will rise to ventricular levels, instead of being at sinus venosus levels as before. We do not know whether these increasing values play a significant role in differentiation of the outer vessel wall layers, but the hypothesis is tempting.

Formation of the coronary vessel media and adventitia is preceded by epithelial-mesenchymal transformation of the epicardium, in particular in the region of the atrioventricular sulcus where the subepicardium is most extensive. The originally squamous epicardium becomes cuboidal and deposits cells in the subepicardium. This condensed area of mesenchymal cells delaminates from the overlying epithelium and joins the vessel plexus in the AV sulcus. Surrounding the presumptive arteries, these mesenchymal cells differentiate into smooth muscle cells or fibroblasts as shown with the antibodies described above. The epicardial cells that acquired a side-to-side position near the vessel endothelium of the most proximal part of the artery, i.e., close to the aortic orifice, express either the procollagen marker M38 or the vascular smooth muscle marker 1E12. This staining distinguishes between future media and future adventitial layer. At HH stage 32 the presumptive coronary veins present in the same region lack both a media and an adventitia around their lumen (Hood and Rosenquist, 1992; Vrancken Peeters et al, 1997a).

A consistent finding in all the chimeras is that if the media/adventitia of the coronary artery is derived from the grafted quail tissue, part of the adjacent epicardial lining is of the same origin. Many cardiac ganglia are present in the vicinity of the differentiating arteries. These ganglia derive from yet another extracardiac cell population, the cardiac neural crest (Kirby et al, 1983; Poelmann et al, 1998; Verberne et al, 1998). In HH stage 43 coronary arterial and venous differentiation is completed. The media and adventitia are formed around the arterial endothelium, reaching from the aorta deep into the muscle of the heart, as many quail cells have been added to the endothelial tubes all the way to the apex. The coronary veins have a very thin media and adventitia, both invaded by quail cells as well.

HUMAN CORONARY DEVELOPMENT

Normal Development

In human embryos from 6 to 19 weeks gestational development, we performed a study on the development of the coronary arteries and veins (Figure 8.2). The specimen were sectioned and stained with antibodies A245 against fibronectin, HHF35 against α-actin, and an anti–α-myosin heavy chain (β-MHC) antibody, using standard immunohistochemistry. One of the main features is that the formation of the media and adventitia is very different in arteries and veins. Whereas the proximal stem of the arteries acquires smooth muscle cells from the epicardium starting at 50 days of gestation, the media from the proximal stems of the veins become populated by β-MHC–positive cells derived from the right atrial and sinus venosus myocardium but not until 77 days of development. The smooth muscle cells present in a single layer around the veins will populate the distal segments only after this time. The same phenomenon, although in a different time scale, has already been described for the quail coronary and pulmonary veins (Vrancken Peeters et al, 1997b), the anterior caval vein (Endo et al, 1996), and the pulmonary veins (Nathan and Gloobe, 1970; de Almeida et al, 1975; Masani, 1986) emptying in the atrium.

Another important feature in these human embryonic hearts was the presence of small vascular shunts between subepicardially located coronary arteries and adjacent veins. These shunts are located in the subepicardial myocardium and were found in the oldest embryos studied.

Anomalous Coronary Vasculature

In the human heart small interarterial and arteriovenous connections remain present after the establishment of the definitive coronary vascular net, as arterial branches feed a labyrinth at the subepicardial-myocardial border that is directly connected to the venous drainage. In human postmortem material these connections are common (Hadziselimovic and Secerov, 1979) and should not be considered as pathologic. In embryonic development the capillary network with its manifold connections provide the opportunity to temporarily overcome the sudden increase in blood flow upon connecting to the aorta. Arteriovenous shunts

CHAPTER 9

The Role of the Extracellular Matrix and Its Receptors in Modulating Cardiac Development

Edie C. Goldsmith, Wayne Carver,
and Thomas K. Borg

INTRODUCTION

Fundamental to the differentiation and morphogenesis of any organ is the interaction between its component cells and the extracellular matrix (ECM). In the early 1980s, the concept of dynamic reciprocity was introduced (Bissell et al, 1982), which proposed that the ECM produced by cells was in turn critical in the regulation of gene expression by these same cells. Studies to date, especially with isolated cells, have shown that the ECM has a profound influence on many cellular processes including the regulation of gene expression (for recent reviews, see Aumailley and Gayraud, 1998; Mauch, 1998). However, it is not yet known how these in vitro results relate to the development of the heart in vivo. The past 20 years have yielded an explosion of papers on this topic, which clearly document the importance of cell-ECM interactions in modulating fundamental cell properties such as differentiation, proliferation, migration, adhesion, and even survival. All of these biologic events are essential to the formation of tissues and organs including those within the cardiovascular system. This chapter focuses on how the components of the ECM influence the formation and function of the heart.

ECM COMPOSITION

The composition and organization of the ECM are very dynamic, particularly during periods of tissue remodeling associated with development and disease (Bishop and Laurent, 1995). Several groups of structural components have been classically described as constituents of the ECM including at least 20 types of collagen; noncollagenous glycoproteins, such as fibronectin and laminin; and proteoglycans. These components have been described in several reviews (Celentano and Frishman, 1997; Ruoslahti and Engvall, 1997; Werb, 1997; Aumailley and Gayraud, 1998; Mauch, 1998), and only a brief overview of these constituents is provided in this chapter.

Collagen

Collagens are a family of macromolecules in the ECM composed of three polypeptide chains and that have at least one domain containing a characteristic triple helical structure. Several types of collagen, including types I, III, and IV, are components of the mature heart, while several others, including types II and VIII, are expressed transiently during embryogenesis (Cheah et al, 1991; Iruela-Arispe and Sage, 1991). Collagen type I is the most abundant fibrillar collagen in the heart and in most other organs. This collagen type is an early marker of mesenchymal cell differentiation in early embryonic valvuloseptal precursors (Sinning et al, 1988), the cardiac cushions, and in the heart, and is the major described component of the mature myocardial interstitium (Borg and Caulfield, 1979; Caulfield and Borg, 1979; Thompson et al, 1979; Carver et al, 1993).

The interstitial collagens form an elaborate, stress-tolerant network in the mature ventricular myocardium (Caulfield and Borg, 1979; Robinson et al, 1987). This network functions to dissipate the mechanical tension associated with heart pumping throughout the ventricular wall. These fibrillar collagens form a weave that also serves to maintain the proper alignment of the cardiac myocytes during contraction. The fact that Mov 13 mice, which lack collagen type I, die during embryonic development due to vascular aneurysms illustrates that this ECM component is critical to cardiovascular morphogenesis and function (Lohler et al, 1984; Kratochwil et al, 1986). That the collagen network is essential to myocardial function was shown by inhibition of collagen cross-linking with β-aminoproprionitrile (BAPN) treatment of neonatal rats (Borg and Caulfield, 1981). Following BAPN treatment, these animals develop severe aneurysms due to the disorganization of the collagen network. Similar studies using animals raised on copper-deficient diets demonstrated abnormalities in collagen distribution (Borg et al, 1985; Capasso et al, 1989). Copper is a cofactor for lysyl oxidase, which is a critical enzyme for collagen cross-linking. Copper-deficient animals showed abnormally shaped myocytes as well as dilated hearts where the lack of collagen cross-linking resulted in a weakened myocardium that was not strong enough to maintain myocardial integrity during neonatal development.

The myocardial interstitium is thought to play a significant role in the contractility and relaxation of the ventricular wall as a whole. Type I and type III collagens form the principal components of the heart interstitium, even colocalizing to the same collagen fibrils (Borg and Terracio, 1990). Because these two interstitial collagens are not structurally or biophysically identical, their relative abundance is an important parameter of cardiovascular function (Weber et al, 1989; Borg and Terracio, 1990). The relative proportions of type I and type III collagen are altered in the heart during embryonic development (Thompson et al, 1979) and in pressure-overload hypertension of the adult (Borg and Terracio, 1990). Changes in the proportions of these collagens may have significant implications for cardiac relaxation, particularly in the adult animal. These studies suggest that the expression and deposition of the collagens are differentially regulated; however, very few studies have been carried out addressing this issue.

Noncollagenous Glycoproteins

A large number of noncollagenous glycoproteins have been described in the heart including fibronectin, laminin, entactin, fibulin, elastin, fibrillin, and tenascin

(Little and Rongish, 1995; Carver et al, 1997). Several of these, including fibronectin and fibulin, have been localized to the developing cardiac cushions and may play a role in mesenchymal cell migration into the valvuloseptal precursors (Linask and Lash, 1988; Spence et al, 1992). Fibronectin has been demonstrated to be present throughout fetal heart development, preceding vessel formation, and might influence migration of endothelial precursors into the myocardium (Rongish et al, 1996). Laminin and entactin are major components of the basement membranes surrounding cardiac myocytes and blood vessels (Little et al, 1989; Price et al, 1992). While much is known concerning the immunohistochemical distribution of these molecules in the basement membrane, little is known about their function. Localization of laminin shows that the myocyte basement membrane forms in a patchy manner and becomes contiguous only in the neonatal heart (Price et al, 1992). It has been speculated that this patchy nature allows for the attachment of collagen on the myocyte surface; however, once the basement membrane is contiguous, no new attachments would be possible. This would present a significant problem in the adult heart in that collagen connections that are sheared or broken would not be able to reform. Depending on the severity, this lack of connections could cause mechanical alterations during contraction.

Proteoglycans

Proteoglycans consist of a core protein covalently linked to carbohydrates termed glycosaminoglycans (Gallagher et al, 1986; Jackson et al, 1991). The biologic functions of proteoglycans are diverse, ranging from structural support to modulation of cell migration and differentiation. Proteoglycans appear to modulate developmental events through their interactions with other matrix macromolecules, cell adhesion proteins, or growth factors (Kjellen and Lindahl, 1991). A number of proteoglycans have been demonstrated to interact with other ECM components including collagen, laminin, and fibronectin (Scholzen et al, 1994). Binding of proteoglycans to other ECM components may alter the three-dimensional organization of the ECM and affect what a cell "sees" in the ECM. For instance, the binding of heparan sulfate proteoglycan to fibronectin modulates cell adhesion by exposing the cell-binding domain of this complex macromolecule (Couchman et al, 1988). However, the role that proteoglycans may play in modulating ECM organization in the developing heart has largely been unexplored.

A number of proteoglycans have been identified in the heart and their expression patterns during development defined, including perlecan-1, versican, N-syndecan, and others (Carey et al, 1992; Murdoch et al, 1994). However, few experimental studies have tested the functional roles of these molecules in cardiogenesis. Inhibition of proteoglycan synthesis by β-xyloside in *Xenopus* embryos results in a range of cardiovascular defects including absence of early heart tube looping (Yost, 1990). Recent work with the *hdf* mutant mouse has implicated versican as playing a critical role in morphogenesis of the embryonic heart (Mjaavedt et al, 1998). While it appears that specific proteoglycans have important effects on cardiogenesis, further tests of their function using in vitro bioassays or knockout mice are warranted.

Other Components

It is now clear that not only are these major structural components critical in the ECM, but also other molecules including matrix-modifying proteases and growth factors play important roles as constituents of the extracellular milieu.

Matrix-Modifying Proteases

Dynamic changes in the ECM are thought to play essential roles in physiologic processes involving tissue remodeling, including morphogenesis and wound healing. The precisely controlled balance between the synthesis and degradation of the ECM is thought to be important in normal embryonic development. Several families of matrix-modifying enzymes including serine proteases and matrix metalloproteases (MMPs) have been identified that are responsible for hydrolysis of various ECM components. The expression and activity of these molecules are postulated to be carefully controlled as abnormal proteolysis may contribute to altered development or pathologic conditions in the adult.

Changes in expression and activation of specific metalloproteases have been associated with several cardiovascular diseases (Tyagi et al, 1996). As mentioned previously, the collagenous network forms an essential structural entity interconnecting myocytes to each other and myocytes to adjacent capillaries (Caulfield and Borg, 1979; Robinson et al, 1987). Overexpression of matrix-modifying proteases is thought to compromise the structural integrity of the collagen network and thereby alter cardiac contractility and relaxation (Tyagi et al, 1996). Elevated levels of activated interstitial collagenase or MMP-1 have been detected in patients with dilated cardiomyopathy (Tyagi et al, 1996). On the other hand, decreased collagenase activation has been found concurrent with myocardial fibrosis in association with hypertension or pressure overload (Bishop et al, 1994). These studies indicate that the level of protease expression and activation is intimately associated with cardiovascular disease; however, much less is known regarding the role of matrix remodeling in heart development.

Expression of several proteases have been documented in the developing cardiac cushions and are postulated to play important roles in the development of the valvuloseptal regions of the heart (McGuire and Orkin, 1992; Nakagawa et al, 1992; Alexander et al, 1997). The expression of urokinase, a serine protease, is elevated in cardiac cushions concurrent with epithelial-mesenchymal transformation (McGuire and Orkin, 1992) and inhibition of urokinase expression has profound inhibitory effects on the ability of cardiac cushion explants to seed on collagen gels (McGuire and Alexander, 1993). Matrix metalloprotease-2 (gelatinase A or 72-kd collagenase) is also abundantly expressed by mesenchymal cells within the cardiac cushions (Nakagawa et al, 1992; Alexander et al, 1997). Studies with isolated heart fibroblasts indicated that MMP-2 is expressed on the leading edge of migratory cells and that this matrix protease is required for remodeling and contraction of collagen gels in vitro (Borg et al, 1997). These studies illustrate that matrix-modifying proteases play critical roles in cell processes, including migration, that are inherent to cardiac morphogenesis. However, a number of questions remain regarding the regulation and function of these proteases in the heart.

Growth Factors

Several growth factors reside as latent forms within the ECM, including basic fibroblast growth factor (bFGF) and transforming growth factor-β (TGF-β). Growth factors bind to the ECM and in many cases to the cell surface through interactions with proteoglycans. Heparan sulfate proteoglycan plays a critical role in binding to bFGF in the ECM and presentation of this growth factor to its receptor (Burgess and Maciag, 1989). Binding of latent growth factors to the ECM is likely to be important in protecting these molecules from degradation and in creating a growth factor reservoir in the extracellular milieu that can be rapidly activated and utilized (Kjellen and Lindahl, 1991). Therefore, matrix-modifying proteases have significant effects on the exposure and activation of these growth factors within the ECM.

CELL SURFACE RECEPTORS

Perception of the ECM by cells requires the presence of specific cell surface receptors that transmit information from the extracellular to the intracellular environment. It is important to remember that information can be transmitted in both directions by transmembrane receptors that connect to the ECM outside of the cell membrane and the cytoskeleton on the internal side. With respect to cardiac development, cell surface receptors can roughly be divided into two groups: those that directly interact with the ECM, and those that maintain cell-cell contact. Receptors involved in the direct recognition of the ECM include integrins, receptor tyrosine kinases (RTKs), and glycosyltransferases, while receptors such as cadherins and cell adhesion molecules (CAMs) are important in maintaining the cell-cell contacts necessary for cardiac development. The signal transducing properties of these cell-cell molecules have been reviewed elsewhere (Ullrich and Schlessinger, 1990; Rosales et al, 1995; Aplin et al, 1998).

ECM Receptors

Integrins

Integrins are the most widely characterized receptors for ECM proteins and can also be involved in cell-cell adhesion. Integrins are heterodimeric, transmembrane proteins composed of an α chain and a β chain that are noncovalently associated (Albelda and Buck, 1990; Baldwin and Buck, 1994). Currently, 15 α chains and eight β chains have been identified. Most α chains have a large extracellular domain, a single transmembrane region, and a relatively short cytoplasmic tail. The α chains have a structure similar to the β chain; however, some α chains have heavy and light subunits that are held together by a disulfide bond (Albelda and Buck, 1990). The extracellular region of the α chains contains seven repeated sequences that are predicted to fold into a β-propeller structure (Springer, 1997).

Ligand binding by integrins requires a divalent cation and for several integrins is mediated by an I domain present in the α chain (see Calderwood et al, 1997, and references therein). Seven of the 15 α chains contain I (inserted) domains in the amino terminal region (Dickcson and Santoro, 1998). The I domain is a region of approximately 200 amino acids located in the amino terminus of the extracellular region of the α chain. Ligand binding to the I domain is dependent on the presence of a divalent cation, either magnesium or manganese but not calcium (Takada et al, 1997). Given that the β chain also participates in ligand binding, there is much speculation as to whether or not this integrin subunit contains an I domain or similar structure. Structural predictions based on comparisons between integrin β chains and von Willebrand factor A domain indicate that an I domain–like structure may be present (Tuckwell and Humphries, 1997) and studies have identified residues predicted to reside in the I domain as critical for ligand binding (Takada et al, 1997). In contrast, additional work demonstrates that β integrin subunits do contain metal ion–dependent ligand binding sites, but the regions of ligand binding are predicted to have little structural similarity to the I domain (Lin et al, 1997).

Integrins are involved in the regulation of a variety of cellular events including differentiation, proliferation, adhesion, and migration (Albelda and Buck, 1990). The importance of integrins in development is illustrated by the fact that many integrin gene knockouts result in embryonic lethality, including several that result in death due to cardiovascular defects (Hynes, 1996). In the myocardium, the majority of integrins belong to the β_1 subfamily (Carver and Terracio, 1993). Analysis of β_1 expression during cardiac development has demonstrated a broad localization early in heart development with a specific localization to valves and endocardial cells late in development (Carver et al, 1994). This pattern of β_1 expression closely mimics myofibrillogenesis in the developing heart. The expression of a β_5 transcript has also been detected in the epicardial region of the heart (Yamada et al, 1995).

Rat cardiac myocytes have been shown to express a number of different α chains associated with β_1 during development. Three α chains, α_1, α_3, and α_5, were associated with neonatal and fetal myocytes, while only α_3 was detected on the adult myocytes. In adult myocytes, β_1 appeared to localize on the cell surface at or near the Z band (Terracio et al, 1991). The expression of α_6 from looping through septation during murine heart development has been determined. This integrin α chain displayed a sustained, high level of expression in atrial myocardium, reduced expression in the ventricular myocardium, and was present in the endocardial cushions (Hierck et al, 1996). The α_7 chain has been detected late in fetal development, predominantly located on myocytes of the ventricular wall (Song et al, 1992). The expression of the α_4 integrin subunit has been detected in the heart during early chick development (Stepp et al, 1994). Overexpression of α_5 in neonatal rat cardiac myocytes resulted in reduced actin expression and abnormalities in sarcomere structure (Simpson et al, 1998).

The differential expression of α chains of the β_1 integrin subfamily during times of increased ECM production and in regions under mechanical stress demonstrates the dynamic relationship between integrins and the ECM during cardiac development. Immunohistochemical staining of tissue sections with antibodies specific to individual integrins or specific ECM components illustrates the codistribution of ECM and its receptors (Figure 9.1). For instance, immunohistochemistry of tissue sections of hearts from 13.5 days of gestation show that β_1 integrin and laminin are both abundant in the developing trabeculae and less abundant in the compact myocardium at this stage (Figure 9.1). These and other data suggest that there is precise expression and accumulation of these molecules in the developing heart.

Receptor Tyrosine Kinases

A collagen specific receptor tyrosine kinase (RTK), discoidin domain receptor 2 (DDR2), has been reported in adult mouse heart (Lai and Lemke, 1994; Shrivastava et al, 1997; Vogel et al, 1997). Receptor tyrosine kinases are a family of transmembrane proteins involved in the conversion of extracellular stimuli into cellular responses. These receptors mediate a variety of cell functions including growth, migration, morphology, and differentiation. All receptor tyrosine kinases share a

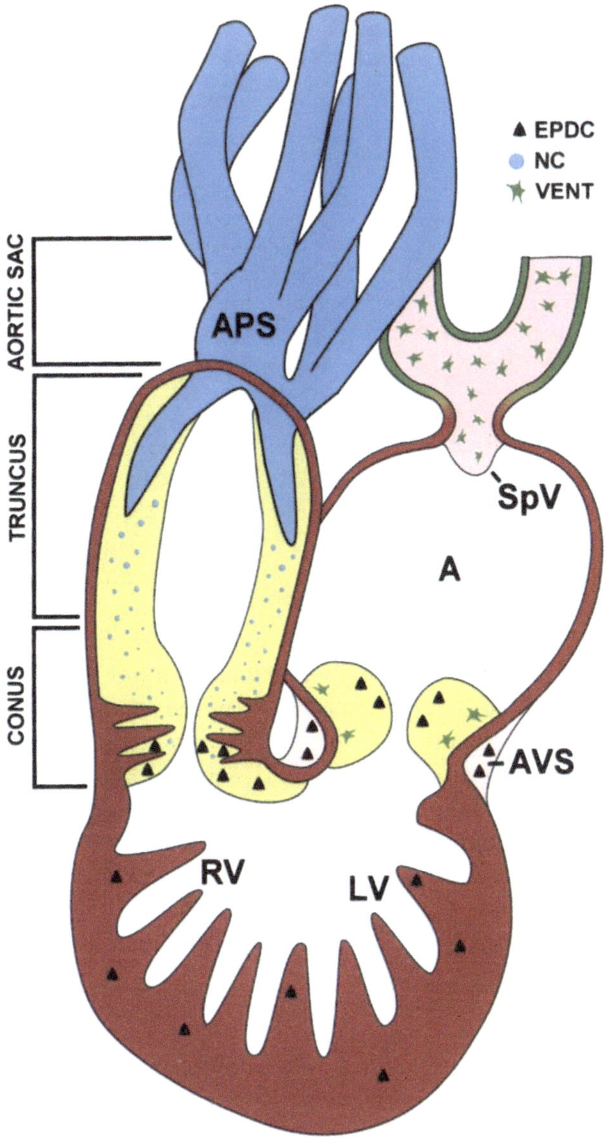

FIGURE 1.2. A stylized heart of a stage-32 chick or day-15 mouse heart. The two limbs of the U-shaped heart are "touching," meaning looping is completed and the conotruncal segment has become "wedged" into the AV and atrial segments of the inflow limb. Yellow represents cushion mesenchyme that has formed in the conus, truncus, and AV canal. Note the migration of extracardiac mesenchymal cells into the cushions. Epicardial-derived cells (EPDCs) migrate from the surface of the heart at sites like the atrioventricular sulcus (AVS) and enter the cushions and subendocardial regions of ventricular trabeculae. Neural crest (NC) cell and tissue (colored blue) enter from the arterial pole as two prongs or columns of cells extending from the aorticopulmonary septum (APS) into the distal end of the definitive tubular heart (i.e., where a myocardial mantle is present). The columns or condensations of crest cells break up into free crest mesenchyme, which migrates into the conus. Correlating with the presence of neural crest, myocardial cells invade the conal cushions, a process called myocardialization. At the venous pole, neural crest derived from the ventral surface of the neural tube, termed "vent cells" migrate into the common atrium as part of a penetrating wedge of dorsal mesocardium called the spina vestibuli (SpV). Ultimately the spina vestibuli will contact the AV cushions during formation of the primary atrial septum, thereby adding a subtype of crest cells to the mesenchyme of the AV cushions. A, atrium.; LV, left ventricle; RV, right ventricle.

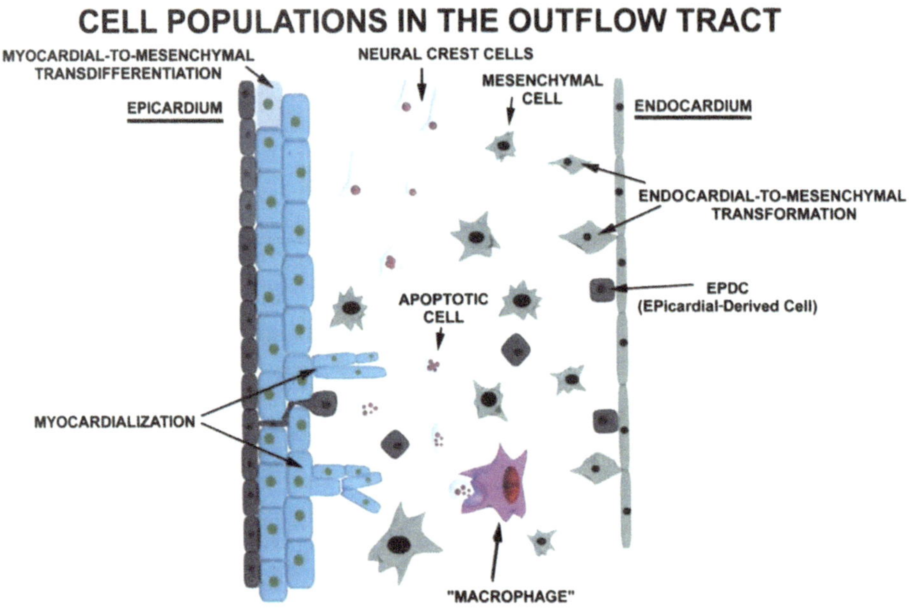

FIGURE 1.3. Diagram of the currently known cell populations among the cushion tissue of the conus region of the outflow tract. Except for the myocardial invasions, all appear as mesenchyme, but of different origins.

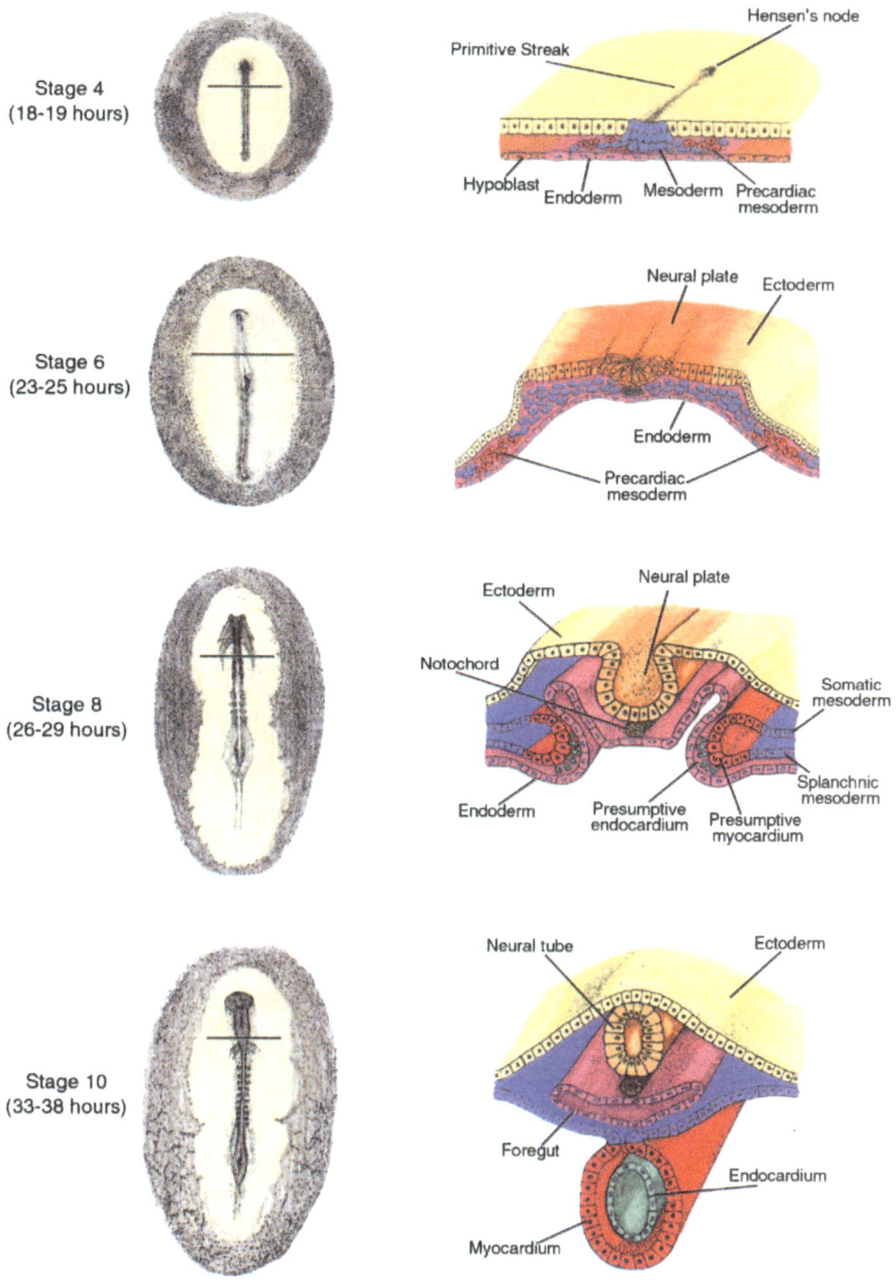

FIGURE 2.1. Overview of heart development in the chick embryo. Dorsal views (left) and sagittal sections (right) at the indicated embryo levels show the origin and movement of cells fated to form the myocardium and endocardium. At stage 4 (mid-gastrulation), cardiogenic progenitors have just involuted and are located within mesoderm on either side of a broad region of the primitive streak. By stage 6, cardiogenic mesoderm is localized to bilateral anterior lateral regions, subtended by cardiogenic endoderm. Separation of somatic and splanchnic layers and folding of the embryo bring the bilateral precardiac regions to toward the midline. By stage 10, bilateral regions have fused to form a single heart tube composed of myocardium and endocardium. Spontaneous rhythmic contractions begin at this stage.

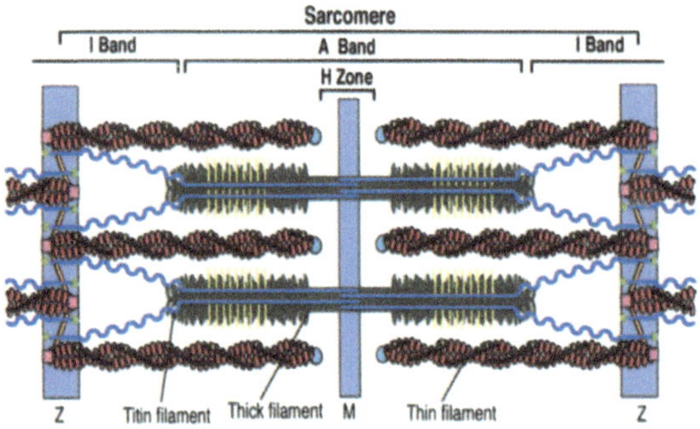

FIGURE 2.2. Schematic of the known organization of the major components of a cardiac muscle sarcomere (a single contractile unit). Z lines form the boundary of the sarcomere, whereas the center of the myosin bipolar, multiheaded, thick filament array is at the M lines, in the middle of the sarcomere. *Thin filaments*: the actin filaments (red) are capped by CapZ (pink squares) at their barbed ends, and by tropomodulin (turquoise) at the pointed ends; the latter end interdigitates with the myosin containing thick filaments. Rodlike tropomyosin molecules (black lines) are associated with each other head to tail, forming two polymers per thin filament. Each tropomyosin molecule binds one troponin complex (yellow). *Thick filaments*: myosin bipolar filaments (green, multiheaded domains) and MyBP-C (yellow transverse lines). The amino termini of the giant titin (blue) molecules, making up the third filament system in striated muscle, fully overlap in the Z disk and interact with T-cap (green). Within the Z disk, α-actinin (gold) crosslinks the actin and titin filaments from opposite sarcomeres. The carboxy-terminal region of titin molecules from opposite half-sarcomeres fully overlap in the M lines. There are three to six titin molecules associated with each thick filament in each half-sarcomere and six thin filaments associated with each thick filament in a hexagonal lattice in vertebrate muscle.

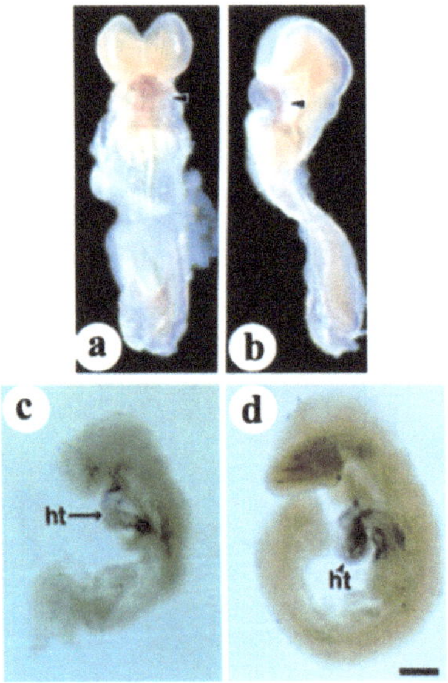

FIGURE 4.1. The expression of *mXin* in mouse embryonic hearts as revealed by in situ hybridization (a,b) Whole-mount in situ hybridization of a day 8.0 p.c. (five somites) embryo; parts a and b are ventral views and left-side views, respectively, of the same embryo under dark-field optics. The *mXin* is expressed specifically in the developing heart tube (arrowheads). (c,d) Whole-mount in situ hybridization of day 8.5 p.c. and day 9.0 p.c. embryos, respectively. Both c and d are left-side views of embryos under bright-field optics. In a manner similar to *cXin* expression in stage 11 chick embryos (Wang et al, 1999), *mXin* is most strongly expressed in posterior and anterior ends of the heart tube. ht, heart; bar = 280 μm.

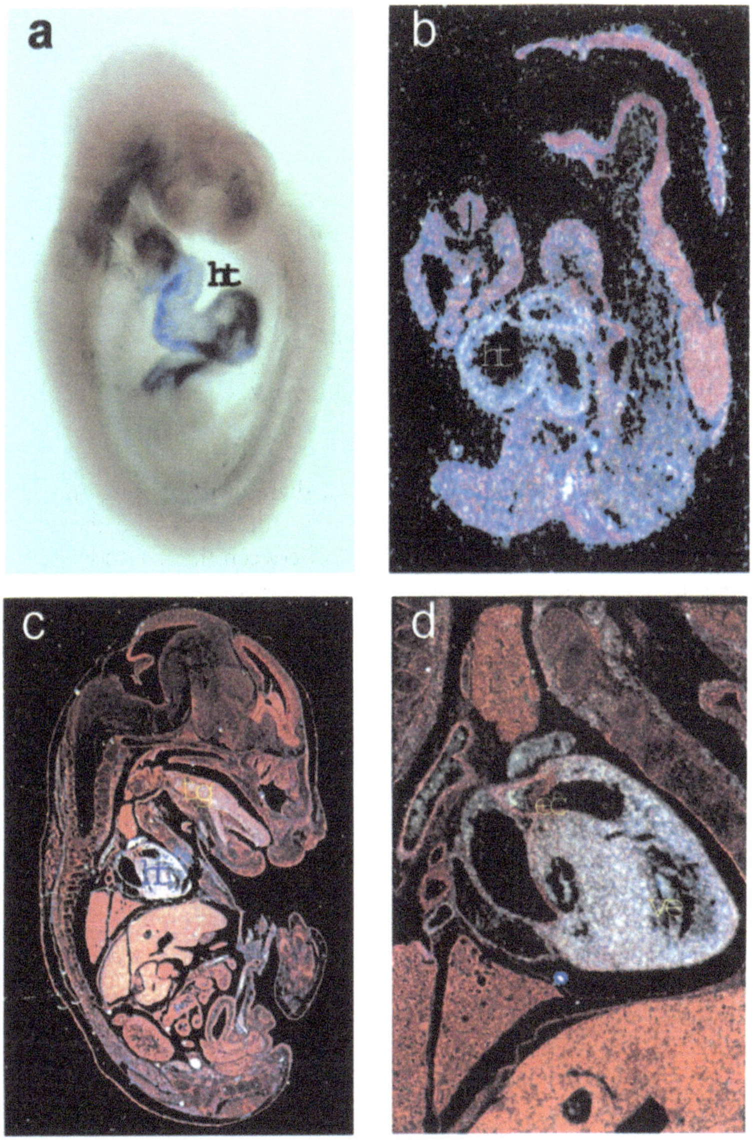

FIGURE 4.2. The expression of *mXin* in mouse embryonic hearts as revealed by in situ hybridization (a) Whole-mount and (b) paraffin-sectioned in situ hybridization of a day 9.5 p.c. embryo. ht, heart. (c,d) paraffin-sectioned in situ hybridization of a day 1.5 p.c. embryo. ht, heart; tg, tongue; ec, endocardial cushion; ve, ventricle.

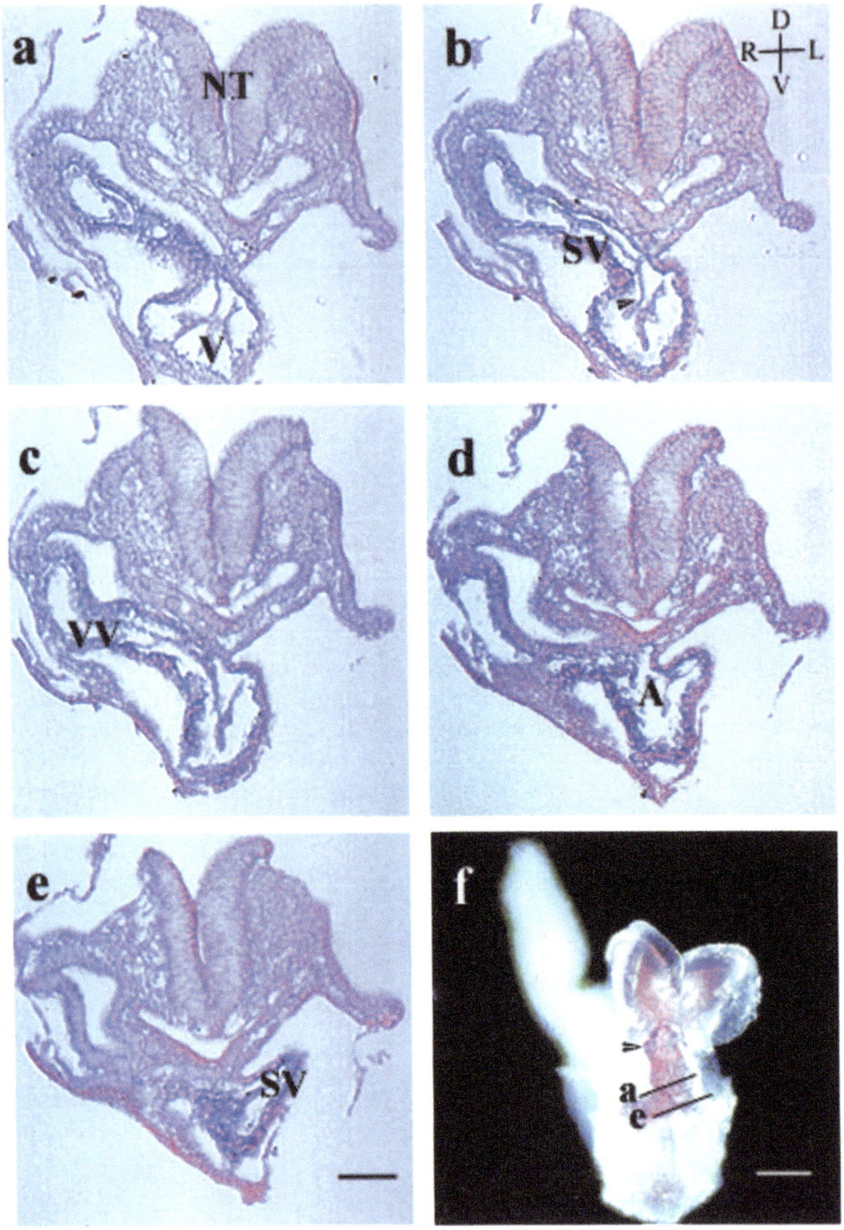

FIGURE 4.3. Expression of *mXin* in the linear heart tube and developing sinus venosus. (a–e) Serial 10-μm-thick cross sections were prepared from a day 8 p.c. (six somites) embryo (f) after whole-mount in situ hybridization. Planes of section for parts a and e from cranial to caudal are indicated by the lines drawn in f. The orientation of sections is depicted in b. The arrowhead in b indicates the endocardial cell layer within the ventricle, which is contiguous to the cell layers (positive for the *mXin* expression) at the sinus venosus and vitelline vein. The arrowhead in f points to a relatively strong hybridization line in the linear heart tube. A, atrium; NT, neural tube; V, ventricle; SV, sinus venosus; VV, vitelline vein. Bars in E and F = 50 μm and 200 μm, respectively.

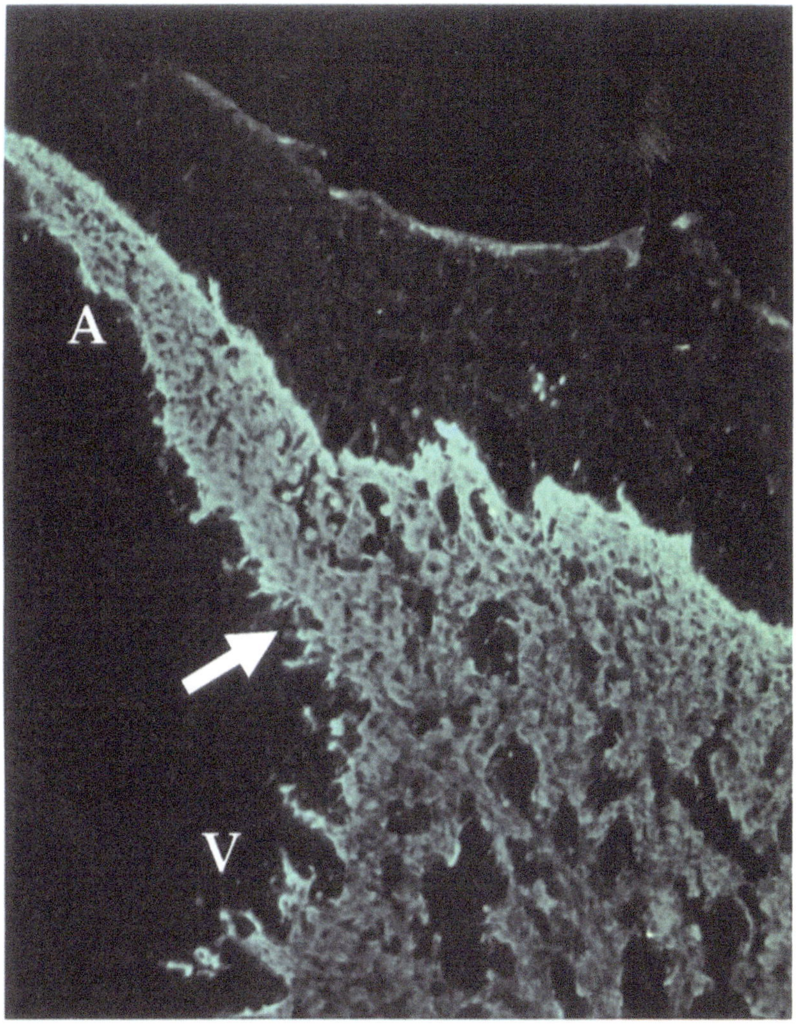

FIGURE 5.1. Morphogenic variation between the atrial and ventricular chambers. Staining of a sectioned embryonic day 7 chicken heart at the atrioventricular junction with MF-20 shows the thick-walled trabeculated ventricle (V) and the thin-walled atrium (A). The valves that separate the atrium and ventricle will later form from the atrioventricular (AV) junction (arrow).

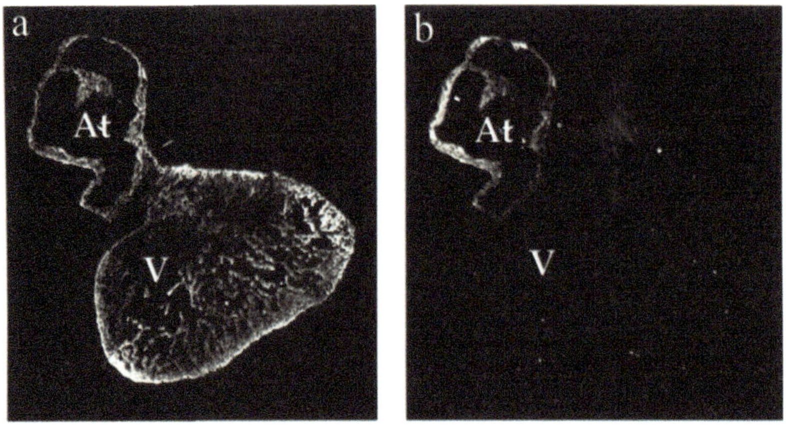

FIGURE 5.2. Chamber-specific myosin heavy chain isoform expression. Sectioned Day 7 chicken embryos were stained with MF-20 (a) and B1 (b) monoclonal antibodies. MF-20 recognizes all sacromeric myosin heavy chain isoforms, whereas B1 antibody is specific to a myosin heavy chain isoform expressed in the atrial chamber.

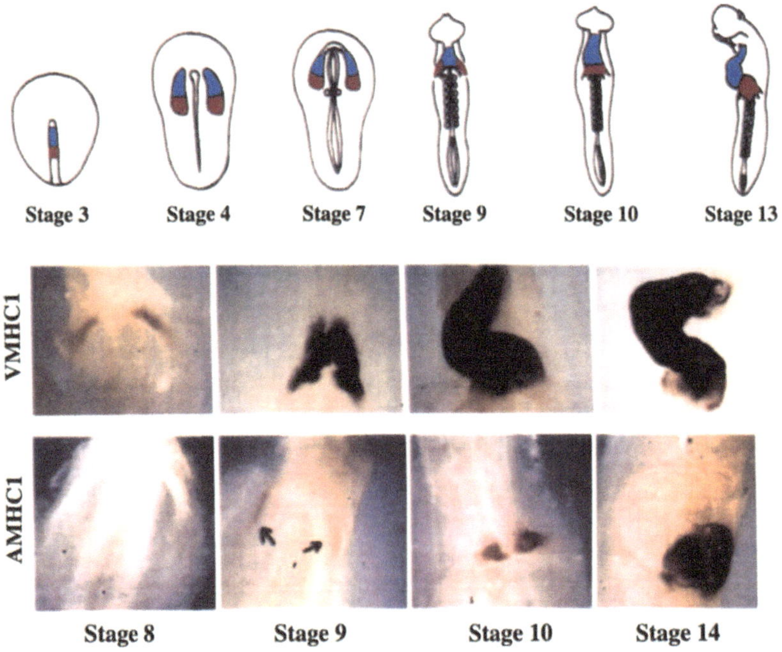

FIGURE 5.3. Schematic diagram of cardiac diversification. Cell fated to an atrial (red) and ventricular (blue) cell lineage are arranged with a distinct anteroposterior polarity within the primitive streak (stage 3). The anteroposterior (AP) polarity of the cardiogenic field is maintained as the cells migrate anterior and laterally within the embryo (stages 4–7). *VMHC1* expression is detected in all cardiogenic cells prior to the fusion of the myotube (stage 10). Posterior cells of the cardiogenic field exhibit an atrial phenotype by stage 9 with the induction of *AMHC1* expression. *AMHC1* expression is detected solely in the atrial chamber through cardiac looping, septation, and compaction. (Adapted from Yutzey et al, 1994.)

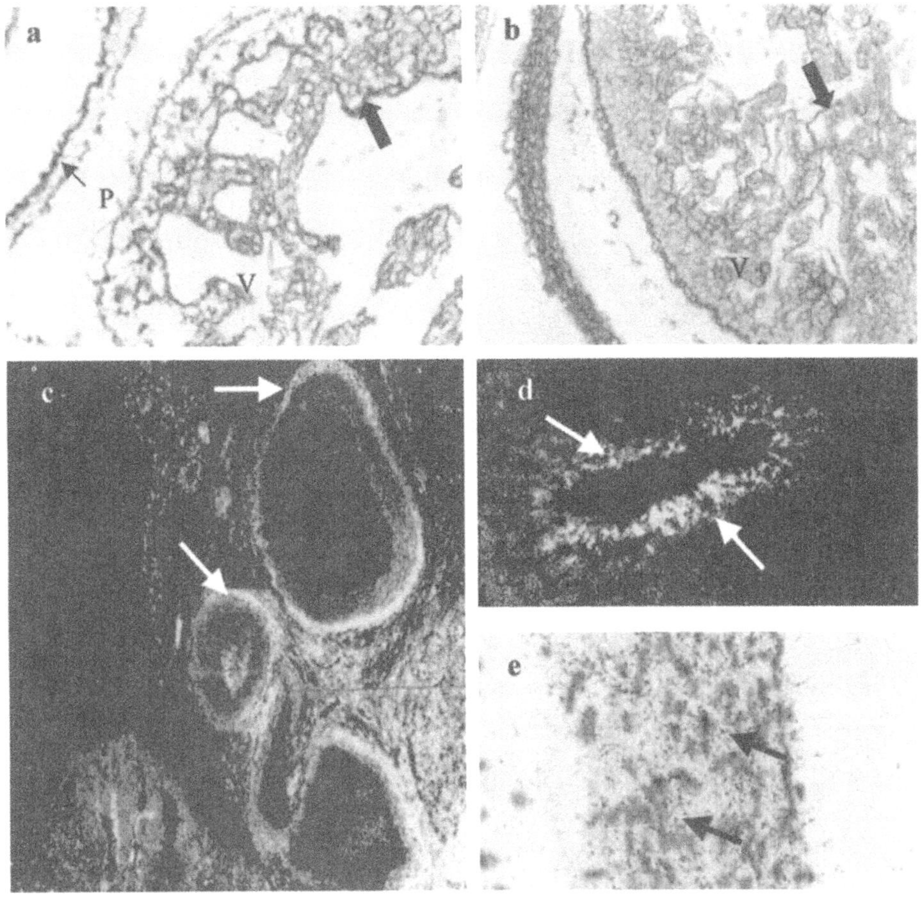

FIGURE 9.1. Photomicrographs of sections of 13.5-day embryonic (a and b) or 3-day neonatal rat hearts illustrating the localization of (a) laminin, (b) α_3 integrin, (c) collagen type I, and (d,e) α_1 integrin. Very similar localization is seen for laminin (a) and the α_3 integrin (b), one of several laminin receptors in the heart. Intense antibody staining is seen in the pericardium (P), outer myocardium of the ventricle (V), and endothelial cells lining the ventricular chamber (arrows). (c–e) Photomicrographs illustrate the colocalization of collagen type I (c) and α_1 integrin, (d) immunohistochemistry, and (e) in situ hybridization in the extracardiac vessels of the neonatal rat. Collagen type I and α_1 integrin, a collagen receptor, are found in the connective tissue regions associated with the vessel walls (arrows).

common structural organization in that every member of this family contains a large extracellular domain responsible for ligand binding, a single membrane spanning segment, and a cytoplasmic region that contains a conserved tyrosine kinase domain (Schlessinger, 1997). In addition, all receptor tyrosine kinases share a common mode of activation involving receptor dimerization as a result of ligand binding. Receptor oligomerization leads to activation of the tyrosine kinase domain, resulting in autophosphorylation of the receptor providing sites for interaction with other molecules involved in signal transduction, which recognize sequences with phosphotyrosine residues such as Src-homology 2 and 3 domains (Lemmon and Schlessinger, 1994).

DDR2 is a transmembrane protein with a molecular weight between 105 and 126 kd (Johnson et al, 1993; Alves et al, 1995) and contains an extracellular region that is responsible for collagen binding. The extracellular domain of DDR2 has been demonstrated to be homologous to the discoidin proteins of *Dictyostelium discoideum* and to the C1 and C2 unit of factor VIII (Johnson et al, 1993; Karn et al, 1993). The cytoplasmic region of DDR2 contains a conserved tyrosine kinase domain with several predicted sites for autophosphorylation and interaction with intracellular proteins (Johnson et al, 1993; Karn et al, 1993; Alves et al, 1995). DDR2 is expressed in a variety of mouse, rat and human tissues, including heart, muscle, brain, and kidney (Karn et al, 1993; Lai and Lemke, 1994).

Specific activation of the tyrosine kinase domain in DDR2 by collagen binding was demonstrated by stimulating the receptor with various types of collagen. Collagen types I and III resulted in the highest degree of receptor phosphorylation, followed by collagen types II and V, which moderately stimulated phosphorylation. No activation of the receptor was observed using collagen type IV (Shrivastava et al, 1997; Vogel et al, 1997). Receptor activation was observed with both soluble and immobilized collagen. Studies have shown that phosphorylation of DDR2 is significantly reduced after deglycosylation of collagen or if the triple helical structure of collagen is altered due to heating (Vogel et al, 1997). Unique to this receptor was the extended period of time required for receptor activation. Activation of DDR2 by type I collagen peaked 90 minutes to 2 hours after stimulation and the receptor remained phosphorylated up to 18 hours later (Vogel et al, 1997). Stimulation of DDR2 by collagen type I over a 4-day period revealed that this receptor is capable of upregulating the expression of MMP-1 (Vogel et al, 1997), an enzyme involved in the degradation of collagen types I, II, III, VII, and X (Cleutjens, 1996).

Glycosyltransferases

Glycosyltransferases are a family of enzymes, normally localized in the Golgi apparatus, that catalyze the transfer of sugar moieties from donor molecules to protein/lipid acceptors and are responsible for the biosynthesis of oligosaccharides found on proteins and lipids (Shur, 1989; Dubois and Shur, 1995). Several members of this protein family have been shown to be expressed on the cell surface with β1,4-galactosyltransferase the best-characterized member of this family. β1,4-galactosyltransferase (GalTase) has been identified on a variety of cells and has been shown to be important in both cell-cell and cell-ECM interactions (Shur, 1989; Dubois and Shur, 1995). GalTase is a type II integral membrane protein with a short, intracellular N-terminal region, single transmembrane domain, and a catalytic, extracellular C-terminal region. The subcellular distribution of GalTase is determined by transcription resulting in either "long" (24 amino acids) or "short" (11 amino acids) forms of the protein, which differ only in their N-terminal regions. The additional 13 amino acids present in the "long" form are responsible for the cell surface expression of β1,4-galactosyltransferase (Lopez et al, 1991). Studies have shown that the "long" form of GalTase is associated with the actin cytoskeleton, possibly through these additional 13 amino acids, on migrating mesenchymal cells (Eckstein and Shur, 1992).

In vitro, GalTase present on migrating fibroblasts was shown to spontaneously galactosylate both glycosaminoglycan (Turley and Roth, 1979) and laminin

(Begovac et al, 1994). Using embryonic cardiac mesenchymal cells, a peptide derived from laminin was able bind to GalTase and inhibit migration in the presence of excess laminin (Loeber and Runyan, 1990). GalTase has been shown to bind N-linked oligosaccharides present primarily in the laminin A chain during cell spreading (Runyan et al, 1988). The presence of laminin in the ECM triggers the cell surface expression of GalTase present intracellularly into developing lamellipodia during migration (Eckstein and Shur, 1989). This enzyme may provide a mechanism by which a cell can recognize and interact with ECM via GalTase binding to N-linked oligosaccharides present on laminin and then release the cell membrane from the ECM substrate after catalytic galactosylation of laminin. Studies of migration rates for GalTase-dependent migration on laminin coated surfaces show that rate of migration is directly dependent on surface expression of GalTase, but that rates are inversely proportional to GalTase association with the cytoskeleton indicating that there are a defined, limited number of binding sites for GalTase on the cytoskeleton (Evans et al, 1993; Appeddu and Shur, 1994).

Several studies have examined the role of GalTase as a signal transducing molecule. The level of cell surface GalTase associated with the cytoskeleton has been shown to directly affect cell growth rate by delivering a growth inhibitory signal through modulation of the EGF receptor (Hinton et al, 1995). Disruptions of GalTase-dependent adhesions have been demonstrated to inhibit cell growth and trigger laminin synthesis in embryonal carcinoma cells (Maillet and Shur, 1994). Together these data support a role for GalTase as a regulator of migration during heart development and as a potential signal transduction molecule.

Cell-Cell Contact Receptors

Cadherins

Cadherins are transmembrane glycoproteins critical for the establishment of the physical cell-cell connections required during embryonic development. This family of proteins, containing three to five repeated domains in the extracellular region and ranging in size from 120 to 140 kd, is characterized by its Ca^{2+}-dependent, homotypic interactions (Geiger and Ayalon, 1992; Takeichi, 1991). The selective, homotypic recognition of cadherins is attributed to the amino terminal (approximately 113 amino acids) regions of these molecules and is believed to occur first by the dimerization of cadherin molecules on the cell surface (called a strand dimer) followed by association with an identical strand dimer on an adjacent cell (Tamura et al, 1998; Uemura, 1998). Cadherins can be grouped into various subclasses, including E-cadherin (epithelial), P-cadherin (placental), liver cell adhesion molecule (L-CAM), and N-cadherin (neural), which have all been well characterized.

N-cadherin is expressed in both the developing and mature myocardium of the heart (Hatta et al, 1987; Duband et al, 1988), where it has been observed both in regions of myocyte-myocyte contact (intercalated disks) and in sarcomere Z-bands associated with α-actinin (Goncharova et al, 1992). The cytoplasmic proteins α, β, and γ (also called plakoglobin) catenins associate with the cytoplasmic region of N-cadherin and provide a bridge between N-cadherin and the actin cytoskeleton. This cadherin-catenin interaction is also required for adhesion events mediated by N-cadherin (Geiger and Ayalon, 1992) as the catenin family is a participant in signal transduction. The association of N-cadherin with an α/β-catenin complex was found to be important in the compartmentalization of heart precursor cells, which potentially involves the Wnt signaling pathway (Linask et al, 1997).

N-cadherin is required for proper development of the myocardium. In a mouse model, a dominant-negative mutation of N-cadherin resulted in failure of heart tube development and lethality by day 10 of development (Radice et al, 1997). Exposure of a developing chick embryo to antibodies that inhibit N-cadherin mediated adhesion resulted in lack of cell differentiation (depending on the time in development during which the embryo was exposed to the N-cadherin antibody) and myofibrillogenesis (Linask et al, 1997). Levels of N-cadherin have been shown to increase during the trabeculation stage of myocardial development in the chick embryo, followed by a constant level of expression when myocytes begin the formation of intercalated disks (Ong et al, 1998). The distribution of N-cadherin on the surface of myocytes changed during development from diffuse throughout the membrane of myocytes in the embryonic heart to a restricted localization in the intercalated disks of mature hearts (Ong et al, 1998).

The role of N-cadherin in cardiac development has been examined using both adult and embryonic myocytes. In adult rat cardiomyocytes, the introduction of a dominant-negative mutant chick N-cadherin resulted in the inability of myocytes to reform cell-cell contacts and loss of myofibril organization (Hertig et al, 1996). N-cadherin antiserum reduced the number of contracting embryonic myocytes (both isolated and clustered, electrically coupled cells) and decreased the amount of plasma membrane involved in cell-cell junctions, including adherens junctions and intercalated disks (Soler and Knudsen, 1994). Myocytes treated with anti–N-cadherin antiserum also contained fewer myofibrils and the fibrils that were present were disorganized and poorly aligned, implicating N-cadherin as a potential mediator of myofibrillogenesis. However, other studies demonstrated that N-cadherin was not essential to the formation of in vivo like myocytes on aligned collagen (Simpson et al, 1994). Together these data indicate that the role of N-cadherin in myofibrillogenesis is controversial and probably involves many steps including formation of the myofiber, signal transduction, and electrical coupling.

Cell Adhesion Molecules

Cell adhesion molecules (CAMs), also referred to as immunoglobulin (Ig) CAMs, belong to a diverse family of cell surface receptors characterized by one or more copies of the Ig domain in their extracellular region. This family of proteins typically possesses large amino terminal extracellular domains, a single transmembrane domain, and a cytoplasmic domain (Rosales et al, 1995). The prototypic member of the CAM family is N-CAM, which contains five Ig folds in its extracellular domain and can be present in multiple forms on the cell surface (Aplin et al, 1998).

Endocardial cushions are essential to cardiac valvulogenesis, and develop and mature in response to a highly coordinated expression of growth factors, cell surface receptors, and other molecules. The transformation of epithelial cells into cushion mesenchyme can be followed by the expression of N-CAM during cushion development. Immunohistochemical examination at stage 8 of chick development indicated that the cardiac primordium contained high levels of N-CAM (Crossin and Hoffman, 1991). At stage 18, the endocardium and myocardium still contained high levels of N-CAM, but once cells began to migrate into the cardiac jelly they no longer expressed N-CAM. This study demonstrates the presence of N-CAM in the myocardium throughout embryonic development and suggests a potential role for this protein in the morphogenetic development of the heart.

REGULATION OF ECM EXPRESSION AND ORGANIZATION

The accumulation, composition, and organization of the ECM all have dramatic effects on cardiovascular development, maintenance, and pathology. Numerous studies have shown that abnormal accumulation of ECM components (fibrosis) deleteriously affects the function of the adult heart and other organ systems (Weber et al, 1989; Borg and Terracio, 1990; Eghbali and Weber, 1990; Chapman et al, 1991). Specific spatial patterns of ECM expression have also been postulated to direct morphogenetic events in the embryonic heart (Linask and Lash, 1988). These studies illustrate that the dynamic expression, deposition, organization, and degradation of the ECM components must be precisely orchestrated. These processes are modulated by a variety of chemical factors including growth factors and cytokines in the cardiovascular system. A number of factors have been shown to affect the expression of diverse ECM components in the heart. Several growth factors, including transforming growth factor-β (TGF-β), platelet-derived growth factor (PDGF), insulin-like growth factor (IGF), and angiotensin II, have been well studied with regard to cardiac hypertrophy and fibrosis. Very few studies have been directed at the developmental roles of these or other factors with regard to ECM production.

It is becoming increasingly clear that the response of cells to environmental factors is cell type as well as age specific. For instance, PDGF has been shown to stimulate laminin and collagen type IV synthesis in kidney mesangial cells (Fleoge et al, 1993) but decreases collagen type IV expression in arterial smooth muscle cells (Okada et al, 1993). The stimulatory effects of TGF-β on ECM expression have been well described (Chua et al, 1991; Eghbali et al, 1991; Katchman et al, 1994). TGF-β has been shown to induce the expression of several collagens and fibronectin, and has been suggested as an important mediator of cardiac hypertrophy and fibrosis. However, the stimulation of collagen type V expression by TGF-β has recently been shown to be cell type specific (Bertelli et al, 1998). Along the same lines, recent studies have indicated that interleukin-1β stimulates the expression of collagen type I by several cell types including dermal fibroblasts (Lertchirakarn et al, 1998). However, this cytokine had conflicting effects on proliferation of different mesenchymal cells in vitro, i.e., treatment of dermal fibroblasts with interleukin-1β (IL-1β) stimulated proliferation of these cells, while interleukin-1β treatment of dental pulp cells resulted in inhibition of proliferation. It is clear, then, that careful examination of specific cell types from particular developmental or physiologic conditions must be carried out before general statements can be made regarding growth factor regulation of ECM expression.

Numerous works have indicated specific patterns of expression of particular ECM components in the developing heart and other organs. These rather unique spatial and temporal patterns of expression suggest that the expression of individual ECM components is modulated differently by various extrinsic factors. PDGF has been shown to stimulate collagen type IV synthesis in arterial smooth muscle cells and to simultaneously inhibit collagen type V expression in these same cells (Okada et al, 1993). As mentioned above, IL-1β has been shown to stimulate collagen type I expression in isolated dermal fibroblasts (Lertchirakarn et al, 1998); however, this same growth factor inhibits collagen type II synthesis and chondrocyte differentiation (Takahashi et al, 1998). TGF-β stimulates both collagen type I and basement membrane collagen type IV expression by glomerular mesangial cells (Silbiger et al, 1998). Recent studies have shown that treatment of isolated mesangial cells with estradiol attenuates the stimulatory effects of TGF-β on collagen type IV expression but not collagen type I expression (Silbiger et al, 1998). The inhibitory effects of estradiol appear to be at the transcriptional level, suggesting that the collagen genes are differentially regulated, in this case, at the DNA level. Subtle changes in the regulation of different ECM components may have profound effects on cellular processes associated with heart development, including migration and proliferation.

Many of these chemical factors have multiple downstream effects. TGF-β, IGF-1, and angiotensin II have all been shown to modulate ECM expression as well as cell proliferation of cardiovascular fibroblasts. The signaling pathways mediating these effects are now being elucidated. It appears that most of these factors stimulate multiple pathways, which may lead to diverse downstream responses. Work from our group indicates that IGF-1 induces collagen expression and that this response is, at least in part, at the transcriptional level (S. Kanekar, unpublished observations). IGF-1 receptor ligation activates several signaling cascades including the mitogen-activated protein kinase (MAPK) pathway and the phosphoinositidyl (PI_3) kinase pathway. Experiments using pharmacologic inhibitors to these different pathways indicate that activation of the PI_3 kinase pathway is critical for IGF-1–induced increase in collagen transcription by heart fibroblasts. Inhibition of the MAPK pathway has no effects on this response. One challenge of future studies is to elucidate the mechanisms through which individual growth factors elicit multiple responses in cardiovascular cells and to determine if one (or more) of these pathways is favored at specific stages of development or disease.

Most studies examining the regulation of ECM expression have been carried out in isolated cells grown in serum-free medium with single growth factors added. While these studies have provided significant insight into ECM modulation by growth factors, it is obvious that this does not mirror the complexity of the in vivo tissue. Several recent studies have begun to address the potential interactions of various chemical factors in ECM modulation. For instance, studies examining collagen type II regulation in chondrocytes illustrated that TGF-β or IGF-1 alone had no effect on gene expression (Yaeger et al, 1997). However, addition of these factors together stimulated collagen type II and aggrecan expression in these cells. It will be particularly important to begin to determine how the complex mixture of chemical factors in the cardiovascular system modulates the ECM during heart development and disease. Do these chemical factors act through some common or overlapping pathway to modulate ECM production?

Mechanical forces play significant roles in the cardiovascular system from early in development throughout adult life (Taber et al, 1992). Experiments using a variety of in vitro systems clearly indicate that mechanical forces alone can initiate a plethora of specific downstream responses in cells, including changes in cell shape, proliferation, and gene expression. For instance, compressive forces stimulate chondrogenesis including the upregulation of collagen type II and aggrecan expression (Takahashi et al, 1998), and mechanical stretch initiates a hypertrophic response in isolated cardiac muscle cells (Sadoshima and Izumo, 1993, 1994). That the synthesis and accumulation of ECM components including collagen type I and type III are increased in the heart concurrent with increases in cardiovascular work load (Chapman et al, 1991) suggested that mechanical stress may modulate ECM production in heart fibroblasts. In vitro studies using a variety of different methods to apply mechanical strain have illustrated that mechanical forces indeed stimulate collagen gene expression by isolated heart fibroblasts and vascular smooth muscle cells (Carver et al, 1991; Bishop, 1998). Work is currently under way to determine how mechanical signals are transduced into specific changes in gene expression.

Mechanical stretch has been shown to induce the expression and secretion of angiotensin II in isolated heart myocytes (Sadoshima et al, 1993; Sadoshima and Izumo, 1995). Angiotensin II was shown in these studies to play a significant paracrine/autocrine role in the hypertrophic response of these cells to mechanical stimulation. Mechanical pressure and angiotensin II have been reported to act synergistically in the regulation of fibronectin expression by vascular cells (Nathalie et al, 1996). At least in kidney glomerular cells, increased expression of collagen in response to cyclic strain is dependent on the paracrine/autocrine actions of TGF-β (Hori et al, 1998). It is obvious that the combinations of chemical and mechanical factors that potentially modulate ECM production by cardiovascular cells are extremely complex. Careful examinations will be required to sort out the contributions of these different stimuli to ECM regulation during development and disease.

MYOFIBRILLOGENESIS

Studies of different stages of development have shown that the ECM together with cell surface receptors plays a critical role in the transduction of mechanical tension and chemical signaling. The changes in mechanical tension and chemical signaling are caused in part by the growth of the heart as well as the increase in blood volume. One of the critical aspects of mechanical/chemical signaling within the ECM is regulation of myocyte phenotype. Initially myocytes are oval shaped and mature into rod-shaped cells. The intrinsic genetic factors result in the formation of myofibrils within the cell; however, the formation of the rod-shaped cell requires mechanical tension, integrins, and attachment to the ECM. In vitro studies have demonstrated that the $\alpha_1\beta_1$ integrin is critical to forming the rod-shaped myocyte (Simpson et al, 1994). In addition to the role of the integrin, collagen must be in the correct orientation for the formation of the rod-shaped myocyte. Along with various chemical factors, mechanical tension has also been shown in ex vivo models of heart growth in the anterior chamber of the eye (Tucker, 1990). In this model, the heart develops in an "unloaded" condition and results in abnormal-shaped myoctes with abnormal myofibrillogenesis. While these in vitro studies demonstrate the role of integrins and collagen in the regulation of phenotype, the regulation of phenotype in vivo is certainly much more complex. Other members of the integrin family and other receptors probably function to regulate myocyte phenotype, as there is little organized collagen in the early developing heart.

During the differentiation of myocytes, the cells are responding to both intrinsic and extrinsic signals to form the rod-shaped phenotype. During this process, several sites of specialization develop in the myocyte including specialized regions of the sarcolemma, cytoskeletal organization, and formation of the contractile apparatus. Several specialized regions of the sarcolemma include the asymmetrical distribution of proteins that form the intercalated disk, gap junctions, basement membrane, and a specialized region that defines the Z band. The formation of these regions involves the coordination of both internal and external signals. While little is known regarding the mechanisms that sort the various protein and lipid moieties into their various compartments, these processes must involve both intra- and extracellular signaling. The development of myofibrils has been demonstrated both in vivo and in vitro, and studies have shown that as the myocyte develops, myofibrils are first developed in the area of the sarcolemma. Attachment of the myofiber to the sarcolemma appears to be regulated in part by the cytoplasmic face of the $\alpha_1\beta_1$ integrin. As the myocyte undergoes growth, it appears that specific integrins sort to specific regions of the cell, which, in part, may dictate where the initial myofibrils form. It is probable that these early myofibrils form the template for additional myofibrillar growth and deposition.

Formation of cell-cell contacts between myocytes is another critical event in the morphogenesis of the heart. The formation of the intercalated disk is an initial step in defining the cellular asymmetry of the myocyte. Studies showing the expression of N-cadherin document that these connections are critical in establishing the muscle phenotype and polarity of the cell. At the time of intercalated disk formation, gap junctional proteins, the connexins, begin to be inserted. The importance of various isoforms of the connexins is not well understood but current molecular and physiologic studies are likely to provide new insights in their function.

Another aspect of muscle cell differentiation is the formation of the basement membrane. The function of this structure in muscle cells is unknown; however, its formation appears to be accomplished in a coordinated manner along with myofibril formation as well as the formation of special sites for attachment of collagen (Price et al, 1992). Studies on laminin deposition both in vivo and in vitro indicate that the basement membrane forms in a coordinated fashion (Hilenski et al, 1991; Price et al, 1992). In vivo studies indicate that laminin deposition occurs initially in a random fashion on the sarcolemma. Immunohistochemical studies show a uniform labeling of the sarcolemma. With maturation of the myocyte, laminin localization begins to coalesce into defined regions with the heaviest staining near the sites of the internal Z band. In vitro studies using neonatal myocytes, indicate that laminin forms into a discrete pattern on the sarcolemma that attaches to the ECM. This pattern also shows that deposition is not uniform but has a patchy nature. Localization studies of other basement membrane components such as collagen type IV have only been done in vitro. The deposition pattern of collagen type IV was both temporally and spatially different from laminin. It is tempting to speculate that the formation of the basement membrane is an important event in defining specialized regions of the sarcolemma, but without more functional data this may be premature.

The third area of asymmetry of the sarcolemma is the formation of a region termed the extracellular-integrin cytoskeleton. As previously described, the association of collagen with the surface of the myocyte occurs in a precise fashion with insertion at or near the Z band. Immunohistochemical studies have shown that a variety of ECM components, cell surface receptors, cytoskeletal components, and signal transduction components have been localized to this region. Such an assembly of molecules, which is important in the interaction of the ECM and cell at one location, would suggest the specialization of such a region. The formation of this region is initiated by the deposition of integrins, which apparently make connections to the forming ECM. With the maturation of the myocyte, additional components are also localized to this site, making it a specialized region for ECM interactions and signal transduction. Disruption of the site by antibodies, antisense oligonucleotides, pharmacologic agents, or other methods of perturbation causes changes in cellular phenotype, myofiber formation, and other dramatic changes in cellular architecture.

Internal specialization of the cardiac myocyte occurs early in development with the formation of the contractile apparatus. The initial myofibrils appear to form along the edge of the sarcolemma. Their terminal association appears to be at the intercalated disk and a lateral attachment with the regions where integrin are inserted. These regions appear to be templates for the forming Z bands. Numerous in vitro models have attempted to predict this formation (Rhee et al, 1990; Imanaka-Yoshida et al, 1993); however, the formation in vitro is different in several critical aspects, making correlation with the in vivo formation difficult. Myocytes grown in planar culture tend to form myofibrils only on the bottom of the culture dish, which results in a unique manner of myofibril formation as opposed to the three-dimensional formation that happens in vivo. Studies on myofibril formation with the aligned myocyte system (Simpson et al, 1994) are more likely to be a better model of myofibril formation.

MIGRATION/ADHESION/DIFFERENTIATION

Critical to the morphogenetic events of the heart are the associated processes of migration and adhesion. Cells in one region must be signaled to (1) migrate to specific regions, (2) stop at specific sites (adherence), and (3) undergo differentiation into specific structures. Frequently forgotten is the fact that it is not until after cells attain their proper position within the heart that the sequential genetic code essential to differentiation is expressed. Cellular interactions with the ECM are likely to play important roles in determining the final destination of a cell within the developing heart and the ability to undergo subsequent differentiation.

Migratory stimulation is a complex event involving perception of a stimulus and alteration of cellular symmetry in order to be able to move in a particular direction. Chemotaxis (response to chemical gradients) and haptotaxis (response to substrate gradients) frequently start in the ECM. The presence of morphogenetic gradients appears to be critical for differentiation in numerous organs, yet little is known about how these gradients are established. A variety of factors can be chemotactic for cell migration including growth factors, cytokines, and fragments of ECM molecules. Clearly, growth factors, which are reviewed elsewhere in this text, provide stimulatory and inhibitory signals to migrating cells. Cyokines such as IL-1 may also be important in the migration of mesenchymal cells of the cardiac cushions (Nakagawa et al, 1992), and it is quite likely that other cytokines will be shown to be important.

Numerous studies have been carried out with developing neural crest cells and regulation of their directional migration during development (Bronner-Fraser,

1994, 1995). These studies have shown that insoluble substrates can play an important role in establishing the direction of migration (haptotaxis). Gradients of different concentrations of ECM components appear to play a role in determining the direction of cellular migration. However, little is know about how these gradients are established, what cells establish the gradients, and how these gradients are perceived by migrating cells.

The signals that dictate stop signals or cellular adhesion are equally as complex as those that signal migration. A variety of models have been speculated for stop signals including the absense or gradual dissipation of the signal in either a soluble or insoluble gradient, and increased expression of receptors that increase adhesion or some combination of these events. Cellular receptors for ECM have been shown to be important regulators of both adhesion and migration (Albelda and Buck, 1990; Baldwin and Buck, 1994). Frequently, only single variables are tested such as integrins or a particular type of ECM substrate, yet the regulation of these events is likely to involve multiple components. Studies of cardiac cushion cells showed that both integrins and sugar transferases were important in regulation of migration, and if one component was blocked, the other could still support migration to some degree (Loeber and Runyan, 1990). These data suggest that relative levels of different ECM receptors may be important in modulating migration and adhesion of heart cells. Perhaps the key will be to understand the genetic regulation and expression of various receptor systems.

Experimental paradigms of cells grown in collagen gels have demonstrated a requirement for a variety of ECM-derived factors (Runyan and Markwald, 1983). These studies showed that cells derived from the endocardium would migrate into the collagen gel in response to signals from the myocardium. As these cells migrated into the gel, they altered the orientation of the collagen. Migration was presumably in response to some gradient involving a variety of cell-ECM interactions including metalloproteases (Alexander et al, 1997), integrins and sugar transferases (Runyan et al, 1988). These data show that a combination of mechanical and chemical signaling are necessary for migration of the endocardially derived mesenchymal cells.

NEW DIRECTIONS OF CELL-ECM STUDIES

Previous studies on the role of ECM components in the formation and function of the heart has provided a wealth of information for future experimentation. Temporal and spatial studies have documented the coordinated regulation of cellular differentiation with changes in the ECM. This dynamic interaction has provided an understanding that the ECM is an integral component in the differentiation of the heart. However, these studies are only the beginning in our attempts to understand the mechanisms of how the components of the ECM are involved in cardiac differentiation and morphogenesis.

Future studies will likely expand on the use of molecular technologies in the production of conditional transgenic animals to examine a wide variety of functions. Important to these investigations will be the development of culture models for both in vitro and *ex vivo* studies. New studies utilizing multicellular systems in vitro will provide experimental tools for the examination of molecular and metabolic processes. For example, it is possible to culture myocytes with an in vivo rod-shaped phenotype in the presence of cardiac fibroblasts. These fibroblasts have been irradiated to prevent cell division, yet retain functions such as collagen synthesis. This system can be combined with adenoviral infection of cells to alter protein expression levels.

Whole-embryo studies have been used by relatively few laboratories but offer great potential for examining differentiation, especially when combined with unique transgenic animals (Price et al, 1992; Baldwin and Solursh, 1994; Baldwin et al, 1994). In addition, these studies can utilize a wide variety of technical innovations, including adenovirus infection, antisense oligonucleotides, pharmacologic interventions, and antibody microinjection to begin to elucidate the functional roles of specific gene products in cardiac morphogenesis. These techniques in conjunction with the above-described model culture systems will permit the analysis of biochemical and molecular mechanisms to determine how the components of the ECM, which are made by the cells themselves, serve to regulate the cells in the developing cardiovascular system.

SUMMARY

The dynamic interactions between cells in the developing heart and the surrounding ECM is critical for proper heart formation and function. The ECM is composed of multiple proteins whose composition and organization vary during times of development and disease. The components of the ECM interact with a variety of cell surface receptors, including integrins and cadherins, which are responsible for the transmission of information from the extracellular environment into the cell. The expression of the ECM is regulated by a variety mechanisms including both mechanical and chemical pathways. The dynamic interactions of ECM, receptors, and proteases provide the necessary signals that regulate adhesion, growth, and development.

REFERENCES

Albelda, S.M., and Buck, C.A. (1990). Integrins and other cell adhesion molecules. *FASEB J* 4:2868–2880.

Alexander, S.M., Jackson, K.J., Bushnell, K.M., and McGuire, P.G. (1997). Spatial and temporal expression of the 72-kDa type IV collagenase (MMP-2) correlates with development and differentiation of valves in the embryonic avian heart. *Dev Dyn* 209:261–268.

Alves, F., Vogel, W., Mossie, K., Millauer, B., Hofler, H., and Ullrich, A. (1995). Distinct structural characteristics of discoidin I subfamily receptor tyrosine kinases and complementary expression in human cancer. *Oncogene* 10:609–618.

Aplin, A.E., Howe, A., Alahari, S.K., and Juliano, R.L. (1998). Signal transduction and signal modulation by cell adhesion receptors: the role of integrins, cadherins, immunoglobulin-cell adhesion molecules, and selectins. *Pharmacol Rev* 50:197–263.

Appeddu, P.A., and Shur, B.D. (1994). Molecular analysis of cell surface $\beta_{1,4}$-galactosyltransferase function during cell migration. *Proc Natl Acad Sci USA* 91:2095–2099.

Aumailley, M., and Gayraud, B. (1998). Structure and biological activity of the extracellular matrix. *J Mol Med* 76:253–265.

Baldwin, H.S., and Buck, C.A. (1994). Integrins and other cell adhesion molecules in cardiac development. *Trends Cardiovasc Med* 4:178–187.

Baldwin, H.S., Lloyd, T.R., and Solursh, M. (1994). Hyaluronate degradation affects ventricular function of the early postlooped embryonic rat heart in situ. *Circ Res* 74:244–252.

Baldwin, H.S., and Solursh, M. (1994). Degradation of hyaluronic acid does not prevent looping of the mammalian heart in situ. *Dev Biol* 136:555–559.

Begovac, P.C., Shi, Y.X., Mansfield, D., and Shur, B.D. (1994). Evidence that cell surface $\beta_{1,4}$-galactosyltransferase spontaneously galactosylates an underlying laminin substrate during fibroblasts migration. *J Biol Chem* 269:31793–31799.

Bertelli, R., Valenti, F., Oleggini, R., et al. (1998). Cell-specific regulation of alpha 1 (III) and alpha 2 (V) collagen by TGF-beta 1 in tubulointerstitial cell models. *Nephrol Dial Transplant* 13:573–579.

Bishop, J.E. (1998). Regulation of cardiovascular collagen deposition by mechanical forces. *Mol Med Today* 4:69–75.

Bishop, J.E., and Laurent, G.J. (1995). Collagen turnover and its regulation in the normal and hypertrophying heart. *Eur Heart J* 16 (suppl C):38–44.

Bishop, J.E., Rhodes, S., Laurent, G.J., Low, R.B., and Stirewalt, W.S. (1994). Increased collagen synthesis and decreased collagen degradation in right ventricular hypertrophy induced by pressure overload. *Cardiovasc Res* 28:1581–1585.

Bissell, M.J., Hall, H.G., and Parry, G. (1982). How does the extracellular matrix direct gene expression? *J Theor Biol* 99:31–68.

Borg, K., Burgess, W., Terracio, L., and Borg, T.K. (1997). Expression of metalloproteases by cardiac myocytes and fibroblasts in vitro. *Cardiovasc Pathol* 6:261–269.

Borg, T.K., and Caulfield, J.B. (1979). Collagen in the heart. *Tex Rep Biol Med* 39:321–333.

Borg, T.K., and Caulfield, J.B. (1981). The collagen matrix of the heart. *Fed Proc* 40:2037–2041.

Borg, T.K., Klevay, L.M., Gay, R.E., Siegel, R., and Bergin, M.E. (1985). Alteration of the connective tissue netwrok of striated muscle in copper deficient rats. *J Mol Cell Cardiol* 17:1173–1183.

Borg, T.K., and Terracio, L. (1990). Interaction of the extracellular matrix with cardiac myocytes during development and disease. In: Robinson, T., ed. *Issues in Biomedicine.* Karger, Basel, pp. 113–129.

Bronner-Fraser, M. (1994). Neural crest cell formation and migration in the developing embryo. *FASEB J* 8:699–706.

Bronner-Fraser, M. (1995). Patterning of the vertebrate neural crest. *Perspect Dev Neurobiol* 3:53–62.

Burgess, W.H., and Maciag, T. (1989). The heparin-binding (fibroblast) growth factor family of proteins. *Annu Rev Biochem* 58:575–606.

Calderwood, D.A., Tuckwell, D.S., Eble, J., Kühn, K., and Humphries, M.J. (1997). The integrin α_1 A-domain is a ligand binding site for collagens and laminin. *J Biol Chem* 272:12311–12317.

Capasso, J.M., Robinson, T.F., and Anversa, P. (1989). Alterations in collagen cross-linking impair myocardial contractility in the mouse heart. *Circ Res* 65:1657–1664.

Carey, D.J., Evans, D.M., Stahl, R.C., et al. (1992). Molecular cloning and characterization of N-syndecan, a novel transmembrane heparan sulfate proteoglycan. *J Cell Biol* 117:191–201.

Carver, W., Nagpal, M., Nachtigal, M., Borg, T.K., and Terracio, L. (1991). Collagen expression in mechanically stimulated cardiac fibroblasts. *Circ Res* 69:113–119.

Carver, W., Price, R.L., Raso, D.S., Terracio, L., and Borg, T.K. (1994). Distribution of β-1 integrin in the developing rat heart. *J Histochem Cytochem* 42:167–175.

Carver, W., and Terracio, L. (1993). Integrin-mediated cell-matrix interactions in heart development and in disease. *Heart Failure* 8:255–263.

Carver, W., Terracio, L., and Borg, T.K. (1993). Collagen expression and accumulation in the neonatal heart. *Anat Rec* 236:511–520.

Carver, W., Terracio, L., and Borg, T.K. (1997). Extracellular matrix maturation and heart formation. In: Burggren, W.W., and Keller, B.B., eds. *Development of Cardiovascular Systems: Molecules to Organisms.* Cambridge University Press, Cambridge, pp. 43–57.

Caulfield, J.B., and Borg, T.K. (1979). The collagen network of the heart. *Lab Invest* 40:354–371.

Celentano, D.C., and Frishman, W.H. (1997). Matrix mettaloproteinases and coronary artery disease: a novel therapeutic approach. *J Clin Pharmacol* 37:991–1000.

Chapman, D., Weber, K.T., and Eghbali, M. (1991). Regulation of fibrillar collagen types I and III and basement membrane type IV collagen gene expression in pressure overloaded rat myocardium. *Circ Res* 67:787–944.

Cheah, K.S., Lau, E.T., and Au, P.K. (1991). Expression of the mouse alpha 1 (II) collagen gene is not restricted to cartilage during development. *Development* 111:945–953.

Chua, C.C., Chua, B.H., Zhoa, Z.Y., Krebs, C., Diglio, C., and Perrin, E. (1991). Effect of growth factors on collagen metabolism in cultured heart fibroblasts. *Connect Tissue Res* 26:271–181.

Cleutjens, J.P.M. (1996). The role of matrix metalloproteinases in heart disease. *Cardiovasc Res* 32:816–821.

Couchman, J.R., Austria, R., Woods, A., and Hughes, R.C. (1988). Adehesion defective BHK cell mutant has cell sruface heparan sulfate proteoglycan of altered properties. *J Cell Physiol* 136:226–236.

Crossin, K.L., and Hoffman, S. (1991). Expression of adhesion molecules during the formation and differentiation of the avian endocardial tissue. *Dev Biol* 145:277–286.

Dickeson, S.K., and Santoro, S.A. (1998). Ligand recognition by the I domain-containing integrins. *Cell Mol Life Sci* 54:556–566.

Duband, J.L., Volberg, T., Sabanay, I., Thiery, J.P., and Geiger, B. (1988). Spatial and temporal distribution of the adherens-junction-associated adhesion molecule A-CAM during avian embryogenesis. *Development* 103:325–344.

Dubois, D.H., and Shur, B.D. (1995). Cell surface $\beta_{1,4}$-galactosyltransferase: a signal transducing receptor? In: Alavi, A., and Axford, J.S., eds. *Glycoimmunology.* Plenum Press, New York, pp. 105–114.

Eckstein, D.J., and Shur, B.D. (1989). Laminin induces the stable expression of surface galactosyltransferase on lamellipodia of migrating cells. *J Cell Biol* 108:2507–2517.

Eckstein, D.J., and Shur, B.D. (1992). Cell surface $\beta_{1,4}$-galactosyltransferase is associated with the detergent-insoluble cytoskeleton on migrating mesenchymal cells. *Exp Cell Res* 201:83–90.

Eghbali, M., Tomek, R., Sukhatme, V.P., Woods, C., and Bahmbi, B. (1991). Differential effects of transforming growth factor-1 and phorbol myristate acetate on cardiac fibroblasts: regulation of collagen mRNAs and expression of early transcription factors. *Circ Res* 69:483–490.

Eghbali, M., and Weber, K.T. (1990). Collagen and the myocardium: fibrillar structure, biosynthesis and degradation in relation to hypertrophy and its regression. *Mol Cell Biochem* 96:1–14.

Evans, S.C., Lopez, L.C., and Shur, B.D. (1993). Dominant negative mutation in cell surface $\beta_{1,4}$-galactosyltransferase inhibits cell-cell and cell-matrix interactions. *J Cell Biol* 120:1045–1057.

Fleoge, J., Eng, E., Young, B.A., et al. (1993). Infusion of platelet-derived growth factor or basic fibroblast growth factor induces selective glomerular mesangial cell proliferation and matric accumulation in rats. *J Clin Invest* 92:2952–2962.

Gallagher, J.T., Lyon, M., Stopak, D., and Steward, W.P. (1986). Structure and function of heparan sulfate proteoglycans. *Biochem J* 236:313–325.

Geiger, B., and Ayalon, O. (1992). Cadherins. *Annu Rev Cell Biol* 8:307–332.

Goncharova, E.J., Kam, Z., and Geiger, B. (1992). The involvement of adherens junction components in myofibrillogenesis in cultured cardiac myocytes. *Development* 114:173–183.

Hatta, K., Takagi, S., Fujisawa, H., and Takeichi, M. (1987). Spatial and temporal expression pattern of N-cadherin cell adhesion molecules correlated with morphogenetic processes of chicken embryos. *Dev Biol* 120:215–227.

Hertig, C.M., Eppenberger-Eberhardt, M., Koch, S., and Eppenberger, H.M. (1996). N-cadherin in adult rat cardiomyocytes in culture. I. Functional role of N-cadherin and impairment of cell-cell contact by a truncated N-cadherin mutant. *J Cell Sci* 109:1–10.

Hierck, B.P., Poelmann, R.E., van Iperen, L., Brouwer, A., and Gittenberger-de Groot, A.C. (1996). Differential expression of alpha-6 and other subunits of laminin binding integrins during development of the murine heart. *Dev Dyn* 206:100–111.

Hilenski, L., Terracio, L., and Borg, T.K. (1991). Myofibrillar and cytoskeletal assembly in neonatal rat cardiac myocytes cultured on laminin and collagen. *Cell Tissue Res* 264:577–587.

Hinton, D.A., Evans, S.C., and Shur, B.D. (1995). Altering the expression of cell surface $\beta_{1,4}$-galactosyltransferase modulates cell growth. *Exp Cell Res* 219:640–649.

Hori, Y., Katoh, T., Hirakata, M., et al. (1998). Anti-latent TGF-beta binding protein-1 antibody or synthetic oligopeptides inhibit extracellular matrix expression induced by stretch in cultured rat mesangial cells. *Kidney Int* 53:1616–1625.

Hynes, R.O. (1996). Targeted mutations in cell adhesion genes: What have we learned from them? *Dev Biol* 180:402–412.

Imanaka-Yoshida, Y., Sanger, J.M., and Sanger, J.W. (1993). Contractile protein dynamics of myofibrils in paired adult rat cardiomyocytes. *Cell Motil Cytoskel* 26:301–312.

Iruela-Arispe, M.L., and Sage, E.H. (1991). Expression of type VIII collagen during morphogenesis of the chicken and mouse heart. *Dev Biol* 144:107–118.

Jackson, R.L., Busch, S.J., and Cardin, A.D. (1991). Glycosaminoglycans: molecular properties, proteins interactions and roles in physiological processes. *Physiol Rev* 71:481–522.

Johnson, J.D., Edman, J.C., and Rutter, W.J. (1993). A receptor tyrosine kinase found in breast carcinoma cells has an extracellular discoidin I-like domain. *Proc Natl Acad Sci USA* 90:5677–5681.

Karn, T., Holtrich, U., Brauninger, A., et al. (1993). Structure, expression and chromosomal mapping of TKT from man and mouse: a new subclass of receptor tyrosine kinases with a factor VIII-like domain. *Oncogene* 8:3433–3440.

Katchman, S.D., Hsu-Wong, S., Ledo, I., Wu, M., and Uitto, J. (1994). Transforming growth factor-beta up-regulates human elastin promoter activity in transgenic mice. *Biochem Biophys Res Commun* 203:485–490.

Kjellen, L., and Lindahl, U. (1991). Proteoglycans: structures and interactions. *Annu Rev Biochem* 60:443–475.

Kratochwil, K., Dziadek, M., Lohler, J., Harbers, K., and Jaenisch, R. (1986). Normal epithelial branching morphogenesis in the absence of collagen I. *Dev Biol* 117:596–606.

Lai, C., and Lemke, G. (1994). Structure and expression of the Tyro 10 receptor tyrosine kinase. *Oncogene* 9:877–883.

Lemmon, M.A., and Schlessinger, J. (1994). Regulation of signal transduction and signal diversity by receptor oligomerization. *Trends Biochem Sci* 19:459–463.

Lertchirakarn, V., Birner, R., and Messer, H.H. (1998). Effects of interleukin-1 beta on human pulpal fibroblast proliferation and collagen synthesis. *J Endodontics* 24:409–413.

Lin, E.C.K., Ratnikov, B.I., Tsai, P.M., et al. (1997). Evidence that the integrin β_3 and β_5 subunits contain a metal ion-dependent adhesion site-like motif but lack and I domain. *J Biol Chem* 272:14236–14243.

Linask, K.K., Knudsen, K.A., and Gui, Y.H. (1997). N-cadherin-catenin interaction: necessary component of cardiac cell compartmentalization during early vertebrate heart development. *Dev Biol* 185:148–164.

Linask, K.K., and Lash, J.W. (1988). A role for fibronectin in the migration of avian precardiac cells. Dose-dependent effects of fibronectin antibody. *Dev Biol* 129:315–323.

Little, C.D., Piquet, D.M., Davis, L.A., Walters, L., and Drake, C.J. (1989). Distribution of laminin, collagen type IV, collagen type I, and fibronectin in chicken cardiac jelly/basement membrane. *Anat Rec* 224:417–425.

Little, C.D., and Rongish, B.J. (1995). The extracellular matrix during heart development. *Experentia* 51:873–882.

Loeber, C.P., and Runyan, R.B. (1990). A comparison of fibronectin, laminin and galactosyltransferase adhesion mechanisms during embryonic cardiac mesenchymal cell migration in vitro. *Devel Biol* 140:401–412.

Lohler, J., Timpl, R., and Jaenisch, R. (1984). Embryonic lethal mutation in mouse collagen I gene causes rupture of blood vessels and is associated with erythropoietic and mesenchymal cell death. *Cell* 38:597–607.

Lopez, L.C., Youakim, A., Evans, S.C., and Shur, B.D. (1991). Evidence for a molecular distinction between Golgi and cell surface forms of $\beta_{1,4}$-galactosyltransferase. *J Biol Chem* 266:15984–15991.

Maillet, C.M., and Shur, B.D. (1994). Perturbing cell surface $\beta_{1,4}$-galactosyltransferase on F9 embryonal carcinoma cells arrests cell growth and induces laminin synthesis. *J Cell Sci* 107:1713–1724.

Mauch, C. (1998). Regulation of connective tissue turnover by cell-matrix interactions. *Arch Dermatol Res* 290:S30–S60.

McGuire, P.G., and Alexander, S.M. (1993). Urokinase production by embryonic endocardial-derived cells: regulation by substrate composition. *Dev Biol* 155:442–451.

McGuire, P.G., and Orkin, R.W. (1992). Urokinase activity in the developing avian heart: a spatial and temporal analysis. *Dev Dyn* 193:24–33.

Mjaatvedt, C.H., Yamamura, H., Capehart, A.A., Turner, D., and Markwald, R. (1998). The Cspg2 gene, disrupted in the hdf mutant, is required for right cardiac chamber and endocardial cushion formation. *Dev Biol* 202:56–66.

Murdoch, A.D., Liu, B., Schwarting, R., Tuan, R.S., and Iozzo, R.V. (1994). Widespread expression of perlecan proteoglycan in basement membranes and extracellular matrices of human tissues as detected by a novel monoclonal antibody against domain III and by in situ hybridization. *J Histochem Cytochem* 42:239–249.

Nakagawa, M., Terracio, L., Carver, W., Birkedal-Hansen, H., and Borg, T.K. (1992). Expression of collagenase and IL-1 alpha in developing rat hearts. *Dev Dyn* 195:87–99.

Nathalie, B., Merval, R., Benessiano, J., Samuel, J.-L., and Tedgui, A. (1996). Pressure and angiotensin II synergistically induce aortic fibronectin expression in organ culture model of rabbit aorta: evidence for a pressure-induced tissue renin-angiotensin system. *Circ Res* 79:70–78.

Okada, Y., Katsuda, S., Watanabe, H., and Kakanisha, I. (1993). Collagen synthesis of human arterial smooth muscle cells: effects of platelet-derived growth factor, transforming growth factor-beta, and interleukin-1. *Acta Pathol Jpn* 43:160–167.

Ong, L.L., Kim, N., Mima, T., Cohen-Gould, L., and Mikawa, T. (1998). Trabecular myocytes of the embryonic heart require N-cadherin for migratory unit identity. *Dev Biol* 193:1–9.

Price, R., Nakagawa, M., Terracio, L., and Borg, T.K. (1992). Ultrastructural localization of laminin on in vivo embryonic, neonatal and adult rat cardiac myocytes and in early rat embryos raised in whole embryo culture. *J Histochem Cytochem* 40:1373–1381.

Radice, G.L., Rayburn, H., Matsunami, H., Knudsen, K.A., Takeichi, M., and Hynes, R.O. (1997). Developmental defects in mouse embryos lacking N-cadherin. *Dev Biol* 181:64–78.

Rhee, D., Sanger, J.M., and Sanger, J.W. (1990). The premyofibril: evidence for its role in myofibrillogenesis. *Cell Motil Cytoskel* 28:1–24.

Robinson, T.F., Factor, S.M., Capasso, J.M., Wittenberg, B.A., Blumenfeld, O.O., and Seifter, S. (1987). Morphology and function of struts between cardiac myocytes of rat and hamster. *Cell Tissue Res* 249:247–255.

Rongish, B.J., Hinchman, G., Doty, M.K., Baldwin, H.S., and Tomanek, R.J. (1996). Relationship of the extracellular matrix to coronary neovascularization during development. *J Mol Cell Cardiol* 28:2203–2215.

Rosales, C., O'Brien, V., Kornberg, L., and Juliano, R. (1995). Signal transduction by cell adhesion receptors. *Biochim Biophys Acta* 1242:77–98.
Runyan, R.B., and Markwald, R.R. (1983). Invasion of mesenchyme into three dimensional collagen gels: a regional and temporal analysis of interaction in embryonic heart tissue. *Dev Biol* 95:108–114.
Runyan, R.B., Versalovic, J., and Shur, B.D. (1988). Functionally distinct laminin receptors mediate cell adhesion and spreading: the requirement for surface galactosyltransferase in cell spreading. *J Cell Biol* 107:1863–1871.
Ruoslahti, E., and Engvall, E. (1997). Integrins and vascular extracellular matrix assembly. *J Clin Invest* 100:S53–S56.
Sadoshima, J., and Izumo, S. (1993). Mechanical stretch rapidly activates multiple signal transduction pathways in cardiac myocytes: potential involvement of an autocrine/paracrine mechanism. *EMBO J* 12:1681–1692.
Sadoshima, J., and Izumo, S. (1994). Roles of integrins in cell swelling-induced tyrosine phosphorylation in cardiac myocytes. *Circulation* 90:291–305.
Sadoshima, J., and Izumo, S. (1995). Rapamycin selectively inhibits angiotensin II-induced increase in protein synthesis in cardiac myocytes *in vitro*. Potential role of 70 kD S6 kinase in angiotensin II-induced cardiac hypertrophy. *Circ Res* 77:1040–1052.
Sadoshima, J., Xu, Y., Slayter, H.S., and Izumo, S. (1993). Autocrine release of angiotensin II mediates stretch-induced hypertrophy of cardiac myocytes *in vitro*. *Cell* 75:972–984.
Schlessinger, J. (1997). Direct binding and activation of receptor tyrosine kinases by collagen. *Cell* 91:869–872.
Scholzen, T., Solursh, M., Suzuki, S., et al. (1994). The murine decorin: complete cDNA cloning, genomic organization, chromosomal assignment, and expression during organogenesis and tissue differentiation. *J Biol Chem* 269:28270–28281.
Shrivastava, A., Radziejewski, C., Campbell, E., et al. (1997). An orphan receptor tyrosine kinase family whose members serve as nonintegrin collagen receptors. *Mol Cell* 1:25–34.
Shur, B.D. (1989). Expression and function of cell surface glactosyltransferase. *Biochim Biophys Acta* 988:389–409.
Silbiger, S., Lei, J., Ziyadeh, F.N., and Neugarten, J. (1998). Estradiol reverses TGF-beta1-stimulated type IV collagen gene transcription in murine mesangial cells. *Am J Physiol* 274:F1113–1118.
Simpson, D.G., Reaves, T.A., Shih, S.T., Burgess, W., Borg, T.K., and Terracio, L. (1998). Cardiac integrins: the ties that bind. *Cardiovasc Pathol* 7:135–143.
Simpson, D.G., Terracio, L., Terracio, M., Price, R.L., Turner, D.C., and Borg, T.K. (1994). Modulation of cardiac myocyte phenotype *in vitro* by the composition and orientation of the extracellular matrix. *J Cell Physiol* 161:89–105.
Sinning, A.R., Lepera, R.C., and Markwald, R.R. (1988). Initial expression of type I procollagen in chick cardiac mesenchyme is dependent upon myocardial stimulation. *Dev Biol* 130:167–174.
Soler, A.P., and Knudsen, K.A. (1994). N-cadherin involvement in cardiac myocyte interaction and myofibrillogenesis. *Dev Biol* 162:9–17.
Song, W.K., Wand, W., Foster, R.F., Biesler, D.A.S., and Kaufman, S.J. (1992). H36-α_7 is a novel integrin a chain that is developmentally regulated during skeletal myogenesis. *J Cell Biol* 117:643–657.
Spence, S.G., Argraves, W.S., Walters, L., Hungerford, J.E., and Little, C.D. (1992). Fibulin is localized at sites of epithelial-mesenchymal transitions in early avian embryos. *Dev Biol* 151:473–484.
Springer, T.A. (1997). Folding of the N-terminal, ligand-binding region of integrin α-subunits into a β-propellor domain. *Proc Natl Acad Sci USA* 94:65–72.
Stepp, M.A., Urry, L.A., and Hynes, R.O. (1994). Expression of α_4 integrin mRNA and protein and fibronectin in the early chicken embryo. *Cell Adhesion Commun* 2:359–375.

Taber, L.A., Keller, B.B., and Clark, E.B. (1992). Cardiac mechanics in the stage 16 chick embryo. *J Biomech Eng* 114:427–434.

Takada, Y., Kamata, T., Irie, A., Puzon-McLaughlin, W., and Zhang, X.P. (1997). Structural basis of integrin-mediated signal transduction. *Matrix Biol* 16:143–151.

Takahashi, I., Nuckolls, G.H., Takahashi, K., et al. (1998). Compressive force promotes sox 9, type II collagen and aggrecan and inhibits IL-1 beta expression resulting in chondrogenesis in mouse embryonic limb bud mesenchymal cells. *J Cell Sci* 111:2067–2076.

Takeichi, M. (1991). Cadherin cell adhesion receptors as a morphogenetic regulator. *Science* 251:1451–1455.

Tamura, K., Shan, W.S., Hendrickson, W.A., Colman, D.R., and Shapiro, L. (1998). Structure-function analysis of cell adhesion by neural (N-) cadherin. *Neuron* 20:1153–1163.

Terracio, L., Rubin, K., Gullberg, D., et al. (1991). Expression of collagen binding integrins during cardiac development and hypertrophy. *Circ Res* 68:734–744.

Thompson, R.P., Fitzharris, T.P., Denslow, S., and LeRoy, E.C. (1979). Collagen synthesis in the developing chick heart. *Tex Rep Biol Med* 39:305–319.

Tucker, D.C. (1990) Genetic, neurohumoral, and hemodynamic influences on spontaneously hypertensive rat heart development in oculo. *Hypertension* 15:247–256.

Tuckwell, D.S., and Humphries, M.J. (1997). A structure prediction for the ligand-binding region of the integrin β subunit: evidence for the presence of a von Willebrand factor A domain. *FEBS Lett* 400:297–303.

Turley, E.A., and Roth, S. (1979). Spontaneous glycosylation of glycosaminoglycan substrates by adherent fibroblasts. *Cell* 17:109–115.

Tyagi, S.C., Campbell, S. E., Reddy, H.K., Tjahja, E., and Voelker, D.J. (1996). Matrix metalloproteinase activity expression in infarcted, noninfarcted and dilated cardiomyopathic human hearts. *Mol Cell Biochem* 155:13–21.

Uemura, T. (1998). The cadherin superfamily at the synapse: more members, more missions. *Cell* 93:1095–1098.

Ullrich, A., and Schlessinger, J. (1990). Signal transduction by receptors with tryosine kinase activity. *Cell* 61:203–212.

Vogel, W., Gish, G.D., Alves, F., and Pawson, T. (1997). The discoidin domain receptor tyrosine kinases are activated by collagen. *Mol Cell* 1:13–23.

Weber, K.T., Jalil, J.E. Janicki, J.S., and Pick, R. (1989). Myocardial collagen remodeling in pressure overload hypertrophy. A case for interstitial heart disease. *Am J Hypertens* 2:931–940.

Werb, Z. (1997). ECM and cell surface proteolysis: regulating cellular ecology. *Cell* 91:439–442.

Yaeger, P.C., Masi, T.L., de Ortiz, J.L., Binette, F., Tubo, R., and McPherson, J.M. (1997). Synergistic action of transforming growth factor-beta and insulin-like growth factor-1 induces expression of type II collagen and aggrecan genes in adult human articular chondrocytes. *Exp Cell Res* 237:318–325.

Yamada, S., Brown, K.E., and Yamada, K.M. (1995). Differential mRNA regulation of integrin subunits α_v, β_1, β_3 and β_5 during mouse embryonic organogenesis. *Cell Adhesion Commun* 3:311–325.

Yost, H.J. (1990). Inhibition of proteoglycan synthesis eliminates left-right assymetry in *Xenopus laevis* cardiac looping. *Development* 110:865–874.

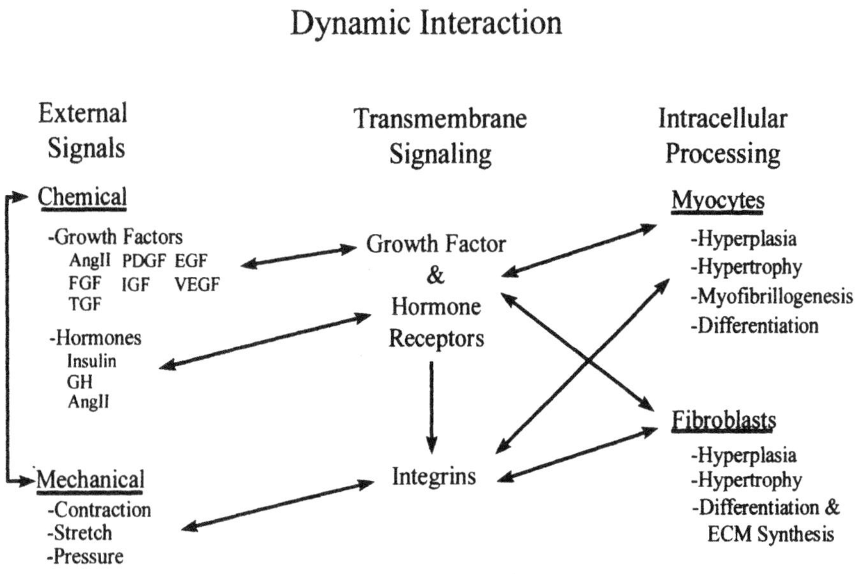

FIGURE 10.1. Normal cardiac development is dependent on numerous interactions and feedback mechanisms between chemical and mechanical factors, transmembrane signaling mechanisms, and cellular responses. Disruption of the normal expression of any component of these feedback mechanisms may result in the abnormal development of the heart in the developing embryo and subsequent death. AngII, angiotensin II; PDGF, platelet-derived growth factor; FGF, fibroblast growth factor; IGF, insulin-like growth factor; VEGF, vascular endothelial growth factor; TGF, transforming growth factor; EGF, epidermal growth factor; GH, growth hormone; ECM, extracellular matrix.

tions that have been determined for angiotensin II (Ang II) in cardiac myocytes and fibroblasts, including a role in cell migration, myofibrillogenesis, ECM production, and having an influence on the expression patterns of other growth factors, no apparent cardiac defects have been described in angiotensinogen knockout animals that survive until the postweaning stages (H.S. Kim et al, 1995). Several possibilities exist to explain this apparent conundrum, including the possibility that maternal Ang II crosses the placenta and provides a source for this growth factor, or that other growth factors may compensate for the absence of Ang II during development. However, it is unknown what factors are involved in the survival of these animals.

Realizing the many difficulties involved in studying the roles of growth factors during embryonic, fetal, and neonatal development, we will summarize much of what is known about the roles of Ang II, platelet-derived growth factor (PDGF), fibroblast growth factor (FGF), vascular endothelial growth factor (VEGF), and insulin-like growth factor (IGF) in cardiac development. Information about transforming growth factor-β (TGF-β) is presented in Chapter 11. While this list does not cover all of the growth factors affecting cardiac development, the data presented do provide an indication of the importance of growth factor regulation and their interactions in heart morphogenesis. Most of our comments will be limited to studies performed utilizing embryonic, fetal, and neonatal cells and animals. In some areas we will also present new data on the role of a growth factor in support of a specific function that has been attributed to it. We have organized the data to present information gathered from receptor localization and cell culture studies first, followed by information gathered from whole-animal studies. Finally, we present information on the interactions of mechanical force and the expression of growth factors and the possibility that several growth factors utilize shared pathways that may explain the survival of some transgenic animals lacking a growth factor.

ANGIOTENSIN II (ANG II)

Molecular Signaling Mechanisms of Ang II

Several reviews have been published concerning the molecular signaling mechanisms of Ang II, which involve several second messenger pathways including phospholipases C, D, and A_2, protein kinase C, tyrosine and mitogen-activated protein (MAP) kinases, as well as the immediate early response genes c-*fos* and Egr-1, late hypertrophic responses for α-skeletal actin and atrial natriuretic factor (ANF), and a host of other genes (Bernstein and Marrero, 1996; Dostal et al, 1997; van Bilsen, 1997; Kodama et al, 1998; Zou et al, 1998). The complexity of the Ang II pathways in the heart increases due to multiple interactions with other growth factors as well as mechanical stretch, different effects of Ang II on the various cell types in the heart, and the possibility that the different Ang II receptor subtypes may act as antagonists. Some information on known pathways for specific cardiac cell responses to Ang II is presented in the next section, but a thorough review of these mechanisms is not included.

In Vivo Expression of Cardiac Ang II Receptors

All of the components for a localized renin-angiotensin system including Ang II, angiotensin-converting enzyme (ACE), and the angiotensin (AT)1 and AT2 receptor subtypes have been identified in embryonic (Price et al, 1997) and neonatal (Dostal et al, 1992a,b) hearts from a variety of species including humans (Schütz et al, 1996). In reports that have distinguished between the in vivo distribution of the AT1a and AT1b receptor subtypes, only the AT1a subtype was found in the developing myocardium (Shanmugam et al, 1994a,b), although Matsubara et al (1994) did find mRNA for the AT1b receptor subtype on isolated cardiac fibroblasts and myocytes. The majority of in vivo studies localizing the Ang II receptors were done in rats with the earliest time point for specific localization at embryonic day (ED) 9.25. At this stage of development and for the next 24 hours, AT1 receptors are localized predominantly on the greater curvature of the developing ventricle. The predominant expression of the AT1 receptor continues to be in the greater curvature of the heart for the next 24 hours of development. During this time period no AT1 localization was found in the developing atria (Price et al, 1997).

By ED 15, diffuse localization of the AT1 receptor occurs in the atrial, ventricular, and septal myocytes (Shanmugam et al, 1994a). Expression of the AT1 receptor in the developing ventricular myocardium continues to increase through late embryonic development and appears to peak sometime between ED 19 and postnatal day 2, after which there is a decline in expression (Sechi et al, 1992; Hunt et al, 1995; Everett et al, 1997). During the peak periods of expression around ED 19, the AT1 receptor is also found in the pericardium and the bundle of His, although minimal labeling is found in the developing cardiac valves (Shanmugam et al, 1994b). Peak expression of the AT1 receptor in the atria occurs at postnatal day 4 and 18, after which there is also a decline in expression in the atria (Hunt et al, 1995).

In general, the temporal and spatial expression patterns of the AT2 receptors follow that of the AT1 receptor density in that expression peaks during late neonatal and early postnatal cardiac development. Using in situ hybridization Shanmugam et al (1996) found mRNA for the AT2 receptors in the annuli of the heart valves starting at ED 21, but found no signal for the AT2 receptor in the valve leaflets or myocardial tissue. However, by utilizing ^{125}I-[Sar^1, Ile^8]angiotensin II binding techniques, equal proportions of the AT1 and AT2 receptors were found to be uniformly distributed throughout the heart by Hunt et al (1995) at ED 14 and Sechi et al (1992) at ED 16 through postnatal day 10. However, Hunt et al (1995) did report a shift in the distribution of the receptors at late peri- and postnatal developmental stages with a twofold higher density of receptors in the atria than in the ventricles. Peak density of the AT2 receptor also occurs during late fetal or early neonatal development and subsequently declines (Grady et al, 1991; Sechi et al, 1992; Everett et al, 1997).

All of the above-cited reports utilized Sprague-Dawley embryonic, fetal, and neonatal rats to determine receptor localization. In one study that utilized Wistar-Kyoto rats, the only Ang II receptor found by autoradiography on slices of neonatal hearts, or by binding studies on membranes isolated from the various chambers of neonatal hearts, was the AT2 receptor in the left atrium. No AT1 or AT2 receptors were localized in the developing ventricles or right atrium by these techniques (Feolde et al, 1993). This study indicates that significant strain and species differences in the expression of the Ang II AT1 and AT2 receptors may exist. This possibility must be accounted for in future studies with transgenic mice in which few cardiac defects have been identified when specific components of the angiotensinogen system have been knocked out. Only two studies have reported Ang II receptor expression in mice. Zemel et al (1989) reported wide spread receptor localization by autoradiographic techniques, but did not discuss receptor localization in the developing heart, and Jones et al (1989) reported a high expression of Ang II receptors in ED 17 to 19 animals. However, they used homogenized embryos, so it was not possible to determine from which organs the receptors originated.

Role of Angiotensin II in Cardiac Myocyte Development, Differentiation, and Function

Several reports have identified the potential for Ang II to induce the hypertrophy of isolated embryonic and neonatal cardiac myocytes independent of cardiac load. In isolated embryonic chick heart cells, addition of 1 μM/day of Ang II resulted in an increase in the rate of protein synthesis of between 14.9% and 21.7% and an increase in the total protein content of between 18.5% and 26.2% following 5, 7, and 9 days of treatment (Aceto and Baker, 1990). In complementary studies Baker and Aceto (1990) also found that a single exposure of cultured embryonic chick heart cells to Ang II for 3 hours resulted in an extended period of protein synthesis for up to 5 to 6 days. Utilizing neonatal rat cardiac myocytes Sadoshima and Izumo (1993) and Miyata and Haneda (1994) also found an increase in protein content in response to Ang II, and constant DNA throughout the culture periods. These studies indicate that hypertrophy rather than hyperplasia occurred in the myocyte cultures. In both studies, losartan, which is a specific inhibitor of the Ang II AT1 receptor subtype, blocked the hypertrophic response to Ang II. While stimulation of the AT1 receptors stimulates myocyte hypertrophy, stimulation of the AT2 receptors may play an antagonistic role to the AT1 receptors in mediating Ang II–stimulated growth. In studies with neontal ventricular myocytes, Booz and Baker (1996) found that blocking the AT2 receptors resulted in a significant increase in protein synthesis and prevented protein degradation.

Associated with the increase in cell size and rates of protein synthesis in cardiac myocytes is the ability of Ang II to rapidly stimulate the formation of premyofibrils (Aoki et al, 1998) and to play a significant role in the maturation of cardiac myofibrils (Sadoshima and Izumo, 1995). In myocyte cultures stimulated with 100 nm of Ang II, thin actin myofibrils form within 5 minutes, and by 30 minutes posttimulation striated actin fibers become prominent (Aoki et al, 1998). The Ang II–stimulated formation of the premyofibrils is mediated by RhoA, a member of the Rho-family of proteins that is known to play a role in the organization of several actin structures (Tapon and Hall, 1997). However, an as yet unidentified Ang II mediated mechanism appears to mediate the maturation process of the myofibrils (Aoki et al, 1998), although the $p70^{S6K}$ S6 kinase plays a central role (Sadoshima and Izumo, 1995). Both the formation of premyofibrils and sarcomere formation are blocked by losartan, which indicates the process of myofibrillogenesis is mediated, at least in part, by the AT1 receptor subtype (Aoki et al, 1998).

The above data indicating that Ang II stimulates an increase in cell size and myofibrillogenesis correlates well with the whole embryo culture studies of Price et al (1997) in which Ang II stimulated an increase in heart size, thickness of the ventricular walls and rate of myofibril maturation. In these studies, blocking the effects of the AT1 receptor with losartan resulted in dilated hearts with a relatively thin ventricular wall and reduced trabeculation. Blocking the AT2 receptor with PD123319 also resulted in reduced growth of the heart and thinning of the ventricular wall; however, the effect of the AT2 receptor block was not as extensive as that observed when the AT1 receptor was blocked.

As we noted above, the initial expression of the Ang II AT1 receptor subtype is associated with the greater curvature of the developing ventricles during the looping stages of heart development (Price et al, 1997). This is also the area where myofibrillar formation is initiated and an area of the heart that is exposed to significant mechanical stress during the looping process (Manasek 1981; Manasek et al, 1984; Taber et al, 1995; Price et al, 1996). Both mechanical stretch and Ang II activate similar pathways involving phospholipases C, D, and A_2, protein kinase C, tyrosine kinases, MAP kinases, the immediate early response gene *c-fos*, late hypertrophic responses for α-skeletal actin, and ANF, and a host of other genes (Sadoshima et al, 1992, 1993; Yamazaki et al, 1993, 1995; Kojima et al, 1994). Using isolated neonatal myocytes, Sadoshima et al (1993) found that mechanically stretching myocytes resulted in an autocrine release of Ang II, and that Ang II can act as the initial mediator of stretch-induced myocyte hypertrophy via the AT1 receptors. Yamazaki et al (1995) and Nyui et al (1997) also found a correlation between the release of Ang II and myocyte hypertrophy, although both groups indicated that mediation of the stretch-induced hypertrophic response was not entirely dependent on Ang II and the AT1 receptors. Nyui et al (1997) also found a stretch-induced MAP kinase response in cardiac myocytes isolated from mice lacking the angiotensinogen gene, and therefore Ang II, indicating that Ang II–independent pathways are available for activation of these cardiac myocyte responses.

In addition to playing a role in the initial formation and maturation of myofibrils, Ang II also mediates a positive inotropic beating response of isolated myocytes when added to the culture media (Rogers et al, 1986; Allen et al, 1988; Abdellatif et al, 1991; Kohout and Rogers, 1995). As little as 5 nM Ang II added to the culture media is capable of increasing the spontaneous beating frequency of myocytes by 100%, although as the rate of contraction increases there is a corresponding decrease in the amount that cells contract per beat and in the shortening and relaxation velocity. Ang II may control contraction frequency by mediating intracellular pH. Acidosis is known to depress the contractility of cardiac myocytes (Orchard and Kentish, 1990) and the application of 100 nM of Ang II to the culture media of beating cardiac myocytes led to a rapid alkalinization of cells and a threefold increase in the recovery rate from an imposed acid load on the myocytes via the Na^+/HCO_3^- mechanism (Allen et al, 1988). Ang II also stimulates the release of arachidonic acid that is known to elicit an HCO_3^- dependent alkalinization that can alter contractility of the heart (Kohout and Rogers, 1995). Both the activation of the Na^+/HCO_3^- and the stimulation of arachidonic acid release are mediated via the AT2 receptor subtype (Lokuta et al, 1994; Kohout and Rogers, 1995).

A final potential role of Ang II in myocyte-related cardiac development is that of structural remodeling of the heart during postnatal development when apoptosis is a major determinant of cardiac restructuring (Kajstura et al, 1995; Cigola et al, 1997). Treatment of cultured neonatal cardiac myocytes with 10^{-9} M Ang II resulted in a 2.5-fold increase in apoptosis, which was mediated by the AT1 receptor subtype and intracellular calcium concentration. Several reports have indicated that Ang II stimulates an increase in intracellular calcium in cardiac myocytes (Allen et al, 1988; Baker et al, 1989; Delisée et al, 1993; Sadoshima et al, 1995; Kudoh et al, 1997). Blocking the activity of the AT1 receptor with losartan or treating cells with the calcium chelator 1,2-bis(O-aminophenoxy)ethane-N,N,N′,N′-tetraacetic acid acetoxymethyl ester (BAPTA-AM) both prevented in vitro myocyte apoptosis (Cigola et al, 1997).

Role of Ang II in Cardiac Fibroblast Development and Function

Cardiac fibroblasts play a critical role in the normal development of the heart through their production of several ECM components. Ang II has the potential to influence this developmental role through a mitogenic capacity on embryonic and neonatal fibroblasts and by stimulating an increased rate of synthesis of ECM components. Both Schorb et al (1993) and Sadoshima and Izumo (1993) found that Ang II stimulated an increase in both DNA and protein content in cultured fibroblasts and an increase in cell number when treated cultures were compared to control cultures. Autocrine production of Ang II by isolated cardiac fibroblasts is similar to that found for cultured cardiac myocytes (Dostal et al, 1992b), although some controversy exists concerning the activity of Ang II in cardiac myocytes. While both Rogers et al (1986) and Matsubara et al (1994) report the presence of

Ang II receptors on cultured myocytes, and numerous reports cited above indicate receptor-mediated activity associated with the AT1 and AT2 receptors on myocytes, N.N. Kim et al (1995) found very little specific binding of Ang II to neonatal cardiac myocytes and determined that the trophic effects of Ang II on the myocytes was mediated via cardiac fibroblasts through an unidentified transferable factor.

Contraction assays of three-dimensional collagen gels have frequently been used as a model to examine ECM remodeling since the basic mechanisms are similar to those that occur during cardiac development. The ability of fibroblasts to remodel cardiac architecture in response to Ang II has been demonstrated by use of collagen gel assays by Burgess et al (1994), Ashizawa et al (1996), and Watson et al (1998). In these studies, the addition of Ang II to the culture media with neonatal cardiac fibroblasts and collagen gels resulted in a significant increase in the contraction rates of the gels. The addition of Ang II to these cultures also resulted in an increase in the expression of integrins, which are important transmembrane signaling molecules between cells and collagen in the ECM [for a review of integrins see Ruoslahti (1991) and Kumar (1998)].

The studies by Ashizawa et al (1996) also indicated that the effects of Ang II on cardiac remodeling may occur indirectly through an increased expression of osteopontin, an acidic phosphoprotein produced by fibroblasts that may play an important role in cardiovascular disease. Ang II stimulation resulted in a fivefold increase in the expression of osteopontin, which was inhibited by blocking the AT1 receptor with losartan. Subsequent antibody blocking of osteopontin prevented collagen gel contraction, indicating osteopontin is a potentially important mediator of Ang II regulation of fibroblast mediated cardiac remodeling (Ahizawa et al, 1996).

Angiotensin and Cardiac Development: In Vivo Studies

Few in vivo studies on the role of Ang II in cardiac development have been done. While not strictly "in vivo," the whole embryo culture studies of Price et al (1997) indicated a role for Ang II in myofibrillogenesis during the looping stages of heart development. Disruption of the normal patterns of myofibrillogenesis, either by inhibiting myofibril formation by blocking the AT1 receptor, or by stimulating myofibrillogenesis with exogenous amounts of Ang II, resulted in abnormal loop formation in a high percentage of embryos in culture. In studies with neonatal pigs, Beinlich et al (1993, 1995) found that normal development of the left ventricle following birth was inhibited by stopping the production of Ang II with angiotensin-converting enzyme inhibitors and by blocking the function of the AT1 receptors.

The remaining in vivo studies manipulating Ang II or receptor expression have been done with transgenic animals. Despite the numerous functions of Ang II described above, no specific cardiac defects have been identified in animals lacking the angiotensinogen gene (H.S. Kim et al, 1995) or angiotensin-converting enzyme (Krege et al, 1995) and thus Ang II at birth. However, animals lacking the angiotensinogen gene die before weaning. Animals lacking the AT1a receptor develop normally (Ito et al, 1995; Harada et al, 1998a,b), although animals that overexpress the AT1 receptor die shortly after birth with myocyte hyperplasia and atrial enlargement (Hein et al, 1997). Animals lacking (Ichiki et al, 1995) and overexpressing (Masaki et al, 1998) the AT2 receptor develop normally, although deletion of the AT2 receptor resulted in an increase in blood pressure in adult animals.

PLATELET-DERIVED GROWTH FACTOR (PDGF)

PDGF is a complex growth factor system that consists of two ligands, PDGF-A and -B, and two receptor subtypes PDGFrα and PDGFrβ. The active forms of PDGF exist as dimers of the PDGF-A and -B chains, which are expressed as PDGF-AA and PDGF-BB homodimers and as PDGF-AB heterodimers. The PDGFrα is capable of binding both the PDGF-A and -B chains, but the PDGFrβ binds only the PDGF-B chain with high affinity (Heldin and Westmark, 1989; Seifert et al, 1993; Ataliotis and Mercola, 1997). Low-affinity binding of the PDGF-AB heterodimer to the PDGFrβ is apparently due to some binding restricted to the PDGF-B ligand (Fretto, 1993). However, activation of PDGFrβ receptors in cells lacking the PDGFrα receptors requires nearly a 100-fold higher concentration of the PDGF-AB ligand, making it unlikely that the PDGF-AB ligand is a physiologically significant activator of the PDGFrβ (Seifert et al, 1993).

The complexity of the PDGF receptor-ligand system is increased since each dimer subtype may elicit different responses in different cell types (Ataliotis and Mercola, 1997) and different signal transduction pathways may be activated by the three ligand dimers (Kondo et al, 1993). Little information is available concerning the activation of signaling pathways by PDGF in developing cardiac cells, although the signal transduction pathways in several other cell types have been reviewed (Ataliotis and Mercola, 1997; Betsholtz and Heldin, 1998). From the studies that have been done, it is probable that PDGF will activate several pathways in common with other growth factors including second messenger systems involving the phospholipases and STAT family (signal transducers and activators of transcription) in the developing heart.

In Vivo Expression of Cardiac PDGF Receptors

In the mouse, PDGFr" mRNA expression has been found in the mesenchymal cells surrounding the developing heart tube epithelium as early as ED 8 (Orr-Urtreger et al, 1992) and by antibody staining in the heart by ED 9 (Takakura et al, 1998). Although specific temporal and spatial expression patterns in various regions of the heart vary by report, expression of α receptors has been found in all regions of the developing heart including the pericardium, cardiac atrioventricular and outflow tract cushions, trabeculae, and myocardium by in situ hybridization and fluorescent antibody labeling by ED 11.5 (Orr-Urtreger and Lonai, 1992; Schatteman et al, 1992, 1995; Takakura et al, 1998; Thielen et al, 1998). Expression in the myocardium is variable and initially in isolated pockets at ED 11.5 (Schatteman et al, 1992), with the highest mRNA signals found in the basal half of the heart (Schatteman et al, 1995). By confocal microscopy we have also found the PDGFrα in close association with differentiating myocytes (unpublished observations).

The earliest expression of the PDGFrβ mRNA has been found in the heart is ED 9.5 in the mouse, although no specific spatial distribution for the receptors was given (Soriano, 1994). By ED 10.5 in the rat, fluorescent staining by confocal microscopy has shown the PDGFrβ heavily localized in the endocardium, cardiac cushions, and myocardium in both the developing ventricular and atrial regions (Thielen et al, 1998). By ED 12.5, strong signal for the PDGFrβ mRNA was found in the pericardium and endocardium of the developing mouse heart, but no localization was found in the myocardium (Shinbrot et al, 1994).

Role of PDGF in the Development of Cardiac Myocytes and Fibroblasts

Even though disruption of the PDGF ligand-receptor system typically results in severe cardiovascular defects, very little is known about the specific roles that either the PDGF-A or -B ligands play in the development of cardiac myocytes and fibroblasts. In studies with isolated chicken cardiac myocytes, Velez et al (1994) found that PDGF stimulated the production of total cytoplasmic and cytoskeletal proteins by 6% and 8% respectively in cells isolated from Hamburger and Hamilton stage 19 embryos, but as myocytes were isolated from progressively older embryos the addition of PDGF to the culture media resulted in a decrease in the total amount of cytoplasmic protein while the total amount of cytoskeletal proteins remained constant. In experiments that examined the effects of PDGF on specific contractile proteins, Velez et al (1994, 1995) found that the addition of PDGF to the culture media resulted in reduced troponin-t, actin, and α-actinin from the cytoplasmic and cytoskeletal protein fractions, suggesting that cardiac myocytes revert to a less differentiated state of development in response to increased levels of PDGF. The specific PDGF ligand used was not given in these reports, and the possibility also exists that the avian PDGF ligand-receptor system differs from the mammalian system since PDGFrβ receptors have not been identified in avian models (Potts and Carrington, 1993).

In unpublished experiments with isolated neonatal rat cardiac myocytes we have found that incubation with 10 ng/mL of PDGF-BB stimulates a significant increase in cell volume (Figure 10.2). In these experiments, isolated myocytes were allowed to grow in culture for 72 hours, after which they were treated with the calcium channel blocker nifedipine for 72 hours. Nifedipine treatment resulted in contractile arrest and the breakdown of the contractile apparatus in the myocytes. This approach resulted in all of the myocytes being in a similar physiologic state and facilitated the process of making morphologic measurements. Following washout of the nifedipine, cells were cultured in serum-free media with PDGF-BB added for up to 4 days. At designated time points, 5-chloromethylfluorescein diacetate (CMFDA) was added to the culture media and myocytes incubated for an additional 30 minutes. CMFDA freely passes through cell membranes and once inside the cell undergoes a glutathione S-transferase–mediated reaction that produces an impermeable product that fluoresces at a wavelength of 490 nm. By confocal microscopy we were then able to optically section the cells at 1 μm intervals and add the areas of each section to determine the volume of the cells. By colabeling the cells with rhodamine phalloidin, we were also able to positively identify all measured cells as myocytes. By this protocol, PDGF-BB induced an increase in cell volume of approximately 15% when compared to control cells.

When added to the culture media of isolated mouse fetal cardiac fibroblasts, PDGF acts as a potent mitogen, increasing cell replication by 128% over that of control cultures. Associated with the increase in cell number was an increase in

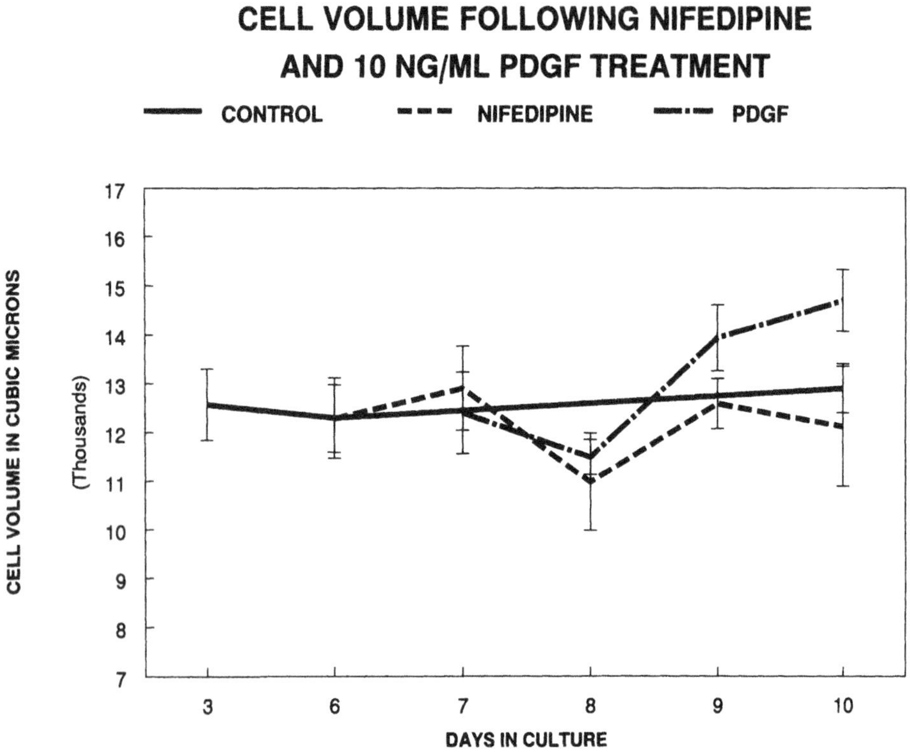

FIGURE 10.2. Treatment of cultured neonatal cardiac myocytes with PDGF-BB results in an increase in the cell volume. See the text for details on the experimental design.

collagen production of 78%, indicating that PDGF may play an important role in the initial modeling of the cardiac ECM (Butt et al, 1995). The specific ligands of PDGF in this study were not identified.

PDGF and Cardiac Development: In Vivo Studies

In all studies in which a ligand or receptor component of the PDGF system is missing, animals die during fetal or perinatal development from kidney, lung, or cardiac defects. In the *Patch* mouse, a naturally occurring mutation in C57Bl6 mice in which the PDGFrα is missing, severe cardiac defects have been observed including dilation, reduced myocardial atrial and ventricular thickness, atrial and ventricular septal defects, distorted formation of the cardiac valves, and persistent truncus arteriosus (Morrison-Graham et al, 1992; Schatteman et al, 1992, 1995). In addition to the reduced ventricular wall thickness, trabeculae fail to develop and myocytes fail to differentiate and form myofibrils (Price et al, 1999; Thielen et al, 1999).

There is also the possibility that not all of these defects are due to the deletion of the PDGFrα since the *Patch* mutation is known to affect the expression of the *c-kit* gene and to include the W^{sh} gene. The inclusions of these genes, which affect melanogenesis, explain the unique coat color of *Patch* animals, although the relationship between the various defects are not entirely clear since the *Patch* mutation has not been totally mapped (Stephenson et al, 1991; Duttlinger et al, 1995).

In complementary studies in which neutralizing antibodies to the PDGF-A ligand were microinjected into mouse deciduas at ED 8.5, 9.5, and 10.5, many of the observed cardiac defects were similar to those seen in the *Patch* mice (Schatteman et al, 1996).

Reported cardiac defects associated with PDGF-A ligand and PDGFrα null mice developed on a 129Sv backround strain do not show the cardiac defects described above in the *Patch* mice and animals in which the function of the ligand has been blocked. In PDGF-A null animals approximately 50% of the embryos die before ED 10, although other animals live through postnatal development for a maximum of 6 weeks and develop secondary right ventricular hypertrophy associated with respiratory defects before death (Boström et al, 1996). PDGFrα-negative animals developed a dilated pericardium and did not survive beyond ED 16, but septation of the heart did occur in some of the surviving ED 15 animals (Soriano, 1997).

The majority of PDGFrβ null mice die shortly before or after birth and exhibit severe hemorrhaging, but no gross developmental defects in the heart (Soriano, 1994). However, PDGF-B ligand negative mice develop severe cardiovascular defects including ventricular hypertrabeculation, thin myocardial walls, and heart failure, and die by ED 19 (Levéen et al, 1994). The lack of heart-related defects in the PDGFrβ null animals may be due to a compensatory effect since PDGFα receptors are present and may be activated by both the PDGF-A and -B ligands. It is also possible that differences in the genetic strains of mice used in the studies may introduce some variations in the development of specific defects and the length of animal survival. In *Patch* mice, crossing the C57Bl6 background in which the original mutation was found with a BalbC mouse strain results in an increase in embryo survival of several days. However, the underlying reasons for the observed differences in cardiac defects and survival times of animals are difficult to determine since little is known about the specific roles the PDGF receptor-ligand system plays in the development and function of cardiac myocytes, fibroblasts, and endothelial cells.

BASIC FIBROBLAST GROWTH FACTOR (bFGF)

Basic fibroblast growth factor is the prototypic member of a large, ever-growing group of angiogenic proteins. To date, there have been some 20 members of the bFGF family identified, most of which are conserved among species; however, both species- and tissue-specific FGF have been reported. Some FGFs lack a signal sequence for export out of the cell, but most possess signal sequences and several of them have been identified as transforming proteins (Galzie et al, 1997). FGFs are potent mitogens for a wide variety of cell types in tissue culture including cardiac endothelial cells, fibroblasts, and myocytes.

The biologic effects of FGF are established as a result of an intracellular signal transduction cascade initiated by the growth factor–bound activated FGF receptors. There are four related FGF tyrosine kinase receptors (FGFR1–4) that are expressed on most types of cells in tissue culture. Different FGFs will activate receptor subtypes to different extents depending on their abilities to bind with high affinity to each receptor type. These receptors initiate a well-studied series of intracellular cascades that eventually result in biologic responses often involving changes in gene transcription. The FGF receptors are very similar in overall structural organization, and with alternative splicing of the extracellular domain create numerous variants that are expressed and activated in differing cell types (Klint and Claesson-Welsh, 1999).

Expression of FGFR-1 is barely detectable at preheart stages, followed by a sequential increase in relative abundance that peaks at stage 24, followed by a decline at days 7 to 14 in the chicken (Sugi et al, 1995). FGFR-1 has long and short isoforms, each of which is also present in embryonic and adult mouse hearts. The function of each isoform is not known, but there appears to be a switch from the long, fetal, form to the short form in adult hearts, leading to the hypothesis that the long form may contribute to the mitogenic capacity of fetal myocytes (Pasumarthi et al, 1995). Using a rat cardiac myoblast cell line devoid of both the long and short isoforms, Sheikh et al (1997) were able to transfect both isoforms into H9c2 cells and demonstrate their ability to activate a mitogenic response.

Immunolocalization of FGF ligands in the developing heart is limited to the myocardium and developing vasculature of the epicardium and is absent in the endocardium (Tomanek et al, 1996; Zhu and Lough, 1996). Early functional studies by Sugi and Lough (1995) showed that the FGF-2 ligand is critical for early cardiac development in the chicken. Subsequent studies from their lab have now demonstrated the presence and function of FGF-1 and FGF-4 ligands in the early endoderm and later in the myocardium. Unlike FGF-2, which was detected at stage 9+, FGF-1 and FGF-4 were not present until stage 11 and 15, respectively. Expression of all FGF isoforms peaked at stages 18 to 24, decreasing thereafter in parallel with reduced myocardial cell proliferation. In addition, both FGF1 and FGF4, like FGF-2, supported the proliferation and differentiation of precardiac myoblasts, resulting in the formation of vesicles containing an adherent multilayer of synchronously contractile cells (Zhu et al, 1996). Endogenous addition of FGF-soaked beads implanted near the primitive ventricle significantly decreased the survival rate and increased the anomaly rate of embryos when compared with controls. The experimental embryos exhibited a wide variety of defects but most were cardiac in nature (Watkins et al, 1998). These data suggest that FGF is critical for cardiac development but that the regulation of FGF must be under strict control.

As mentioned above, many FGFs and FGFRs are conserved among species, a case in point is the heartless (*htl*) gene in drosophila. Heartless is a FGFR homolog that has been shown to be involved with early directional migration of *Drosophila* mesoderm. A second role for the Htl protein in promoting the specification of the precursors to certain cardiac and somatic muscle cells in the embryo has now been demonstrated (Michelson et al, 1998). These results suggest that the requirement for FGFR signaling in *Drosophila* and vertebrate mesoderm development and early cardiac programming may be similar.

Role of FGF in Cardiac Myocyte Development and Function

There is considerable evidence for the effects of FGF on cardiac endothelial cells, and recent work has shown that cardiac myocytes also respond to FGF. Cardiac hypertrophy is characterized by an increase in cell size in the absence of cell division and is accompanied by a number of qualitative and quantitative changes in gene expression (Schaub et al, 1997). Upregulation of so-called fetal genes, i.e., β-myosin heavy chain, α-skeletal and α-smooth muscle actin and ANF has been observed concomitant with the downregulation of α-myosin heavy chain and the

Ca^{2+} pump of the sarcoplasmic reticulum (Schaub et al, 1997). FGF appears to produce both of these effects in cardiac myocytes. In addition, FGF increases the accumulation of cytoplasmic troponin-T (TnT) at the subcellular level (Velez et al, 1994). Because expression of thin filament molecules is considered a good marker of differentiation in muscle cell cultures, increased expression of TnT by FGF at stage 19 but not 39 in the chicken, supports the proposed role of FGF in early heart development. In vitro cell culture studies have shown that FGF upregulates α-smooth muscle actin expression and restricts myofibrillar growth with a sharp boundary in the perinuclear region (Harder et al, 1996). Addition of IGF, which is known to work through the same downstream signaling pathway as FGF, produces the opposite effects on both gene expression and morphology. Surprisingly, triiodothyronine (T_3), which uses a completely different signaling pathway, produces a similar response to that of FGF, and both IGF and FGF appear dependent on T_3 to bring about their typical responses (Schaub et al, 1997).

The other major response of cardiac myocytes to FGF is the alteration in the calcium current and Ca^{2+} channels expression. Patch clamp experiments using ventricular myocytes have shown that addition of FGF induces an increase of fast burst openings, mediated by Ca^{2+} -permeable channels. Confocal microscopy of intracellular free Ca^{2+} has provided direct evidence that FGF induced an increase of cytoplasmic and nucleoplasmic free Ca^{2+} concentrations that were generated in part by Ca^{2+} influx through the plasma membrane (Merle et al, 1997). Whole-cell patch clamp experiments demonstrated that the transient outward current (Ito) in cardiomyocytes is modulated by FGF. Treatment of cardiomyocytes with FGF from day 6 to 15 exhibited a dose-dependent increase in the Ito. This increase was produced by 72-hour treatments and not by the prolonged administration from day 6 to 15. Addition of a specific tyrosine kinase inhibitor (methyl-2,5-dihydroxycinnamate) showed preferential inhibition of the FGF-stimulated increase in Ito density. This is most likely due to the parallel change in the ionic selectivity of the channel (Guo et al, 1995, 1996).

Role in Cardiac Fibroblast Development and Function

Cardiac fibroblasts constitute greater than 90% of the nonmyocyte cells that contribute to the structure of the heart (Booz and Baker, 1995). It is well established that cardiac fibroblasts are predisposed to transform into a phenotype with muscle-specific features and that TGF-β1 is a specific inducer of this event but does not by itself induce proliferation (Sigel et al, 1996). FGF is known to be a powerful mitogen for cardiac fibroblasts and it appears that TGF-β1 controls this FGF-induced mitogenic program (Butt et al, 1995; Sigel et al, 1996). In addition, FGF promotes cardiac fibroblast production of collagen in vitro, albeit at a much lower potency than TGF-β, PDGF, and IGF. The collagen increase by FGF is most likely due to both the proliferative effect on fibroblasts and an increased synthesis (Butt et al, 1995). Finally, FGF also upregulates connexin43 (Cx43), the major connexin in the rat heart. Cultured cardiac fibroblasts show a significant increase in the amount of Cx43 mRNA and protein after a 6-hour incubation with FGF. This increase was not seen at 30 minutes, nor was the increased cellular communication that was present after 6 hours. Thus, it appears that FGF increased Cx43 expression caused increased coupling between cardiac fibroblasts (Doble and Kardami, 1995).

Genetic Manipulation of FGF Ligands and Receptors

With a family as large as the FGFs, one of the best ways to determine individual gene function is by the use of gene targeting knockout mice. This has been done for a number of the family members for both ligands and receptors. In the case of FGF receptor knockouts, both FGFR-1 and FGFR-2 null mutants are embryonic lethal. FGFR-2 –/– mice die just a few days postimplantation, whereas the FGFR-1 null mice make it to the second week of gestation before dying (Ciruna et al, 1997; Arman et al, 1998).

Mice lacking the FGF-2 ligand gene are hypotensive but respond normally to Ang II–induced hypertension (Dono et al, 1998). FGF-3 and FGF-6 ligand knockouts both appear to have no cardiovascular abnormalities (McKay et al, 1996; Fiore et al, 1997). Mice lacking the FGF-10 gene die shortly after birth due to abnormal lung development but show normal cardiovascular development (Sekine et al, 1999). One could view these results as disappointing given the overwhelming evidence that FGF influences heart development. However, a careful examination of the heart has not been done in all cases and it is quite possible that compensatory mechanisms by other FGFs or other factors may allow cardiovascular development to continue as normal in these mice.

INSULIN-LIKE GROWTH FACTORS I AND II (IGF-I AND IGF-II)

Ligand and Receptor Characteristics

Insulin, IGF-I, and IGF-II are structurally and functionally related with insulin mediating acute metabolic actions, while the insulin-like growth factors primarily mediate physiologic roles regulating growth (Fuller et al, 1992; Louvi et al, 1997). The emphasis of this discussion is on IGF-I and IGF-II even though insulin may activate the IGF-I receptor under some circumstances (Sklar et al, 1989, Hudson et al, 1996) and IGF-II will bind to and activate the insulin receptor (Louvi et al, 1997). The IGF-I and IGF-II receptors are distinctly different in structure. The IGF-I receptor is in the tyrosine kinase receptor family and consists of a heterotetramer with two β subunits that span the cell membrane and two α subunits that bind the IGF ligands. The IGF-II receptor consists of a single membrane spanning domain mannose 6-phosphate receptor that has distinct binding sites for IGF-II and mannose (Sklar et al, 1989; 1992; Haskell and Tucker, 1994). This receptor is only functional for IGF-II in mammalian systems with no activity reported in amphibians or avians, indicating a recent divergence in the evolution of the IGF-II growth factor ligand/receptor system (Sklar et al, 1992; Antin et al, 1996).

The IGF-I ligand only binds to the IGF-I receptor with high affinity but will also bind to the IGF-II/mannose 6-phosphate receptor with reduced affinity (Ito et al, 1993; Haskell and Tucker, 1994). The IGF-II ligand binds to both receptor types with equal affinity (Ito et al, 1993; Louvi et al, 1997). Binding of IGF-II to the IGF-1 receptor results in signal transduction while binding to the IGF-II/mannose 6-phosphate receptor results in degradation of excess IGF-II by receptor-mediated endocytosis (Matzner et al, 1992; Eggenschwiler et al, 1997). As discussed further below, regulation of circulating and locally produced quantities of both IGF ligands by IGF-binding proteins (Carr et al, 1995) and of IGF-II by receptor-mediated endocytosis is critical in the normal development of the heart. IGF-I receptor-mediated signaling in the heart has been studied by Foncea et al (1997) and Lavandero et al (1998). IGF-I receptor activation results in the phosphorylation of the two β subunits of the receptor and an increase in the phosphotyrosine content of Extracellular Signal Related Kinase (ERKs) Insulin Receptor Substrate-I (IRS-1), phospholipase C-γ_1 and Pl3-K and the sequential activation of protein Kinase C (PKC) isozymes, the ERK cascade, and p90 RSK.

Expression of IGF Ligands and Receptors

Although some species variation occurs in the time of expression, all of the available studies indicate that IGF-I ligand and receptors function primarily during the peri- and postnatal periods of development and that IGF-II ligand and receptors function primarily during fetal and perinatal development. Both IGF ligands and receptors have been found in all four chambers of the heart as well as in the developing cardiac cushions and outflow tract (Stylianopoulou et al, 1988; Bondy et al, 1990; Cheung et al, 1996). IGF-1 mRNA levels from isolated rat cardiac myocytes was greatest from late fetal (ED 18–19) and 1 day postnatal myocytes and decreased significantly in myocytes isolated from animals after 21 days of postnatal development. The downregulation of IGF-I closely parallels the reduced capacity of cardiac myocytes to divide as the heart matures, indicating that IGF-I may play a role in the proliferative capacity of myocytes (Cheng et al, 1995; Anversa et al, 1996).

A similar decrease in the expression of IGF-I receptor mRNA was found in fibroblasts isolated from rat hearts. In these studies Reiss et al (1995) found a decrease of 77% in the expression of IGF-I receptor mRNA when comparing 21-day postnatal isolated fibroblasts with fetal fibroblasts. The major decrease in the expression of IGF-I receptor mRNA in fibroblasts occurred during the first 5 days of postnatal development. Even though there was a large reduction in the expression levels of IGF-I receptor during early postnatal development, IGF-I ligand expression did not change in isolated fibroblasts.

IGF-II receptor expression is at least twofold higher in the heart than in any other tissue during fetal development (Sklar et al, 1992). Expression of the IGF-II ligand has been detected from the earliest stages of precardiac mesoderm development and increases through fetal development in avian, rat, and ovine experimental models (Cheng et al, 1995; Antin et al, 1996; Cheung et al, 1996). Expression of the IGF-II ligand peaks during mid- to late gestation and rapidly decreases after birth to barely detectable levels by 2 months of age in the rat (Engelmann et al, 1989; Senior et al, 1990; Cheng et al, 1995; Liu et al, 1996). Decreases in IGF-II ligand and receptor expression in fibroblasts mirrored that seen in myocytes with a rapid reduction in the levels of mRNA immediately after birth (Reiss et al, 1995).

Role of Insulin-Like Growth Factors in Myocyte Development, Differentiation, and Function

As indicated above, the apparent role of the IGF-II/mannose 6-phosphate receptor is a reduction in the amount of circulating and local IGF-II. This indicates that signaling by both IGF ligands is primarily mediated by the IGF-I receptor. Some controversy exists concerning the principal role of IGF-I stimulation of cardiac myoyctes. While several studies have indicated an increase in DNA synthesis and myocyte number following IGF-I stimulation, others have found that myocytes hypertrophy without a concomitant increase in number after IGF-I stimulation. Using serum-free cultures, Kajstura et al (1994) found a fivefold increase in Bromodeoxy Uridine (BrdU) labeling and an increase in the number of myocytes exhibiting DNA synthesis without an increase in myocyte volume after IGF-I stimulation. Supplemental quantities of IGF-I added to myocyte culture media containing serum induced a 34% increase in cell number and had no effect on myocyte hypertrophy over a 6-day period when compared with control cultures. Antisense oligodeoxynucleotides to the IGF-I receptor almost abolished the DNA synthesis and had no effect on ANF, an indicator of cardiac myocyte hypertrophy. These results support the conclusions of Anversa et al (1996) in their review on myocyte growth that the IGF-I ligand/receptor system is not involved in myocyte hypertrophy.

However, Lavandero et al (1998) found that exposure of isolated neonatal myocytes to 10 nM of IGF-I for 1, 6, or 24 hours resulted in an increase in protein synthesis of 1.9-, 1.7-, and 2.2-fold and an increase in cell size without a concomitant increase in [^{3}H]thymidine incorporation. Ito et al (1993) also found a doubling in cell size following treatment of cultured myocytes with 10^{-7} M IGF-I as well as a dose-dependent increase in the synthesis of myosin light chain-2 (MLC-2), troponin 1, and skeletal α-actin. Increased mRNA levels for MLC-2 and troponin 1 occurred within 60 minutes and reached maximum levels at 6 hours. Additional studies by Guo et al (1997) and Harder et al (1998) have also indicated that treatment of myocyte cultures with IGF-I induces myocyte hypertrophy and enhanced myofibrillar growth without an increase in cell number.

The potential for IGF-I to perform additional roles beyond those of myocyte proliferation and hypertrophy also exists. Using isolated neonatal myocytes Guo et al (1997) found that IGF-I can regulate potassium channel expression in myocytes and may modulate cardiac excitability. In a recent study, Wang et al (1998) also found that IGF-I may increase cardiac myocyte viability by reducing the expression of Bax and Caspase 3 activation, two factors that stimulate the rate of DNA fragmentation and apoptosis, a process that may function in the normal remodeling of the heart during development.

IGF-II stimulates both the proliferation and differentiation of fetal myocytes. Treatment of quail precardiac mesodermal cells with concentrations of IGF-II above 1 ng/mL resulted in a threefold increase in the proliferation of cells as well as the enhanced differentiation of myocytes as indicated by sarcomere organization and contractility (Antin et al, 1996; Eisenberg and Bader, 1996). Similar results have been observed with primary cultures of fetal and neonatal rat cardiac myocytes. Liu et al (1996) found that treatment of fetal myocytes with 50 ng/mL of IGF-II induced DNA synthesis as well as an increase in several hypertrophic markers including MLC-2 mRNA, ANF, β-myosin heavy chain, and skeletal α-actin. Associated with the increase in the expression of the hypertrophic markers was an increase in myofibrillar organization.

Neonatal myocytes in cultures treated with IGF-II expressed an increase in the hypertrophic markers and myofibrillar organization, but did not show an increase in DNA synthesis or cell number. In experiments with neonatal cardiac myocytes treated with 10^{-7} M IGF-II for 48 hours, Adachi et al (1994) found a 131% increase in cell size and protein content and an increase in the expression of mRNA for several hypertrophic markers including MLC-2, troponin 1, and skeletal α-actin. Increases in cell size and protein content in these experiments were mediated via calcium channels and blocked by the calcium channel blocker nicardipine.

IGFs and Cardiac Development: In Vivo Studies

In targeted mutation studies, each IGF ligand and receptor has been removed singly and in combination with varying results on cardiac development and animal viability (reviewed by Louvi et al, 1997). The survival of IGF-I null animals depends on the genetic strain, but in those that do survive, body weight is reduced by 60%. The initial phenotypic effects of IGF-I null mutations occur around ED 13.5 in the mouse and involve a reduction in body size, although overall proportions of the body do not change (Baker et al, 1993). IGF-I receptor mutants (–/–) die immediately after birth due to respiratory failure (Louvi et al, 1997). No references concerning developmental cardiac defects in animals in which the IGF-I ligand/receptor system has been disrupted by targeted mutations were found.

In experiments in which IGF-I was subcutaneously injected into 3- and 15-day-old rat pups, there was a significant increase in heart weights (Phillips et al, 1988). Similar increases in heart weight (26%) were found when IGF-I levels were supplemented in late gestation sheep by intravascular infusion (Lok et al, 1996). No cardiac defects beyond the increased heart weights were described in these studies. In transgenic mice in which the α-myosin heavy chain promoter was used to overexpress IGF-I, myocyte proliferation occurred in juvenile and young adult mice without an increase in myocyte size. No changes in myocyte number or size were noted in fetal or neonatal animals with this model since the α-myosin heavy chain transgenic model does not increase the expression of the targeted gene until after birth (Reiss et al, 1996).

In targeted gene studies resulting in animals heterozygous for the IGF-II allele there was a 10-fold decrease in the level of IGF-II transcripts that resulted in reduced embryonic size and the death of 10 of 77 embryos. In surviving animals growth deficiencies remained, but animals reached maturity and were fertile (DeChiara et al, 1990). No indication of cardiac defects were given for the embryos that died. In mice lacking both the IGF-II and IGF-II/mannose 6-phosphate receptor alleles, survival does not extend beyond the first few postnatal weeks (Ludwig et al, 1996).

In homozygous and heterozygous animals in which the normal maternal inheritance (maternally imprinted gene) of the IGF-II/mannose 6-phosphate receptor has been disrupted, the levels of circulating and tissue IGF-II are not properly controlled resulting in the increased stimulation of the IGF-I receptors. These animals die at birth and exhibit a two- to fourfold increase in heart size. Additional cardiac defects associated with disruption of the expression of the IGF II receptor include hyperplasia of the ventricular myocardium, septal defects involving large blood-filled sinusoids and a reduction in the number of myocyte layers in the septum, and poorly developed tricuspid valves (Lau et al, 1994; Wang et al, 1994). Animals that lack the IGF-II receptor are rescued from perinatal lethality by also knocking out the IGF-1 receptor allele. Double receptor null animals are smaller than wild-type littermates but survive with no apparent defects (Ludwig et al, 1996) as a result of IGF-II stimulation of insulin receptors (Louvi et al, 1997).

Further increasing the level of circulating IGF-II by creating animals with a double mutation involving the removal of the IGF-II/mannose 6-phosphate receptor and the overexpression of IGF-II resulted in hyperplasia of endothelial cells in addition to the above-noted defects. Ventricular septal defects, in which the communication between the ventricles due to fenestration of the septum, and enlargement of the aortic and pulmonary valves and accumulation of pericardial fluids were also noted. These defects, as well as several other described defects in these mutants are also characteristic of the Beckwith-Wiedemann and Simpson-Golabi-Behmel syndromes that occur during human development and have been mapped to similar genes (Eggenschwiler et al, 1997).

VASCULAR ENDOTHELIAL GROWTH FACTOR (VEGF)

Since a chapter in the second volume of this series will address the several different VEGF ligand isoforms and receptors in detail, only a brief synopsis is presented here. The primary role of VEGF is to facilitate the cascade of angiogenic events. However, evidence indicates that stretch induces the expression of VEGF in adult hearts (Li et al, 1997), and in isolated neonatal cardiac myocytes and fibroblasts VEGF activates Raf-1, MAP kinase, and S6 kinase (p90rsk) (Seko et al, 1998). Through the activation of these pathways in cardiac myocytes, it is possible that VEGF plays an important role in the paracrine stimulation of endothelial cells and the development of the required vasculature and ventricular sinusoids during cardiac development (Shifren et al, 1994; Tomanek et al, 1998, 1999).

The various VEGF isoforms primarily bind to two tyrosine kinase receptors, VEGFR-1 (*flt-1*) and VEGFR-2 (*flk-1*). Gene knockout experiments for both types of receptors result in normal development of heterozygous animals and embryonic lethality of null animals at midgestation. Null *flt-1* animals have abnormally large vascular structures, dorsal aortae and ventricles filled with disorganized endothelial cells, and an increased number of endothelial cells lining the ventricles (Fong et al, 1995). In contrast, in *flk-1* null embryos endothelial cells and blood islands failed to develop, resulting in an absence of blood cells and vessels (Shalaby, 1995).

In VEGF ligand-deficient animals the hearts showed a developmentally delayed common atrium, primitive thin-walled ventricle, and rudimentary aortae. Heterozygous VEGF animals failed to develop a connection between the heart and vasculature and died at ED 9 to 10 (Carmeliet et al, 1996; Ferrara et al, 1996). Addition of exogenous quantities of VEGF on methylcellulose carriers inserted into the midbrain or on the forelimb during the ex ovo culture of quail embryos resulted in enlarged atria, altered septation patterns, thin ventricular walls, and death following overnight incubation (Feucht et al, 1997). In other studies in which VEGF was injected into developing quail and chicken embryos, numerous vascular defects have also been described (Drake and Little, 1995; Tomanek et al, 1998).

GROWTH FACTOR INTERACTIONS AND COMBINED SIGNALING MECHANISMS

One of the most difficult aspects of determining the specific role of a growth factor in cardiac development is the potential for numerous interactions to occur not only between a growth factor and its receptors, but also between a growth factor and other signaling mechanisms. These interactions may include an increased expression due to mechanical load, shared signaling mechanisms with integrins, and/or the increased expression of ECM proteins and other growth factors in response to a single stimulus. For example, an increase in mechanical load on cardiac myocytes is known to induce an increase in the local expression of Ang II (Sadoshima et al, 1993; Yamazaki et al, 1995). In turn, the increased expression of Ang II has the potential to affect the activity of the ligands and receptors of PDGF-BB (Bobik et al, 1990; Linesman et al, 1995), TGF-β (Sharma et al, 1994; Lee et al, 1995), IGF-I (Brink et al, 1996), and VEGF (Chua et al, 1998). Mechanical activity has also been shown to mediate the function of other cell types and integrins (Burgess et al, 1994; Ashizawa et al, 1996; Watson et al, 1998). Once the Ang II signal transduction pathways have been stimulated, there is also the potential for significant cross-talk between the pathways mediating the effects of mechanical force, integrins, and growth factors. Several reviews illustrating possible interactions between these and other signaling mechanisms have been published (Sastry and Horwitz, 1996; Hefti et al, 1997; Schaub et al, 1997; van Bilsen, 1997).

FUTURE DIRECTIONS

While studies describing the expression patterns of ligands and receptors have provided significant information concerning the role of growth factors in cardiac development, as noted by Hyer and Mikawa (1997), to correctly assess the role of various growth factors in development it is not enough to conclude they are important based solely on expression patterns. To fully understand the role of growth factors in cardiac development, it will be necessary to further develop techniques that permit the molecular dissection of the numerous signal transduction pathways stimulated by growth factors and the various interactions they elicit. Although at times cost prohibitive in generation and maintenance, animals that overexpress or lack receptors and ligands have become valuable tools in studying the effects of growth factors on cardiac development, although they do not always generate the expected results. For example, as noted above, despite the numerous functions of the cardiac angiotensin system in development, animals lacking angiotensinogen develop normally to the postweaning stages. In addition, animals lacking the AT1 receptor do not develop any obvious morphologic defects and survive to adulthood (Ito et al, 1995; H.S. Kim et al, 1995; Krege et al, 1995). However, animals that overexpress the AT1 receptor develop myocyte hyperplasia and die shortly after birth (Hein et al, 1997).

Mammalian whole embryo and cell culture techniques, in combination with viral transfection and antisense oligonucleotide protocols, are also becoming valuable mechanisms to dissect the interactions that occur between growth factors and various signaling mechanisms (Slavkin, 1995; Hyer and Mikawa, 1997). Whole-embryo culture techniques provide the opportunity to treat embyros in the absence of maternal influences without disrupting the normal cell/cell and cell/ECM interactions that occur in the heart. However, this technique is limited to a narrow developmental window of approximately 48 hours during the looping stages of cardiac development so the effects of growth factors on late fetal and neonatal development cannot be studied. While cell culture techniques typically remove cells from their normal interactions with other cells, ECM components, and other factors affecting cell growth and differentiation, they typically provide experiments with the tightest controls in determining the effects of a growth factor. With the numerous antisense oligonucleotide, viral, and signal transduction probes that are becoming available to block and enhance specific actions of growth factors, the opportunity to determine their functional interactions in cardiac development should soon be possible.

SUMMARY

Cardiac development is dependent on the normal expression of several growth factors and their receptors. This chapter reviewed much of what is currently known about the roles of some of these growth factors in the embryonic, fetal, and neonatal development of the heart. However, our understanding of the specific roles of these growth factors and their complex interactions with other chemical and mechanical signaling mechanisms is incomplete. As new probes and techniques are developed, it should be possible to further our understanding of these interactions and to determine the mechanisms by which growth factors affect heart development.

ACKNOWLEDGMENTS

We would like to thank Benny Davidson for his help in preparation of the manuscript. This work was funded in part by National Institutes of Health grants 5RO3HD35253-02, 2RO1HL4224910A1, and 5RO1HL58893-02.

REFERENCES

Abdellatif, M.M., Neubauer, C.F., Lederer, W.J., and Rogers, T.B. (1991). Angiotensin-induced desensitization of the phosphoinositide pathway in cardiac cells occurs at the level of the receptor. *Circ Res* 69:800–809.

Aceto, J.F., and Baker, K.M. (1990). [Sar[1]]angiotensin II receptor-mediated stimulation of protein synthesis in chick heart cells. *Am J Physiol* 258(*Heart Circ Physiol* 27):H806–H813.

Adachi, S., Ito, H., Akimoto, H., et al. (1994). Insulin-like growth factor-II induces hypertrophy with increased expression of muscle specific genes in cultured rat cardiomyocytes. *J Mol Cell Cardiol* 26:789–795.

Allen, I.S., Cohen, N.M., Dhallan, R.S., Gaa, S.T., Lederer, W.J., and Rogers, T.B. (1988). Angiotensin II increases spontaneous contractile frequency and stimulates calcium current in cultured neonatal rat heart myocytes: insights into the underlying biochemical mechanisms. *Circ Res* 62:524–534.

Antin, P.B., Yatskievych, T., Dominguez, J.L., and Chieffi, P. (1996). Regulation of avian precardiac mesoderm development by insulin and insulin-like growth factors. *J Cell Physiol* 168:42–50.

Anversa, P., Kajstura, J., Cheng, W., Reiss, K., Cigola, E., and Olivetti, G. (1996). Insulin-like growth factor-1 and myocyte growth: the danger of a dogma. Part I. Postnatal myocardial development: nomal growth. *Cardiovasc Res* 32:219–225.

Aoki, H. Izumo, S., and Sadoshima, J. (1998). Angiotensin II activates RhoA in cardiac myocytes. A critical role of RhoA in angiotensin II-induced premyofibril formation. *Circ Res* 82:666–676.

Arman, E., Haffner-Krausz, R., Chen, Y., Heath, J.K., and Lonai, P. (1998). Targeted disruption of fibroblast growth factor (FGF) receptor 2 suggests a role for FGF signaling in pregastrulation mammalian development. *Proc Natl Acad Sci USA* 95:5082–5087.

Ashizawa, N., Graf, K., Do, Y.S., et al. (1996). Osteopontin is produced by rat cardiac fibroblasts and mediates A_{II}-induced DNA synthesis and collagen gel contraction. *J Clin Invest* 98:2218–2227.

Ataliotis, P., and Mercola, M. (1997). Distribution and functions of platelet-derived growth factors and their receptors during embryogenesis. *Annu Rev Cytol* 172:95–127.

Baker, J., Liu, J., Robertson, E.J., and Efstratiadis, A. (1993). Role of insulin-like growth factors in embryonic and postnatal growth. *Cell* 75:73–82.

Baker, K.M., and Aceto, J.F. (1990). Angiotensin II stimulation of protein synthesis and cell growth in chick heart cells. *Am J Physiol* 259(*Heart Circ Physiol* 28):H610–H618.

Baker, K.M., Singer, H.A., and Aceto, J.F. (1989). Angiotensin II receptor-mediated stimulation of cytosolic-free calcium and inositol phosphates in chick myocytes. *J Pharmacol Exp Ther* 251:578–585.

Beinlich, C.J., Baker, K.M., White, G.J., and Morgan, H. (1993). Control of growth in neonatal pig hearts. *Mol Cell Biochem* 119:3–9.

Beinlich, C.J., Rissinger, C.J., and Morgan, H.E. (1995). Mechanisms of rapid growth in the neonatal pig heart. *J Mol Cell Cardiol* 27:273–281.

Bernstein, K.E., and Marrero, M.B. (1996). The importance of tyrosine phosphorylation in angiotensin II signaling. *Trends Cardiovasc Med* 6:179–187.

Betsholtz, C., and Heldin, C.-H. (1998). Platelet derived growth factor: a regulator of connective tissue development and reaction. In: Reed, R.K., and Rubin, K., eds. *Connective Tissue Biology: Integration and Reductionism*. Portland Press, London, pp. 179–193.

Bobik, A., Grinpukel, S., Little, P.J., Grooms, A., and Jackman, G. (1990). Angiotensin II and noradrenaline increase PDGF-BB receptors and potentiate PDGF-B stimulated DNA synthesis in vascular smooth muscle. *Biochem Biophys Res Commun* 166:580–588.

Bondy, C.A., Werner, H., Roberts C.T. Jr., and LeRoith, D. (1990). Cellular pattern of insulin-like growth factor-I (IGF-I) and type IIGF receptor gene expression in early organogenesis: comparison with IGF-II gene expression. *Mol Endocrinol* 4:1386–1398.

Booz, G.W., and Baker, K.M. (1995). Molecular signaling mechanisms controlling growth and function of cardiac fibroblasts. *Cardiovasc Res* 30:537–543.

Booz, G.W., and Baker, K.M. (1996). Role of type 1 and type 2 angiotensin receptors in angiotensin II-induced cardiomyocyte hypertrophy. *Hypertension* 28:635–640.

Boström, H., Willetts, K., Pekny, M., et al. (1996). PDGF-A signaling is a critical event in lung alveolar myofibroblast development and alveogenesis. *Cell* 85:863–873.

Brink, M., Wellen, J., and Delafontaine, P. (1996). Angiotensin II causes weight loss and decreased circulating insulin-like growth factor I in rats through a pressor-independent mechanism. *J Clin Invest* 97:2509–2516.

Burgess, M.L., Carver, W.E., Terracio, L., Wilson, S.P., Wilson, M.A., and Borg, T.K. (1994). Integrin-mediated collagen gel contraction by cardiac fibroblasts. Effects of angiotensin II. *Circ Res* 74:291–298.

Butt, R.P., Laurent, G.J., and Bishop, J.E. (1995). Collagen production and replication by cardiac fibroblasts is enhanced in response to diverse classes of growth factors. *Eur J Cell Biol* 68:330–335.

Carmeliet, P.I., Ferreira, V., Breier, G., et al. (1996). Abnormal blood vessel development and lethality in embryos lacking a single VEGF allele. *Nature (Lond)* 380:435–439.

Carr, J.M., Owens, J.A., Grant, P.A., Walton, P.E., Ownes, P.C., and Wallace, J.C. (1995). Circulating insulin-like growth factors (IGFs), IGF-binding proteins (IGFBPs) and tissue mRNA levels of IGFBP-2 and IGFBP-4 in the ovine fetus. *J Endocrinol* 145:545–557.

Cheng, W. Reiss, K., Kajstura, J., Kowal, K., Quaini, F., and Anversa, P. (1995). Down-regulation of the IGF-1 system parallels the attenuation in the proliferative capacity of rat ventricular myocytes during postnatal development. *Lab Invest* 72:646–655.

Cheung, C.Y., Johnson, D.D., and Reyes, V. (1996). Ontogeny of insulin-like growth factor-I and -II gene expression in ovine fetal heart. *J Soc Gynecol Invest* 3:309–315.
Chua, C.C., Hamdy, R.C., and Chua, B.H. (1998). Upregulation of vascular endothelial growth factor by angiotensin II in rat heart endothelial cells. *Biochim Biophys Acta* 1401:187–194.
Cigola, E., Kajstura, J., Baosheng, L., Meggs, L.G., and Anversa, P. (1997). Angiotenisn II activates programmed myocyte cell death in vitro. *Exp Cell Res* 231:363–371.
Ciruna, B.G., Schwartz, L., Harpal, K., Yamaguchi, T.P., and Rossant, J. (1997). Chimeric analysis of fibroblast growth factor receptor-1 (Fgfr1) function: a role for FGFR1 in morphogenetic movement through the primitive streak. *Development* 124:2829–2841.
DeChiara, T.M., Efstratiadis, A., and Robertson, E.J. (1990). A growth-deficiency phenotype in heterozygous mice carrying an insulin-like growth factor II gene disrupted by targeting. *Nature* 345:78–80.
Delisée, C., Schaeffer, P., Cazaubon, C., and Chatelain, P. (1993). Characterization of cardiac angiotensin AT_1 receptors by [^{3}H]SR47436. *Eur J Pharmacol* 247:139–144.
Doble, B.W., and Kardami, E. (1995). Basic fibroblast growth factor stimulates connexin-43 expression and intercellular communication of cardiac fibroblasts. *Mol Cell Biochem* 143:81–87.
Dono, R., Texido, G., Dussel, R., Ehmke, H., and Zeller R. (1998). Impaired cerebral cortex development and blood pressure regulation in FGF-2-deficient mice. *EMBO J* 17:4213–4225.
Dostal, D.E., Hunt, R.A., Kule, C.E., et al. (1997). Molecular mechanisms of angiotensin II in modulating cardiac function: intracardiac effects and signal transduction pathways. *J Mol Cell Cardiol* 29:2893–2902.
Dostal, D.E., Rothblum, K.N., Chernin M.I., Cooper, G.R., and Baker, K.M. (1992a). Intracardiac detection of angiotensinogen and renin: a localized renin-angiotensin system in neonatal rat heart. *Am J Physiol* 263:C838–C850.
Dostal, D.E., Rothblum, K.N., Conrad, K.M., Cooper, G.R., and Baker, K.M. (1992b). Detection of angiotensin I and II in cultured rat cardiac myocytes and fibroblasts. *Am J Physiol* 263:C851–863.
Drake, C.J., and Little, C.D. (1995). Exogenous vascular endothelial growth factor induces malformed and hyperfused vessels during embyronic neovascularization. *Proc Natl Acad Sci USA* 92:7657–7661.
Duttlinger, R., Manova, K., Berrozpe, G., et al. (1995). The W^{SH} and *Ph* mutations affect the c-*kit* expression profile: c-*kit* misexpression in embryogenesis impairs melanogenesis in W^{SH} and *Ph* mutant mice. *Proc Natl Acad Sci USA* 92:3754–3758.
Eggenschwiler, J., Ludwig, T., Fisher, P., Leighton, P.A., Tilghman, S.M., and Efstratiadis, A. (1997). Mouse mutant embryos overexpressing IGF-II exhibit phenotypic features of the Beckwith-Wiedemann and Simpson-Golabi-Behmel syndromes. *Genes Dev* 11:3128–3142.
Eisenberg, C.A., and Bader, D.M. (1996). Establishment of the mesodermal cell line QCE-6. A model system for cardiac cell differentiation. *Circ Res* 78:205–216.
Engelmann G.L., Boehm, K.D., Haskell, J.F., Khairallah, P.A., and Ilan, J. (1989). Insulin-like growth factors and neonatal cardiomyocyte development: ventricular gene expression and membrane receptor variations in normotensive and hypertensive rats. *Mol Cell Endocrinol* 63:1–14.
Everett, A.D., Fisher, A. Tufro-McReddie, A., and Harris, M. (1997). Developmental regulation of angiotensin type 1 and 2 receptor gene expression and heart growth. *J Mol Cell Cardiol* 29:141–148.
Feolde, E., Vigne, P., and Frelin, C. (1993). Angiotensin II receptor subtypes and biological responses in the rat heart. *J Mol Cell Cardiol* 25:1359–1367.
Ferrara, N., Carvermoore, K., Chen, H., et al. (1996). Heterozygous embryonic lethality induced by targeted inactivation of the VEGF gene. *Nature (Lond)* 380:439–442.

Feucht, M., Christ, B., and Wilting, J. (1997). VEGF induces cardiovascular malformation and embryonic lethality. *Am J Pathol* 151:1407–1416.

Fiore, F., Planche, J., Gibier, P., Sebille, A., deLapeyriere, O., and Birnbaum, D. (1997). Apparent normal phenotype of Fgf6–/– mice. *Int J Dev Biol* 41:639–642.

Foncea, R., Andersson, M., Ketterman, A., et al. (1997). Insulin-like growth factor-I rapidly activates multiple signal transduction pathways in cultured rat cardiac myocytes. *J Biol Chem* 272:19115–19124.

Fong, G.H., Rossant, J., Gertsenstein, M., and Breitmen, M.I. (1995). Role of the Flt-1 receptor tyrosine kinase in regulating the assembly of vascular endothelium. *Nature (Lond)* 376:66–70.

Fretto, L.J., Snape, A.J., Tomlinson, J.E., et al. (1993). Mechanism of platelet-derived growth factor (PDGF) AA, AB, and BB binding to α and β PDGF receptors. *J Biol Chem* 268:3625–3631.

Fuller, S.J., Mynett, J.R., and Sugden, P.H. (1992). Stimulation of cardiac protein synthesis by insulin-like growth factors. *Biochem J* 282:85–90.

Galzie, Z., Kinsella, A., and Smith, J.A. (1997). Fibroblast growth factors and their receptors. *Biochem Cell Biol* 75:669–685.

Grady, E.F., Sechi, L.A., Grifin. C.A., Schambelan, M., and Kalinyak, J.E. (1991). Expression of AT_2 receptors in the developing rat fetus. *J Clin Invest* 88:921–933.

Guo, W., Kada, K., Kamiya, K., and Toyama, J. (1997). IGF-I regulates K^+-channel expression of cultured neonatal rat ventricular myocytes. *Am J Physiol (Heart Circ Physiol* 41):H2599–H2606.

Guo, W., Kamiya, K., and Toyama, J. (1995). bFGF promotes functional expression of transient outward currents in cultured neonatal rat ventricular cells. *Pflugers Arch* 430:1015–1017.

Guo, W., Kamiya, K., and Toyama, J. (1996). Modulated expression of transient outward current in cultured neonatal rat ventricular myocytes: comparison with development in situ. *Cardiovasc Res* 32:524–533.

Harada, K., Komuro, I., Shiojima, I., et al. (1998a). Pressure overload induces hypertrophy in angiotensin II type 1A receptor knockout mice. *Circulation* 97:1952–1959.

Harada, K., Komuro, I., Zou, Y., et al. (1998b). Acute pressure overload could induce hypertrophic responses in the heart of angiotensin II type 1a knockout mice. *Circ Res* 82:779–785.

Harder, B.A., Hefti, M.A., Eppenberger, H.M., and Schaub, M.C. (1998). Differential localization in sarcomeric and nonsarcomeric contractile structures of cultured cardiomyocytes. *J Struct Biol* 122:162–175.

Harder, B.A., Schaub, M.C., Eppenberger, H.M., and Eppenberger-Eberhardt, M. (1996). Influence of fibroblast growth factor (bFGF) and insulin-like growth factor (IGF-I) on cytoskeletal and contractile structures and on atrial natriuretic factor (ANF) expression in adult rat ventricular cardiomyocytes in culture. *J Mol Cell Cardiol* 28:19–31.

Haskell, J.F., and Tucker, D.C. (1994). Binding of insulin-like growth factors (IGF-I and IGF-II) to the IGF-II/mannose 6-phosphate receptor in fetal rat myocardium. *Endocrinology* 135:231–239.

Hefti, M.A., Harder, B.A., Eppenberger, H.M., and Schaub, M.C. (1997). Signaling pathways in cardiac myocyte hypertrophy. *J Mol Cell Cardiol* 29:2873–2892.

Hein, L., Stevens, M.E., Barsh, G.S., Pratt, R.E., Kobilka, B.K., and Dzau, V. (1997). Overexpression of angiotensin AT1 receptor transgene in the mouse myocardium produces a lethal phenotype associated with myocyte hyperplasia and heart block. *Proc Natl Acad Sci USA* 94:6391–6396.

Heldin, C.H., and Westmark, B. (1989). Platelet-derived growth factor: three isoforms and two receptor subtypes. *Trends Genet* 5:108–111.

Hudson, E.K., Wang, D., Bieber, L.L., Buja, L.M., and McMillin, J.B. (1996). Increased carnitine palmitoyltransferase in cardiac myocytes is mediated by insulin growth factor I. *Am J Physiol (Heart Circ Physiol* 40):H422–H427.

Hunt, R.A., Ciuffo, G.M., Saavedra, J.M., and Tucker, D.C. (1995). Quantification and localisation of angiotensin II receptors and angiotensin converting enzyme in the developing rat heart. *Cardivasc Res* 29:834–840.

Hyer, J., and Mikawa, T. (1997). Retroviral techniques for studying organogenesis with a focus on heart development. *Mol Cell Biochem* 172:23–35.

Ichiki, T., Labosky, P.A., Shiota, C., et al. (1995). Effects on blood pressure and exploratory behaviour of mice lacking angiotensin II type-2 receptor. *Nature* 377:748–750.

Ito, H., Hiroe, M., Hirata, Y., et al. (1993). Insulinlike growth factor-I induces hypertrophy with enhanced expression of muscle specific genes in cultured rat cardiomyocytes. *Circulation* 87:1715–1721.

Ito, M., Oliverio, M.I., Mannon, P.J., et al. (1995). Regulation of blood pressure by the type 1A angiotensin II receptor gene. *Proc Natl Acad Sci USA* 92:3521–3525.

Jones, C., Millan, M.A., Naftolin, F., and Aguilera, G. (1989). Characterization of angiotensin II receptors in the rat fetus. *Peptides* 10:459–463.

Kajstura, J., Cheng, W., Reiss, K., and Anversa, P. (1994). The IGF-1 receptor system modulates myocyte proliferation but not myocyte cellular hypertrophy *in vitro*. *Exp Cell Res* 215:273–283.

Kajstura, J., Mansukhani, M., Cheng, W., et al. (1995). Programmed cell death and expression of the protooncogene bcl-2 in myocytes during postnatal maturation of the heart. *Exp Cell Res* 219:110–121.

Kim, H.S., Krege, J.H., Kluckman, K.D., et al. (1995). Genetic control of blood pressure and the angiotensinogen locus. *Proc Natl Acad Sci USA* 92:2735–2739.

Kim, N.N., Villarreal, F.J., Printz, M.P., Lee, A.A., and Dillman, W.H. (1995). Trophic effects of angiotensin II on neonatal rat cardiac myocytes are mediated by cardiac fibroblasts. *Am J Physiol* 269(*Endocirinol Metab* 32):E426–E437.

Klint, P., and Claesson-Welsh, L. (1999). Signal transduction by fibroblast growth factor receptors. *Front Biosci* 4:165–177.

Kodama, H., Fukuda, K., Pan, J., et al. (1998). Biphasic activation of the JAK/STAT pathway by angiotensin II in rat cardiomyocytes. *Circ Res* 82:244–250.

Kohout, T.A., and Rogers, T.B. (1995). Angiotensin II activates the Na^+/HCO^-_3 symport through a phosphoinositide-independent mechanism in cardiac cells. *J Biol Chem* 270:20432–20438.

Kojima, M., Shiojima, I., Yamazaki, T., et al. (1994). Angiotensin II receptor antagonist TCV-116 induces regresson of hypertensive left ventricular hypertrophy in vivo and inhibits the intracellular signaling pathway of stretch-mediated cardiomyocyte hypertrophy in vitro. *Circulation* 89:2204–2211.

Kondo, T., Konishi, F., Inui, H., and Inagami T. (1993). Differing signal transductions elicited by three isoforms of platelet-derived growth factor in vascular smooth muscle cells. *J Biol Chem* 268:4458–4464.

Krege, J.H., John, S.W.M., Langenback, L.L., et al. (1995). Male-female differences in fertility and blood pressure in ACE-deficient mice. *Nature* 375:146–148.

Kudoh, S., Komuro, I., Mizuno, T., et al. (1997). Angiotensin II stimulates c-jun NH_2-terminal kinase in cultured cardiac myocytes of neonatal rats. *Circ Res* 80:139–146.

Kumar, C.C. (1998). Signaling by integrin receptors. *Oncogene* 17:1365–1373.

Lau, M.M.H., Stewart, C.E.H., Liu, Z., Bhatt, H., Rotwein, P., and Stewart, C.L. (1994). Loss of the imprinted IGF2/cation-independent mannose 6-phosphate receptor results in fetal overgrowth and perinatal lethality. *Genes Dev* 8:2953–2963.

Lavandero, S., Foncea, R., Pérez, V., and Sapag-Hagar, M. (1998). Effect of inhibitors of signal transduction on IGF-I-induced protein synthesis associated with hypertrophy in cultured neonatal rat ventricular myocytes. *FEBS Lett* 422:193–196.

Lee, A.A., Dillmann, W.H., McCulloch, A.D., and Villarreal, F.J. (1995). Angiotensin II stimulates the autocrine production of transforming growth factor-β1 in adult rat cardiac fibroblasts. *J Mol Cell Cardiol* 27:2347–2357.

Levéen, P., Pekny, M., Gebre-Medhin, S., Swolin, B., Larsson, E., and Betsholtz, E. (1994). Mice deficient for PDGF B show renal, cardiovascular, and hematological disorders. *Genes Dev* 8:1875–1887.

Li, J., Hampton, T., Morgan, J.P., and Simons, M. (1997). Stretch-induced VEGF expression in the heart. *J Clin Invest* 100:18–24.

Linseman, D.A., Benjamin, C.W., and Jones, D.A. (1995). Convergence of angiotensin II and platelet-derived growth factor receptor signaling cascades in vascular smooth muscle cells. *J Biol Chem* 270:12563–12568.

Liu, Q., Yan, H., Dawes, N.J., Mottino, G.A., Frank, J.S., and Zhu, H. (1996). Insulin-like growth factor II induces DNA synthesis in fetal ventricular myocytes in vitro. *Circ Res* 79:716–726.

Lok, F., Owens, J.A., Mundy, L., Robinson, J.S., and Owens, P.C. (1996). Insulin-like growth factor I promotes growth selectively in fetal sheep in late gestation. *Am J Physiol* 270(*Regulatory Integrative Comp Physiol* 39):R1148–R1155.

Lokuta, A.J., Cooper, C., Gaa, S.T., Wang, H.E., and Rogers, T.B. (1994). Angiotensin II stimulates the release of phospholipid-derived second messengers through multiple receptor subtypes in heart cells. *J Biol Chem* 269:4832–4838.

Louvi, A., Accili, D., and Efstratiadis, A. (1997). Growth-promoting interaction of IGF-II with the insulin receptor during mouse embryonic development. *Dev Biol* 189:33–48.

Ludwig, T., Eggenschwiler, J., Fisher, P., D'Ercole, A.J., Davenport, M.L., and Efstratiadis, A. (1996). Mouse mutants lacking the type 2 IGF receptor (IGF2R) are rescued from perinatal lethality in *IGF2* and *IGF1r* null backrounds. *Dev Biol* 177:517–535.

Manasek, F.J. (1981). Determinants of heart shape in early embryos. *FASEB J* 40:2011–2016.

Manasek, F.J., Isobe, Y. Shimada, Y., and Hopkins, W. (1984). The embryonic myocardial cytoskeleton, interstitial pressure, and the control of morphogenesis. In: Nora, J.J., and Takao, A., eds. *Congenital Heart Disease: Causes and Processes.* Futura, New York, pp. 359–376.

Masaki, H., Kurihara, T., Yamaki, A., et al. (1998). Cardiac-specific overexpression of angiotensin II AT2 receptor causes attenuated response to AT1 receptor-mediated pressor and chronotropic effects. *J Clin Invest* 101:527–535.

Matsubara, H., Kanasaki, M., Mursawa, S., Tsukaguchi, Y., Nio, Y., and Inada, M. (1994). Differential gene expression and regulation of angiotensin II receptor subtypes in rat cardiac fibroblasts and cardiomyocytes in culture. *J Clin Invest* 93:1592–1601.

Matzner, U., Von Figura, K., and Pohlmann, R. (1992). Expression of the two mannose 6-phosphate receptors is spatially and temporally different during mouse embryogenesis. *Development* 114:965–972.

McKay, I.J., Lewis, J., and Lumsden, A. (1996). The role of FGF-3 in early inner ear development: an analysis in normal and Kreisler mutant mice. *Dev Biol* 174:370–378.

Merle, P.L., Usson, Y., Robert-Nicoud, M., and Verdetti, J. (1997). Basic FGF enhances calcium permeable channel openings in adult rat cardiac myocytes: implication in the bFGF-induced increase of free Ca^{2+} content. *J Mol Cell Cardiol* 29:2687–2698.

Michelson, A.M., Gisselbrecht, S., Zhou, Y., Baek, K.H., and Buff, E.M. (1998). Dual functions of the heartless fibroblast growth factor receptor in development of the *Drosophila* embryonic mesoderm. *Dev Genet* 22:212–229.

Miyata, S., and Haneda, T. (1994). Hypertrophic growth of cultured neonatal rat heart cells mediated by type 1 angiotenisn II receptor. *Am J Physiol* 266(*Heart Circ Physiol*):H2443–H2451.

Morrison-Graham, K., Schatteman, G.C., Bork, T., Bowen-Pope, D.F., and Weston, J.A. (1992). PDGF receptor mutation in the mouse (*Patch*) perturbs the development of a non-neuronal subset of neural crest-derived cells. *Development* 115:133–142.

Nyui, N., Tamura, K., Mizuno, K., et al. (1997). Stretch-induced map kinase activation in cardiomyocytes of angiotensinogen-deficient mice. *Biochem Biophys Res Commun* 235:35–41.

Orchard, C.H., and Kentish, J.C. (1990). Effects of changes in pH on the contractile function of cardiac muscle. *Am J Physiol* 258:C967–C981.

Orr-Urtreger, A., Bedford, M.T., Do, M.-S., Eisenbach, L., and Lonai, P. (1992). Developmental expression of the α receptor for platelet-derived growth factor, which is deleted in the embryonic lethal *Patch* mutation. *Development* 115:289–303.

Orr-Urtreger, A., and Lonai, P. (1992). Platelet-derived growth factor-A and its receptor are expressed in separate, but adjacent cell layers of the mouse embryo. *Development* 115:1045–1058.

Pasumarthi, K.B., Jin, Y., Bock, M.E., Lytras, A., Kardami, E., and Cattini, P.A. (1995). Characterization of fibroblast growth factor receptor 1 RNA expression in the embryonic mouse heart. *Ann NY Acad Sci* 752:406–416.

Phillips, A.F., Persson, B., Hall, K., Skottner, A., Sanegen, T., and Sara, U.R. (1988). The effects of biosynthetic insulin-like growth factor-I supplementation on somatic growth, maturation, and erythropoiesis in the neonatal rat. *Pediatr Res* 23:298–305.

Price, R.L., Carver, W., Simpson, D.G., et al. (1997). The effects of angiotensin II and specific receptor blockers on embryonic cardiac development and looping patterns. *Dev Biol* 192:572–584.

Price, R.L., Chintanowonges, C., Shiraishi, I., Borg, T.K., and Terracio, L. (1996). Local and regional variations in myofibrillar patterns in looping rat hearts. *Anat Rec* 245:83–93.

Price, R.L., Thielen, T.E., Borg, T.K., and Terracio, L. (1999). Microscopy as a tool in understanding the role of growth factors in cardiac development. *Microsc Microanal.*

Potts, J.D., and Carrington, J.L. (1993). Selective expression of the chicken platelet-derived growth factor α (PDGFα) receptor during limb bud development. *Deve Dyn* 198:14–21.

Reiss, K., Cheng, W., Ferber, A., et al. (1996). Overexpression of insulin-like growth factor-1 in the heart is coupled with myocyte proliferation in transgenic mice. *Proc Natl Acad Sci USA* 93:8630–8635.

Reiss, K., Cheng, W., Kajstura, J., Sonnenblick, E.H., Meggs, L.G., and Anversa, P. (1995). Fibroblast proliferation during myocardial development in rats is regulated by IGF-1 receptors. *Am J Physiol* 269(*Heart Circ Physiol* 38):H943–H951.

Rogers, T.B., Gaa, S.T., and Allen, I.S. (1986). Identification and characterization of functional angiotensin II receptors on cultured heart myocytes. *J Pharmacol Exp Ther* 236:438–444.

Ruoslahti, E. (1991). Integrins. *J Clin Invest* 87:1.

Sadoshima, J., and Izumo, S. (1993). Molecular characterization of angiotensin II-induced hypertrophy of cardiac myocytes and hyperplasia of cardiac fibroblasts. Critical role of the AT_1 receptor subtype. *Circ Res* 73:413–423.

Sadoshima, J., and Izumo, S. (1995). Rapamycin selectively inhibits angiotensin II-induced increase in protein synthesis in cardiac myocytes in vitro. Potential role of 70-kD S6 kinase in angiotensin II-induced cardiac hypertrophy. *Circ Res* 77:1040–1052.

Sadoshima, J., Jahn, L., Takahashi, T., Kulik, T.J., and Izumo, S. (1992). Molecular characterization of the stretch-induced adaptation of cultured cardiac cells. *J Biol Chem* 267:10551–10560.

Sadoshima, J., Qui, Z., Morgan, J.P., and Izumo, S. (1995). Angiotensin II and other hypertrophic stimuli mediated by G protein-coupled receptors activate tyrosine kinase, mitogen-activated protein kinase, and 90-kD S6 kinase in cardiac myocytes. The critical role of Ca^{2+}-dependent signaling. *Circ Res* 76:1–15.

Sadoshima, J., Xu, Y., Slayter, H.S., and Izumo, S. (1993). Autocrine release of angiotensin II mediates stretch-induced hypertrophy of cardiac myocytes in vitro. *Cell* 75:977–984.

Sastry, S.K., and Horwitz, A.F. (1996). Adhesion-growth factor interactions during differentiation: an integrated biological response. *Dev Biol* 180:455–467.

Schatteman, G.C., Loushin, C., Li, T., and Hart, C.E. (1996). PDGF-A is required for normal murine cardiovascular development. *Dev Biol* 176:133–142.

Schatteman, G.C., Morrison-Graham, K., Van Koppen, A., Weston, J.A., and Bowen-Pope, D.F. (1992). Regulation and role of PDGF receptor α-subunit expression during embryogenesis. *Development* 115:123–131.

Schatteman, G.C., Motley, S.T., Effmann, E.L., and Bowen-Pope, D.F. (1995). Platelet-derived growth factor receptor alpha subunit deleted *Patch* mouse exhibits severe cardiovascular dysmorphogenesis. Teratology 51:351–366.

Schaub, M.C., Hefti, M.A., Harder, B.A., and Eppenberger, H.M. (1997). Various hypertrophic stimuli induce distinct phenotypes in cardiomyocytes. *J Mol Med* 75:901–920.

Schorb, W., Booz, G.W., Dostal, D.E., Conrad, K.M., Chang, K.C., and Baker, K.M. (1993). Angiotensin II is mitogenic in neonatal rat cardiac fibroblasts. *Circ Res* 72:1245–1254.

Schütz, S., Le Moullec, J.M., Corvol, P., and Gasc, J.M. (1996). Early expression of all the components of the renin-angiotensin-system in human development. *Am J Pathol* 149:2067–2079.

Sechi, L.A., Griffin, C.A., Grady, E.F., Kalinyak, J.E., and Schambelan, M. (1992). Characterization of angiotensin II receptor subtypes in rat heart. *Circ Res* 71:1482–1489.

Seifert, R.A., Van Koppen, A., and Bowen-Pope, D.F. (1993). PDGF-AB requires PDGF receptor α-subunits for high-affinity, but not for low-affinity, binding and signal transduction. *J Biol Chem* 268:4473–4480.

Sekine, K., Ohuchi, H., Fujiwara, M., et al. (1999). Fgf10 is essential for limb and lung formation. *Nat Genet* 21:138–141.

Seko, Y., Takahashi, N., Tobe, K., Ueki, K., Kadowki, T., and Yazaki, Y. (1998). Vascular endothelial growth factor (VEGF) activates raf-1, mitogen-activated protein (MAP) kinases, and S6 kinase (p90rsk) in cultured rat cardiac myocytes. *J Cell Physiol* 75:239–246.

Senior, P.V., Byrne, S., Brammar, W.J., and Beck, F. (1990). Expression of the IGF-II/mannose-6-phosphate receptor mRNA and protein in the developing rat. *Development* 109:67–73.

Shalaby, F., Rossant, J., Yamaguchi, T.P., et al. (1995). Failure of blood island formation and vasculogenesis in Flk-1-deficient mice. *Nature (Lond)* 376:62–66.

Shanmugam, S., Corvol, P., and Gasc, J.-M. (1994a). Ontogeny of the two angiotensin II type 1 receptor subtypes in rats. *Am J Physiol* (*Endocrinol Metab* 30):E828–E836.

Shanmugam, S., Corvol, P., and Gasc, J.-M. (1996). Angiotensin II type 2 receptor mRNA expression in the developing cardiopulmonary system of the rat. *Hypertension* 28:91–97.

Shanmugam, S., Monnot, C., Corvol, P., and Gasc, J.-M. (1994b). Distribution of type 1 angiotensin II receptor subtype messenger RNAs in the rat fetus. Hypertension 23:137–141.

Sharma, H.S., van Heugten, H.A.A., Goedbloed, M.A., Verdouw, P.D., and Lamers, J.M.J. (1994). Angiotensin II induced expression of transcription factors precedes increase in transforming growth factor-β1 mRNA in neonatal cardiac fibroblasts. *Biochem Biophys Res Commun* 205:105–112.

Sheikh, F., Jin, Y., Pasumarthi, K.B., Kardami, E., and Cattini, P.A. (1997). Expression of fibroblast growth factor receptor-1 in rat heart H9c2 myoblasts increases cell proliferation. *Mol Cell Biochem* 176:89–97.

Shifren, J.L., Doldi, N., Ferrara, N., Mesiano, S., and Jaffe, R.B. (1994). In the human fetus, vascular endothelial growth factor is expressed in epithelial cells and myocytes, but not vascular endothelium: implications for mode of action. *J Clin Endocrinol Metab* 79:316–322.

Shinbrot, E., Peters, K.G., and Williams, L.T. (1994). Expression of the platelet-derived growth factor β receptor during organogenesis and tissue differentiation in the mouse embryo. *Dev Dyn* 199:169–175.

Sigel, A.V., Centrella, M., and Eghbali-Webb, M. (1996). Regulation of proliferative response of cardiac fibroblasts by transforming growth factor-beta 1. *J Mol Cell Cardiol* 28:1921–1929.

Sklar, M.M., Kiess, W., Thomas, C.L., and Nissley, S.P. (1989). Developmental expression of the insulin-like growth factor II/mannose 6-phosphate receptor in the rat. *J Biol Chem* 264:16733–16738.

Sklar, M.M., Thomas, C.L., Municchi, G., et al. (1992). Developmental expression of rat insulin-like growth factor-II/mannose 6-phosphate receptor messenger ribonucleic acid. *Endocrinology* 130:3483–3491.

Slavkin, H.C. (1995). Antisense oligonucleotides—an experimental strategy to advance a causal-analysis of development. *Int J Deve Biol* 39:123–126.

Soriano, P. (1994). Abnormal kidney development and hematological disorders in PDGF β-receptor mutant mice. *Genes Dev* 8:1888–1896.

Soriano, P. (1997). The PDGFα receptor is required for neural crest cell development and for normal patterning of the somites. *Development* 124:2691–2700.

Stephenson, D.A., Mercola, M., Anderson, E., et al. (1991). Platelet-derived growth factor receptor α-subunit gene (*Pdgfra*) is deleted in the mouse patch (*Ph*) mutation. *Proc Natl Acad Sci USA* 88:6–10.

Stylianopoulou, F., Efstratiadis, A., Herbert, J., and Pintar, J. (1988). Pattern of the insulin-like growth factor II gene expression during rat embryogenesis. *Development* 103:497–506.

Sugi, Y., and Lough, J. (1995). Activin-A and FGF-2 mimic the inductive effects of anterior endoderm on terminal cardiac myogenesis in vitro. *Dev Biol* 168:567–574.

Sugi, Y., Sasse, J., Barron, M., and Lough, J. (1995). Developmental expression of fibroblast growth factor receptor-1 (cek-1; flg) during heart development. *Dev Dyn* 202:115–125.

Taber, L.A., En Lin, I., and Clark, E.B. (1995). Mechanics of cardiac looping. *Dev Dyn* 203:42–50.

Takakura, N., Yoshida, H., Ogura, Y., Kataoka, H. Nishikawa, S., and Nishikawa, S.-I. (1998). PDGFRα expression during mouse embryogenesis: immunolocalization analyzed by whole-mount immunohistostaining using the monoclonal anti-mouse PDGFRα antibody APA5. *J Histochem Cytochem* 45:883–893.

Tapon, N., and Hall, A. (1997). Rho, Rac and Cdc42 GTPases regulate the organization of the actin cytoskeleton. *Curr Opin Cell Biol* 9:86–92.

Thielen, T., Carver, W., Simpson, D.G., Borg, T.K., Terracio, L., and Price, R.L. (1998). Temporal and spatial expression patterns of PDGF receptors in embryonic rat heart as detected through confocal scanning laser microscopy. *Microsc Microanal* 4(suppl 2):1166–1167.

Thielen, T.E., Terracio, L., Borg, T.K., and Price, R.L. (1999). In vivo characterization of cardiac tissue defects associated with PDGFRα in the *patch* mouse. *Microsc Microanal.*

Tomanek, R.J., Haung, L., Suvarna, P.R., O'Brien, L.C. Tatajska, A., and Sandra, A. (1996). Coronary vascularization during development in the rat and its relationship to basic fibroblast growth factor. *Cardiovasc Res* 31:E116–126.

Tomanek, R.J., Lotun, K., Clark, E.B., Suvarna, P.R., and Hu, N. (1998). VEGF and bFGF stimulate myocardial vascularization in embryonic chick. *Am J Physiol* 274(5 pt 2):J1620–H1626.

Tomanek, R.J., Ratajska, A., Kitten, G.T., Yue, X., and Sandra, A. (1999). Vascular endothelial growth factor expression coincides with coronary vasculogenesis and angiogenesis. *Dev Dyn* 215:54–61.

Van Bilsen, M. (1997). Signal transduction revisited: recent developments in angiotensin II signaling in the cardiovascular system. *Cardiovasc Res* 36:310–322.

Velez, C., Aranega, E., Melguizo, C., Fernandez, J.E., Prados, J., and Aranega, A. (1994). Modulation of contractile protein troponin-T in chick myocardial cells by basic fibroblast growth factor and platelet-derived growth factor during development. *J Cardiovasc Pharmacol* 24:906–913.

Velez, C., Aranega, E., Prados, J.C., Melguizo, C., Alvarez, L., and Aranega, A. (1995). Basic fibroblast and platelet-derived growth factors as modulators of actin and α-actinin in chick myocardiocytes during development. *Proc Soc Exp Biol Med* 210:57–63.

Wang, L., Ma, W., Markovich, R., Chen, J., and Wang, P.H. (1998). Regulation of cardiomyocyte apoptotic signaling by insulin-like growth factor 1. *Circ Res* 83:516–522.

Wang, Z.-Q., Fung, M., Barlow, D.P., and Wagner, E.F. (1994). Regulation of embryonic growth and lysosomal targeting by the imprinted *IGF2/Mpr* gene. *Nature* 372:464–467.

Watkins, B.P., Bolender, D.L., Lough, J., and Kolesari, G.L. (1998). Teratogenic effects of implanting fibroblast growth factor-2 soaked beads in the cardiac region of the stage 24 chick embryo. *Teratology* 57:140–145.

Watson, S., Burnside, T., and Carver, W. (1998). Angitoensin II-stimulated collagen gel contraction by heart fibroblasts: role of the AT1 receptor and tyrosine kinase activity. *J Cell Physiol* 177:224–231.

Yamazaki, T., Komuro, I., Kudoh, S., et al. (1995). Angiotensin II partly mediates mechanical stress-induced cardiac hypertrophy. *Circ Res* 77:258–265.

Yamazaki, T., Tobe, K., Hoh, E., et al. (1993). Mechanical loading activates mitogen-activated protein kinase and S6 peptide kinase in cultured rat cardiac myocytes. *J Biol Chem* 268:12069–12076.

Zemel, S., Millan, M.A., and Aguilera, G. (1989). Distribution of angiotensin receptors and renin in the mouse fetus. *Endocrinology* 124:1774–1780.

Zhu, X., and Lough, J. (1996). Expression of alternatively spliced and canonical basic fibroblast growth factor mRNAs in early embryo and developing heart. *Dev Dyn* 206:139–145.

Zhu, X., Sasse, J., McAllister, D., and Lough, J. (1996). Evidence that fibroblast growth factors 1 and 4 participate in regulation of cardiogenesis. *Dev Dyn* 207:429–438.

Zou, Y., Komuro, I., Yamazaki, T., et al. (1998). Cell type-specific angiotensin II-evoked signal transduction pathways. Critical roles of $G_{\beta\gamma}$ subunit, Src family, and Ras in cardiac fibroblasts. *Circ Res* 82:337–345.

CHAPTER **11**

Transforming Growth Factor-β Signal Transduction in the Atrioventricular Canal During Heart Development

Angelique S. Boyer and Raymond B. Runyan

During embryonic development of the heart, one critical component of the process is the formation of the mesenchymal cells of the cardiac cushions. Cells within the cushions arise by an epithelial-mesenchymal cell transformation of the overlying endothelium and provide the progenitors of the valves and septum intermedium. In the last several years, significant progress has been made toward understanding this process at the cellular and molecular level. Particularly, the function and localization of transforming growth factor-β (TGF-β) and its receptors during epithelial-mesenchymal cell transformation have been studied. Several TGF-β isoforms are expressed throughout the heart (Akhurst et al, 1990; Potts et al, 1992; Barnett et al, 1994; Boyer et al, 1999b) and have been implicated in activities such as angiogenisis (Battegay, 1995) and myocyte hypertrophy (Schneider and Parker, 1990). However, the only cardiac-specific defects produced by loss of the various TGF-βs in null mice were valvular and septal defects seen in TGF-β_2 knockout animals (Sanford et al, 1997). Together, mouse and chicken studies point to critical roles for members of this growth factor family in the normal formation of cardiac valves.

This chapter reviews research on the role of TGF-β and TGF-β receptors during epithelial-mesenchymal cell transformation and signal transduction pathways that mediate this transformation. The distribution of TGF-β isoforms during epithelial-mesenchymal cell transformation is discussed in both mouse and chicken species. The function of TGF-β isoforms and TGF-β receptors is discussed in regard to functional data obtained largely in the chicken. Based on this evidence, we propose a model of TGF-β signal transduction during epithelial-mesenchymal cell transformation, in which two TGF-β isoforms (β_2 and β_3) play separate and sequential roles in the continuum of tissue interaction, endothelial activation, endothelial to mesenchymal cell transformation, and finally mesenchymal cell migration. The two TGF-β receptors (types II and III) studied to date also play distinct roles in signal propagation from TGF-βs. The TGF-β type II receptor propagates a signal from TGF-β_3 and the type III receptor propagates a signal from TGF-β_2.

TGF-β AND TGF-β RECEPTORS

Transforming Growth Factor Beta

Transforming growth factor-β (TGF-β) is a low molecular weight protein of approximately 25,000 daltons composed of two identical subunits. Since the original isolation of TGF-β_1 from human platelets, it was found to be the prototype of a gene family characterized by a conserved pattern of cysteines (Massagué, 1990). The TGF-β gene family includes five numbered members (TGF-β_{1-5}) that share at least 70% sequence identity at the nucleic acid level. No species has more than three numbered TGF-β isoforms. Chicken TGF-β_4 is the avian homologue of mammalian TGF-β_1. The published chicken TGF-β_1 sequence has a homology with mammalian isoforms that suggests a misidentification of this sequence (Burt, 1992). There are only two identified TGF-β isoforms (TGF-β_2 and TGF-β_5) in frogs.

During the past 15 years, many proteins with lesser but significant homology with TGF-β have been identified in species ranging from *Caenorhabditis elegans* and *Drosophila* to humans. These related molecules are identified as members of the TGF-β superfamily. This superfamily consists of a large number of related growth factors divided into several subfamilies that include TGF-βs, activins, bone morphogenetic proteins (BMPs), Vg1-related proteins, Growth/differentration factors (GDFs), and numerous other proteins (Massagué, 1998). TGF-β was originally named for its ability to enable cells to form colonies in soft agar. Additional functions of TGF-β molecules include growth stimulation, growth inhibition, cell differentiation, stimulation of mobility, and the stimulation of extracellular matrix (ECM) and ECM receptor production (Massagué, 1998). TGF-β is expressed in many cell types, and has been reported to both stimulate and inhibit cell growth depending on physiologic conditions and cell type. In general, TGF-β is classified as an inhibitor of cell growth that interacts with specific TGF-β receptors on cell surfaces and produces an effect on cell cycle components (Moses et al, 1990; Moses, 1992; Alexandrow and Moses, 1995).

TGF-β is secreted in a latent (dimer) complex that contains the mature, functional TGF-β domain linked with a latency associated protein (LAP). This latent form of TGF-β is activated by proteases such as plasmin (Lyons et al, 1988; Antonelli-Olridge et al, 1989). Mature (active) TGF-β remains associated with LAP via noncovalent interactions. Two LAP molecules within the latent complex are coupled by a disulfide bond (Miyazono et al, 1988). The TGF-β–LAP complex, in turn, binds to ECM (Taipale et al, 1994). Plasminogen activators (PAs) are the major mediators of proteolysis. At least two types of PAs exist in tissues: the urokinase- (u-PA) and tissue-type plasmingen activators (t-PA). u-PA has been implicated in the role of epithelial-mesenchymal transformation in the heart (McGuire and Alexander, 1992, 1993a,b).

The role of the TGF-β family of molecules in embryogenesis is particularly interesting. In the *Xenopus* embryo, several members of the TGF-β family, such as TGF-β_2, -β_3, and activin, have been demonstrated to induce mesoderm formation (Rosa et al, 1988). A member of the TGF-β superfamily, Vg1, was found in the maternal messenger RNA (mRNA) of the early embryo (Weeks and Melton, 1987). Another TGF-β–related molecule, müllerian inhibiting substance, promotes the regression of the ducts during embryogenesis (Trelstad et al, 1982; Tsuji et al,

1992). Homozygous null mutations of a TGF-β gene family member in *Drosophila* (decapentaplegic, Dpp) cause early embryonic death by producing completely ventralized embryos (Panganiban et al, 1990; Sekelsky et al, 1995). Studies in heart development demonstrate that cardiac atrioventricular (AV) canal endothelia respond to a combination of TGF-β and another, unknown ECM component by transforming into mesenchyme (Runyan et al, 1992). Together, these data demonstrate a widespread and important role for members of the TGF-β family in critical processes of embryonic development.

TGF-β Receptors

Although other TGF-β receptors (TβRs) were identified, the major focus of research has been on three that are widespread, TβRs I, II, and III (Massagué et al, 1992). TβRI and TβRII are glycoproteins of approximately 55 kd and 70 kd, respectively. The extracellular domains of TβRI and TβRII are both relatively short. The larger cytoplasmic sequence for both receptors contains serine/threonine kinase domain. Based on growth inhibition studies, TβRII binds the TGF-β molecule and forms a heterodimer with TβRI (Wrana et al, 1992). The TβRII kinase then phophorylates the bound TβRI and then TβRI initiates signal transduction within the cell (Wrana et al, 1994). TGF-β_1 and -β_3 bind to TβRII with much greater affinity than does TGF-β_2.

TGF-β receptor III is a large proteoglycan also known as betaglycan (Wang et al, 1991). β-Glycan is composed of a 120 kd core protein and glycosaminoglycan (GAG) side chains that contain heparan sulfate and chondronitin sulfate (Segarini and Seyedin, 1988). It has a very small cytoplasmic domain that does not appear to have any enzymatic activity. TβRIII has a 10-fold higher affinity for TGF-β_1 and TGF-β_3 than TβRII. TβRIII is proposed to bind TGF-β_2 and present it to the TβRII receptor since it has a greater affinity for TGF-β_2 than TβRII (Lopez-Casillas et al, 1993). Although the prevailing dogma is that TβRIII functions solely through its presentation of TGF-β to the TβRII receptor, recent evidence suggests that there may be other mechanisms of signal transduction, such as the interaction of TβRIII with TβRI (Taniguchi et al, 1998).

Another type III receptor-like protein is endoglin. Endoglin is a cell surface glycoprotein that forms a homodimer of 180 kd and shares significant cytoplasmic domain homology with betaglycan (TβRIII receptor) (Gougos and Letarte, 1988a,b). Endoglin, in contrast to TβRIII, has a high affinity for TGF-β_1 and -β_3 but low affinity for TGF-β_2 (Cheifetz et al, 1992). A proposed role for endoglin is to present TGF-β_3 to the type II receptor, although recent research suggests that endoglin is an accessory protein during the binding of TGF-β to the type II receptor rather than presenting TGF-β to the type II receptor (Barbara et al, 1999).

Smad Proteins

Although there is evidence for additional signal transduction pathways, recent progress in TGF-β signaling has been largely on a family of molecules known as the Smad proteins. Smad proteins are a substrate for type I receptor kinase. The first member of Smad family identified was a *Drosophila* signaling molecule known as Mad (Mothers against Dpp) (Sekelsky et al, 1995). Three molecules were identified in *C. elegans* (from mutations that produced a small phenotype) with a similarity to Mad and were named Sma2, 3, and 4 (Savage et al, 1996). From Sma and Mad came the family name Smad. Smads 1 to 8 were subsequently cloned from human, mouse, and frog (Chen et al, 1996; Eppert et al, 1996; Graff et al, 1996; Hoodless et al, 1996; Lechleider et al, 1996; Liu et al, 1996; Yingling et al, 1996; Imamura et al, 1997; Nakao et al, 1997a; Watanabe et al, 1997; Labbe et al, 1998; Zhang et al, 1998). The sequence for several additional unpublished Smads can be found in GenBank.

The first evidence that Smad is downstream of a TGF-β receptor came from the inhibition of TGFβ signaling by Mad mutations (Hoodless et al, 1996; Wiersdorff et al, 1996). Smad proteins are phosphorylated by TβRI in response to TGF-β signaling (Eppert et al, 1996; Hoodless et al, 1996; Lagna et al, 1996; Lechleider et al, 1996; Yingling et al, 1996; Zhang et al, 1996). Phosphorylated Smads then translocate into the nucleus (Hoodless et al, 1996; Liu et al, 1996), and activate transcription (Liu et al, 1996; Zhang et al, 1998).

Although details are still being worked out, there are three types of Smads: Smads that are phosphorylated by type I receptors, Smads that help transduce signal by binding to receptor-phosporylated Smads, and Smads that inhibit the signaling function of the other Smads. Among the receptor phosphorylated Smads, Smad1, Smad5, and Smad8 are substrates for the BMP receptor and mediate BMP signaling (Hoodless et al, 1996; Liu et al, 1996; Kretzschmar et al, 1997; Yamamoto et al, 1997; Macias-Silva et al, 1998). Smads 2 and 3 are substrates of the TβRI and respond to TGF-β and activin (Lagna et al, 1996; Nakao et al, 1997b; Dennler et al, 1998; Lebrun et al, 1999). In vertebrates, Smad4 has been shown to bind the receptor-phosphorylated Smads and then act as a transcription factor (Zhang et al, 1998). Smads 6 and 7 are negative regulators of the Smad signaling pathway and prevent the phosphorylation and nuclear translocation of the other Smads (Imamura et al, 1997; Nakao et al, 1997a; Ishisaki et al, 1998; Nakayama et al, 1998).

VALVE AND SEPTUM DEVELOPMENT

It is clear from other chapters in this volume that the heart is the first functional organ in the developing embryo. At the stage when valvular and septal development begins, the heart is a hollow tube-like structure with two cell layers in the midst of looping. The outer surface is a thin myocardial cell layer, while the inner, luminal surface is an endothelium. Between the two cell layers lies an expanse of cell-free ECM. The ECM of the heart is known as the "cardiac jelly" and is described as a specialized, expanded basement membrane (Kitten et al, 1987; Icardo and Manasek, 1992). The constituents of the cardiac jelly include type I collagen, hyaluronan, chondroitin sulfates, and a number of extracellular glycoproteins (Markwald et al, 1984; Krug et al, 1985; Kitten et al, 1987). More traditional components of a basement membrane including laminin and type IV collagen can be found near the myocardial cell surfaces (Kitten et al, 1987). In the region of the AV canal, the ECM is expanded into two opposing projections, the cardiac cushions. Autoradiographic evidence suggests that the ECM that fills the cushions is predominately produced by the myocardial cell layer (Krug et al, 1985).

At a specific stage in development, a population of endothelial cells lining the AV canal and outflow tract detach from adjacent cells, extend filopodia, and invade the underlying ECM (Markwald et al, 1984). This event is termed an epithelial-mesenchymal cell transformation. The resultant mesenchymal cells migrate toward the surrounding myocardium and proliferate to populate the entire AV canal. Once the opposing cushion pads fuse in the midline, the mesenchyme provides the first cellular constituents of the septum intermedium and the valvular leaflets of the mitral and tricuspid valves. The septum intermedium subsequently contributes to the lower portion of the atrial septum and the membranous portion of the ventricular septum (Markwald et al, 1984). There are indications that additional cells may also contribute to the final valvular and septal structures of the adult heart (see Chapters 1 and 8).

EPITHELIAL-MESENCHYMAL CELL TRANSFORMATION

Comparison of Heart Development Between Mouse and Chicken

At least three distinct stages occur during epithelial-mesenchymal cell transformation: endothelial cell activation, mesenchymal cell formation, and cell invasion into the ECM. Endothelial cells begin to receive the activation signal from the myocardium as early as stage 13+ in the chicken (Ramsdell and Markwald, 1997). Endothelial activation is characterized by a loss of cell-cell contacts, cellular hypertrophy and polarization, and an increased expression of ECM molecules (Bolender et al, 1980; Krug et al, 1985; Crossin and Hoffman, 1991). Activated endothelial cells undergo morphologic changes to become mesenchymal cells that invade the ECM. Several studies suggest that a majority of AV canal endothelial cells are activated, while only a small subset of cells continue to differentiate into mesenchyme (Krug et al, 1985; Runyan et al, 1990; Wunsch et al, 1994). Concomitant with mesenchymal cell formation is the acquisition of cell motility. Although cell migration is required for mesenchymal cell invasion, it can be distinguished from mesenchymal cell formation by treatments that inhibit migration but not cellular transformation (Loeber and Runyan, 1990).

The majority of studies on the role of TGF-βs during epithelial-mesenchymal cell transformation have been performed in mouse and chicken. It is important to clarify the relationship between chicken and mouse stages as defined by the criteria of Hamburger and Hamilton (1951) for chicken and the "house mouse" (Theiler, 1989) for the mouse. The detailed process of epithelial-mesenchymal cell transformation is based on the works of the Markwald and Runyan laboratories in the chicken (Runyan and Markwald, 1983; Markwald et al, 1984; Runyan et al, 1990), and the Akhurst laboratory in the mouse (Lehnert and Akhurst, 1988; Akhurst et al, 1990; Millan et al, 1991; Dickson et al, 1993). Table 11.1 displays the equivalent developmental stages of chicken and mouse hearts during steps in the epithelial-mesenchymal cell transformation.

Table 11.1. Stages of chicken and mouse heart development

	Endothelial activation	Endothelial to mesenchymal transformation	Mesenchymal cell migration
Chick	HH 14; (45–50 hours)	HH 16+–18; (50–60 hours)	HH 17–19; (52–64 hours)
Mouse	E 8–9	E 9–9.5	E 10–11

HH, Hamburger-Hamilton stage; E, embryonic day.

the heart during epithelial mesenchymal cell transformation. TGF-β_2 transcripts were present in both the AV canal and ventricular regions of the heart with a higher level of expression in the ventricle, while TGF-β_3 RNA was largely expressed in the AV canal (Potts et al, 1992). Since RNase protection assays could not be used to identify the cells in the heart that express TGF-β_2 and TGF-β_3, further experiments using in situ hybridization and immunostaining techniques at different stages of development were undertaken to localize TGF-β_2 and TGF-β_3 in the heart.

At the time of transformation, chick embryonic heart contains three cell types in the AV canal region: myocardial, endothelial, and mesenchymal. In stage 14 chick embryo sections (before cell transformation), TGF-β_2 RNA and protein are localized in both the myocardium and endothelium (Boyer et al, 1999a). TGF-β_2 mRNA is detected in the AV cushion and outflow tract region in stage 18 hearts after transformation (Barnett et al, 1994). In stage 19 (posttransformation stage) hearts, TGF-β_2 mRNA continues to be present throughout the entire myocardium and endothelium (Boyer et al, 1999a). TGF-β_2 mRNA is localized in both the forming trabeculae and the AV canal. There is distinctly higher TGF-β_2 expression at the luminal surface in both the atrial and ventricular areas, which reflects localization to the endothelial lining (Boyer et al, 1999a).

TGF-β_3 mRNA was found to be more abundant in the AV canal than in the ventricle via RNase protection assays (Potts et al, 1992). Immunolocalization studies show that TGF-β_3 protein is expressed only in the myocardium in stage 14 chick AV canal (Huang et al, 1995) and in the activated endothelium of stage 16 chick heart (Nakajima et al, 1994). Using in situ hybridization, TGF-β_3 mRNA was found in the myocardium throughout stages 14 to 19 (Boyer et al, 1999a). However, TGF-β_3 mRNA was not detected in the endothelium of the stage 14 heart (Boyer et al, 1999a). In the stage 17 chick AV canal, both TGF-β_3 mRNA and protein are present in all cells (endothelium, mesenchyme, and myocardium). In contrast to the expression of TGF-β_2 mRNA, TGF-β_3 mRNA is absent around the lumen in stage 19 heart, indicating that TGF-β_3 is not expressed in endothelial cells of nontransforming regions (Boyer et al, 1999a). TGF-β_3 mRNA expression was stronger in the epicardial portion than the endocardial and trabecular portions of the myocardium.

Together, in situ and immunolocalization data demonstrate that (1) expression of TGF-β_3 within the myocardium precedes transformation, and (2) TGF-β_3 was expressed in the endothelial and mesenchymal cells only during epithelial mesenchymal cell transformation.

FUNCTION OF TGF-β ISOFORMS DURING HEART DEVELOPMENT

Mouse

The evidence for TGF-β function in embryonic heart development in the mouse comes principally from gene targeting. Knockout mice for all three TGF-β isoforms were generated (Shull et al, 1992; Proetzel et al, 1995; Sanford et al, 1997). Mice null for TGF-β_1 show no gross developmental abnormality at birth. The main effect of loss of TGF-β_1 is an inflammation of the heart (Shull et al, 1992). When the null mutation was shifted to a severe combined immunodeficiency disease

(SCID) background to suppress the inflammatory response, no structural defects were observed. As a null mutation for TGF-β_1 in a different genetic background produced vasculogenic defects prior to heart formation, a role for modifier genes in TGF-β_1 function is clear (Bonyadi et al, 1997).

The TGF-β_3 null mouse showed only a cleft palate phenotype with no obvious gross malformation in the heart (Proetzel et al, 1995). Of the three isoform knockouts, only the TGF-β_2 null mouse demonstrates severe cardiac defects (Sanford et al, 1997). These include both AV canal valvular structures and the great vessels. In these mice, defects in a number of cardiac structures were described. There was some variation between littermates, but the general phenotype appeared similar to a tetralogy of Fallot with an atrial septal defect (Sanford et al, 1997). Although malformed, valvular and septal strutures did develop in these mice, suggesting that the defects observed did not develop from a complete loss of epithelial-mesenchymal cell transformation. Such defects might occur by a reduction in mesenchymal cell numbers. Interestingly, experiments with the in vitro collagen gel assay using murine explants showed that some mesenchymal cell formation could take place in each of the three TGF-β null mice (Runyan and Doetschman, unpublished data). Thus, in the mouse, there may be a redundancy in isoform activities that permits some cell transformation. A requirement for TGF-β activity in epithelial-mesenchymal cell transformation in the mouse was demonstrated by the effects of an antibody toward the TGF-β–binding protein, which completely blocked mesenchymal cell formation by murine AV canal explants (Nakajima et al, 1997).

Chicken

The most detailed research on the function of TGF-β isoforms during cushion formation has been conducted in the chicken system using the in vitro collagen culture system described above. Runyan and Markwald (1983) first demonstrated that epithelial-mesenchymal cell transformation in the endothelium was induced by a stimulus from adjacent myocardium. Experiments with exogenous growth factors showed that while exogenous TGF-β alone does not cause cell transformation, any isoform of TGF-β, in combination with ventricular myocardium (which does not induce transformation by itself), will produce epithelial mesenchymal cell transformation by competent AV canal endothelium in vitro (Potts and Runyan, 1989). Further, studies with TGF-β_3 antisense oligonucleotides inhibited epithelial-mesenchymal cell transformation concomitant with the loss of TGF-β_3 mRNA, demonstrating a functional role for TGF-β_3 (Potts et al, 1991). In those experiments, TGF-β_2 antisense oligonucleotides did not inhibit cell transformation. In light of the results seen in the mouse, the efficacy of oligonucleotide-mediated degradation of TGF-β_2 mRNA in the avian explants was not analyzed, and the oligonucleotide sequence used in that study may not have been effective.

Recently, inhibition of stage 14 AV canal explants with TGF-β isoform specific antibodies revealed separate and sequential roles for the two isoforms found in the chick heart (Boyer et al, 1999a). Anti–TGF-β_2 antibody treatment produced a polygonal endothelium with retained cell-cell adhesions and largely prevented mesenchymal cell formation. Conversely, anti–TGF-β_3 antibodies enabled normal endothelial cell-cell separation (activation) but blocked the separated cells from invading the gel matrix. This appearance of separated endothelial cells is consistent with previous observations using antisense oligonucleotides and pan-specific anti–TGF-β antibodies (Potts and Runyan, 1989; Potts et al, 1991).

As described above, TGF-β_3 becomes expressed in the AV canal endothelium concomitant with activation (Nakajima et al, 1994; Huang et al, 1995; Boyer et al, 1999a). There is debate as to whether TGF-β_3 is a component of the inducing stimulus from the myocardium or an autocrine response by the endothelium to other components of the signal. Since Nakajima and colleagues (1994) noted that TGF-β_3 was not found, they favored an autocrine hypothesis. Similarly, Ramsdell and Markwald (1997) showed that TGF-β_3 sustains transformation of an "activated endothelium." TGF-β_3 is present in the ECM of the heart, where it is capable of contributing to the transforming signal (Ghosh and Brauer, 1996; Boyer et al, 1999a). Since TGF-β_3 is made by the myocardium well before transformation occurs and can be found in the ECM, the simplest explanation is that both myocardial-derived and endothelial-derived TGF-β_3 molecules may participate in cell transformation. This mechanism provides a potential for amplification of the inductive signal.

It is important to note that TGF-βs require a proteolytic activation step to become functional (Massagué, 1990). Therefore, urokinase-type plasminogen activator (u-PA) or another protease that can activate TGF-β isoforms may be a critical component of the transforming stimulus. u-PA is expressed in the heart, and its perturbation inhibits mesenchymal cell migration (McGuire and Alexander, 1992, 1993b). Expression of u-PA by AV canal endothelial cells is induced by a stimulus from the myocardium (Romano and Runyan, unpublished data). Thus, the inductive component that regulates u-PA in the heart could be a critical element in the regulation of TGF-β activities in the AV canal.

TGF-β_3 antibodies produced a partial inhibition of transformation in stage 16 AV explants and had no effect on the transformation of stage 18 explants (Boyer et al, 1999a). These data suggest that there is a stage in the transformation process where transforming cells no longer require TGF-β_3. In contrast, the finding that TGF-β_2 antibody inhibited transformation more effectively than TGF-β_3 antibodies at both earlier and later stages (Boyer et al, 1999a) suggests that at distinct stages of cell transformation TGF-β_2 plays a more important role than TGF-β_3. Consistent with this premise, the expression of two protein markers of the mesenchymal cell phenotype, cell surface β-1,4-galactosyltransferase (GalTase) and fibrillin 2, are regulated differently by TGF-β isoforms. Both isoforms increased expression of fibrillin 2 and both antibodies decreased fibrillin 2 expression in cultures, indicating that both TGF-β isoforms are required for ECM production. Increased GalTase expression in TGF-β_2–treated cultures alone suggests that TGF-β_2 may play a role in cell motility.

LOCALIZATION OF TGF-β RECEPTORS DURING HEART DEVELOPMENT

Obviously critical to the role of TGF-β in the heart is the distribution and function of the TGF-β receptors. TGF-β type II and III receptors were cloned from chick by Barnett and colleagues (1994). The expression of protein and mRNA for both receptors was at a relatively high level in stage 18 hearts (Barnett et al, 1994). Whole mount immunolocalization of type II receptor shows that It is mainly expressed in the endothelial cells throughout the heart and the rest of the embryo

(Brown et al, 1996). TGF-β type II receptor was also expressed in the migrating mesenchymal cells (Brown et al, 1996).

In contrast to the localization of TGF-β type II receptor, immunolocalization of the TβRIII (betaglycan) in the chicken embryo showed a unique expression only in the endothelium overlying the transforming regions of the heart, the AV canal, and the outflow tract (Brown et al, 1999). Mesenchymal cells derived from the cardiac endothelia continued to express TβRIII. This pattern of TβRIII expression suggests the important role for the receptor in mediating epithelial-mesenchymal cell transformation.

The distribution of endoglin in the heart and AV canal also was explored (Qu et al, 1998; Vincent et al, 1998). In chicken, endoglin is expressed in the myocardium and endocardium at early stages of heart development (stage 10) (Vincent et al, 1998). However, as development proceeds, endoglin expression is restricted. In stage 14, endoglin is expressed on both myocardium and endothelial cells, but begins to decrease after stage 14, particularly in the myocardium. At stages 15 to 16, there is very little endoglin expressed in the heart. By the time of cell transformation in the AV canal, endoglin is expressed only on endothelium and newly formed mesenchyme (Vincent et al, 1998). In human tissues, endoglin was detected in the valvular tissues at the earliest stage examined (4 weeks) (Qu et al, 1998). However, these tissue samples were collected after mesenchymal cell formation had taken place, and the lack of myocardial staining in the myocardium is consistent with the studies in the chick.

FUNCTION OF TGF-β RECEPTORS DURING HEART DEVELOPMENT

Functional analysis of TβRII and TβRIII has been aided by the production of blocking antibodies (Brown et al, 1996, 1999). Treatment of AV explants with blocking antibodies to avian TβRII produced the expected blockade of epithelial-mesenchymal cell transformation (Brown et al, 1996). Consistent with the reported greater affinity of this receptor for TGF-β_1 or -β_3 (Lopez-Casillas et al, 1994), the morphology of the blocked cultures was similar to that of cultures treated with antisense oligonucleotide against TGF-β_3 or antibodies against this isoform. Under all three stages (14, 16, and 19), endothelial cells showed the separation associated with activation but were unable to invade the collagen gel as mesenchyme. Interestingly, the TβRIII blocking antibody was more effective than the TβRII antibody (Brown et al, 1999). Antibody blockade of TβRIII prevents cell-cell separation similar to that observed with an anti–TGF-β_2 antibody (Boyer et al, 1999a; Brown et al, 1999; Boyer and Runyan, unpublished data). Both type II and type III receptor antibodies inhibited mesenchymal cell migration (Brown et al, 1996, 1999). Based on numerous significant differences in marker protein expression by treated cells (Boyer and Runyan, unpublished data), as well as the differences in blockade of epithelial-mesenchymal cell transformation, it is clear that the two receptors are acting in different signal transduction pathways.

In addition to the loss-of-function experiments in culture summarized in the previous paragraph, a critical role for TβRIII in cell transformation was recently shown by gain-of-function experiments. It was shown that ventricular endothelium is refractory to the myocardial inductive signal and incapable of producing mesenchymal cells (Runyan and Markwald, 1983; Potts and Runyan, 1989;

Runyan et al 1990). When TβRIII is expressed by a retroviral construct within ventricular explants, endothelial cells will respond to exogenous TGF-β_2 ligand and transform into mesenchyme (Brown et al, 1999). Although there are reasons for caution concerning the identity of the invaded cells, these experiments clearly demonstrate that TβRIII and exogenous TGF-β_2 ligand will produce mesenchymal cells. Clearly, TβRIII plays a much more significant role in the heart than presenting TGF-β_2 to the TβRII.

SUMMARY AND CONCLUSIONS

Based on the data summarized here, we argue that the two isoforms of TGF-β and the two TGF-β receptors have complementary roles in the formation of mesenchyme in the avian heart AV canal. Acute loss of TGF-βs by either antibody or antisense oligonucleotide-mediated degradation of message in explant tissues produces distinguishable differences in activity between the two isoforms.

The story in the mouse is much less clear. There are differences not only in isoform types present but also in the distribution of the isoforms between mouse and chicken. For instance, although TGF-β_1 activity is similar to TGF-β_3 in receptor binding, it does not directly replace the avian TGF-β_3 distribution. Murine TGF-β_1 localization is largely endothelial while avian TGF-β_3 is prominent in the myocardium. TGF-β_2 appears to have a similar distribution in the AV canal of mouse and chicken hearts after transformation. Since there is a paucity of data from mouse hearts at times prior to the onset of cell transformation, some of our confusion may be produced by this lack of information. However, results from the knockout mouse point to TGF-β_2 as the uniquely critical isoform. Our demonstration that TGF-β_2 is critical to avian cardiac cell transformation does reconcile a role for this growth factor but does not explain the differences in TGF-β_1 and -β_3.

The observed differences in isoform distribution and function between the species can be directly contrasted to an apparent common role for TGF-β_3 in palate formation in both species. The TGF-β_3 null mouse has a cleft palate (Sanford et al, 1997), while the normal chicken has a cleft palate and a concomitant lack of TGF-β_3 within the medial edge epithelia of the palatal shelves. Delivery of TGF-β_3 to the chicken palate produces a fused palate and epithelial-mesenchymal cell transformation of the medial edge epithelium (Sun et al, 1998). Though the chicken does not normally produce a fused palate and the mouse does, the tissue responses to the growth factor in both species are the same.

Due to the similarity in the inhibition pattern between antibodies to TGF-β_3 and TβRII, and between TGF-β_2 and TβRIII antibodies, and because of the known affinities between these TGF-βs and the same receptors, we propose the following TGF-β signal transduction model during epithelial-mesenchymal transformation in the chicken (Figure 11.1). At an early stage, TGF-β_3 is one of the inducing signals from the myocardium. After the endothelial cells receive the inducing signal, these endothelial cells need a TGF-β_2 signal to activate. This signal is propagated through the type III receptor. Since TGF-β_2 is already expressed by the endothelium, expression of u-PA or another protease may be all that is needed to produce active TGF-β_2 at the endothelial surface. Once endothelial cells are fully activated, they require TGF-β_3 to transform into mesenchymal cells, and this process is propagated by type II receptor. The transformation stimulus may be

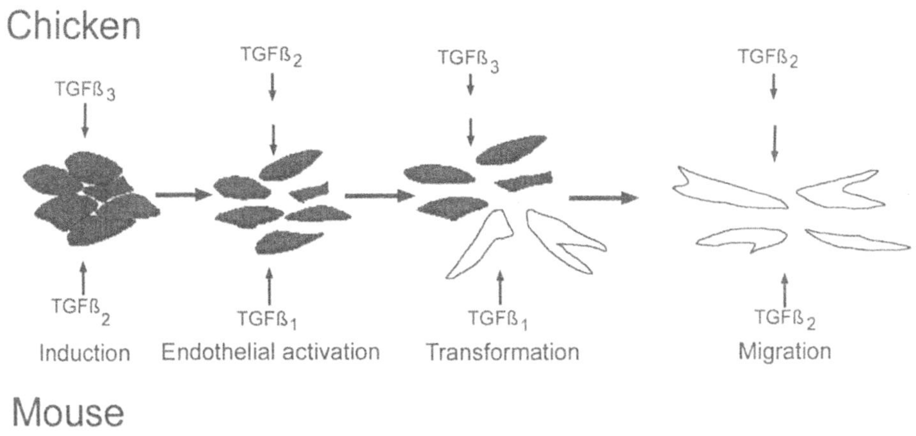

FIGURE 11.1. The transforming growth factor-β (TGF-β) signal transduction model during epithelial-mesenchymal cell transformation. The shaded cells are endothelium. The clear cells are mesenchyme. The upper half of the graph presents proposed TGF-β signal pathways in chicken; the lower half presents proposed TGF-β signal pathways in mouse.

amplified by the expression and activation of TGF-β_3 within the endothelium. Once the mesenchymal cells are formed from endothelial cells, they again utilize TGF-β_2 during migration into the extracellular matrix.

Based on the comparative chicken data, we can suggest common and differing roles for TGF-β isoforms in the mouse (Figure 11.1). In the mouse, TGF-β_1 is expressed first in the endothelial cells when cushion formation begins. As cushion formation proceeds, TGF-β_1 expression is restricted to the endothelial cells overlying the AV canal and outflow tract regions. After mesenchymal cell formation, TGF-β_1 is also detected in the mesenchymal cells. In contrast, TGF-β_2 expression starts in the myocardium and is restricted in the myocardium throughout the transformation process. Later, TGF-β_2 is also detected in the forming mesenchymal cells. Based on these observations, we propose that TGF-β_2 is a component of the inducing signal from myocardium that initiates epithelial mesenchymal cell transformation in the mouse. TGF-β_1 is required for endothelial cell activation and transformation. Both TGF-β_1 and TGF-β_2 are likely to be involved in mesenchymal cell migration and valve remodeling. Comparison of null mouse phenotypes would suggest that TGF-β_2 provides sufficient redundancy for TGF-β_1 function, but that TGF-β_1 cannot adequately compensate for the loss of the β_2 isoform. This conjecture concerning the mouse will have to be explored experimentally. More importantly, we will need to explore the genetics of human congenital heart disease to resolve whether human heart development utilizes TGF-β isoforms in a manner analogous to either of the experimental models.

REMAINING ISSUES

Despite the progress made on TGF-β signaling during epithelial mesenchymal cell transformation, there are still many remaining issues that need to be resolved. One such issue is the role of the TβRI receptor during cell transformation. From the published mechanisms of TGF-β signaling, we would expect that the type I receptor would mediate signals initiated from the TβRII during cell transformation. However, it is not clear whether the TβRI propagates a signal from the TGF-β_2/TβRIII interaction as documented in other systems (Taniguchi et al, 1998). Alternatively, TβRIII may act via a previously undescribed signal transduction pathway.

The function of Smad proteins during epithelial-mesenchymal transformation also needs to be explored. Smad proteins are a known bridge between TGF-βs and gene regulation. Examination of Smads in the heart will provide new insights as to how signals from the different TGF-βs are propagated and, more importantly, how downstream events may be regulated.

We previously showed that G-protein signal transduction pathways and TGF-β signal transduction pathways are distinct during epithelial-mesenchymal cell transformation (Boyer et al, 1999b). However, there may be an interaction of TGF-β signaling pathways with additional pathways.

Finally, signal transduction, initiated through the various TGF-βs, begins the process of cell phenotype change. We have begun the study of downstream genes or proteins regulated by TGF-βs during heart development. There are several transcription factors associated with embryonic cellular transformations such as Brachyury, Mox-1, and Slug (Huang et al, 1995; Romano and Runyan, 1999). It will be instructive to explore the regulation of these genes in the heart and relate them both to the TGF-βs and the downstream markers of cell phenotype.

ACKNOWLEDGMENTS

Dr. Boyer was supported by a fellowship from the Arizona Affiliate of the American Heart Association. Research support was provided by National Institutes of Health grants HL20984 and HL54986 (Runyan). Dr. Runyan was an established investigator of the American Heart Association during much of the progress of this work.

REFERENCES

Akhurst, R.J., Lehnert, S., Faissner, A., and Duffie, E. (1990). TGF beta in murine morphogenetic processes: the early embryo and cardiogenesis. *Development* 108(4):645–656.

Alexandrow, M.G., and Moses, H.L. (1995). Transforming growth factor beta and cell cycle regulation. *Cancer Res* 55:1452–1457.

Antonelli-Olridge, A., Saunders, K.B., Smith, S.R., and D'Amore, P.A. (1989). An activated form of transforming growth factor-β is produced by cocultures of endothelial cells and pericytes. *Proc Natl Acad Sci USA* 86:4544–4548.

Barbara, N.P., Wrana, J.L., and Letarte, M. (1999). Endoglin is an accessory protein that interacts with the signaling receptor complex of multiple members of the transforming growth factor-beta superfamily. *J Biol Chem* 274:584–594.

Barnett, J.V., Moustakas, A., Lin, W., et al. (1994). Cloning and developmental expression of the chick type II and type III TGFβ receptors. *Dev Dyn* 199:12–27.

Battegay, E.J. (1995). Angiogenesis: mechanistic insights, neovascular diseases and therapeutic prospects. *J Mol Med* 73:333–346.

Bernanke, D.H., and Markwald, R.R. (1982). Migratory behavior of cardiac cushion tissue cells in a collagen lattice system. *Dev Biol* 91:235–245.

Bolender, D.L., Seliger, W.G., and Markwald, R.R. (1980). A histochemical analysis of polyanionic compounds found in the extracellular matrix encountered by migrating cephalic neural crest cells. *Anat Rec* 196:401–412.

Bonyadi, M., Rusholme, S.A., Cousins, F.M., et al. (1997). Mapping of a major genetic modifier of embryonic lethality in TGF beta 1 knockout mice. *Nat Genet* 15:207–211.

Boyer, A.S., Ayerinskas, I.I., Vincent, E.B., McKinney, L., Weeks, D.L., and Runyan, R.B. (1999a). TGFβ2 and TGFβ3 have separate and sequential activities during epithelial mesenchymal cell transformation in the embryonic heart. *Dev Biol* 208:530–545.

Boyer, A.S., Erickson, C.P., and Runyan, R.B. (1999b). Epithelial-mesenchymal transformation in the embryonic heart is mediated through distinct pertussis toxin-sensitive and TGFβ signal transduction mechanisms. *Dev Dyn* 214:81–91.

Brown, C.B., Boyer, A.S., Runyan, R.B., and Barnett, J.V. (1996). Antibodies to the type II TGF beta receptor block cell activation and migration during atrioventricular cushion transformation in the heart. *Dev Biol* 174:248–257.

Brown, C.B., Boyer, A.S., Runyan, R.B., and Barnett, J.V. (1999). Requirement of the type III TGFβ receptors for endocardial cell transformation in the heart. *Science* 283:2080–2082.

Burt, D.W. (1992). Evolutionary grouping of the transforming growth factor-β superfamily. *Biochem Biophys Res Commun* 184:590–595.

Cheifetz, S., Bellon, T., Cales, C., et al. (1992). Endoglin is a component of the transforming growth factor-beta receptor system in human endothelial cells. *J Biol Chem* 267:19027–19030.

Chen, Y., Lebrun, J.J., and Vale, W. (1996). Regulation of transforming growth factor beta- and activin-induced transcription by mammalian Mad proteins. *Proc Natl Acad Sci USA* 93:12992–12997.

Crossin, K.L., and Hoffman, S. (1991). Expression of adhesion molecules during the formation and differentiation of the avian endocardial cushion tissue. *Dev Biol* 145:277–286.

Dennler, S., Itoh, S., Vivien, D., ten Dijke, P., Huet, S., and Gauthier, J.M. (1998). Direct binding of Smad3 and Smad4 to critical TGF beta-inducible elements in the promoter of human plasminogen activator inhibitor-type 1 gene. *EMBO J* 17:3091–3100.

Dickson, M.C., Slager, H.G., Duffie, E., Mummery, C.L., and Akhurst, R.J. (1993). RNA and protein localisations of TFG beta 2 in the early mouse embryo. *Development* 117:625–639.

Elsdale, T., and Bard, J. (1972). Collagen substrata for studies on cell behavior. *J Cell Biol* 54:626–637.

Eppert, K., Scherer, S.W., Ozcelik, H., et al. (1996). MADR2 maps to 18q21 and encodes a TGFbeta-regulated MAD-related protein that is functionally mutated in colorectal carcinoma. *Cell* 86:543–552.

Fananapazir, K., and Kaufman, M.H. (1988). Observations on the development of the aortico-pulmonary spiral septum in the mouse. *J Anat* 158:157–172.

Ghosh, S., and Brauer, P.R. (1996). Latent transforming growth factor-β is present in the extracellular matrix of embroyonic hearts in situ. *Dev Dyn* 205:126–134.

Gougos, A., and Letarte, M. (1988a). Biochemical characterization of the 44G4 antigen from the HOON pre-B leukemic cell line. *J Immunol* 141:1934–1940.

Gougos, A., and Letarte, M. (1988b). Identification of a human endothelial cell antigen with monoclonal antibody 44G4 produced against a pre-B leukemic cell line. *J Immunol* 141:1925–1933.

Graff, J.M., Bansal, A., and Melton, D.A. (1996). *Xenopus* Mad proteins transduce distinct subsets of signals for the TGF beta superfamily. *Cell* 85:479–487.

Hamburger, V., and Hamilton, H.L. (1951). A series of normal stages in the development of the chick embryo. *J Morphol* 88:49–92.

Heine, U.I., Munoz, E.F., Flanders, K.C., et al. (1987). Role of transforming growth factor-β1 in the development of the mouse embryo. *J Cell Biol* 105:2861–2876.

Hoodless, P.A., Haerry, T., Abdollah, S., et al. (1996). MADR1, a MAD-related protein that functions in BMP2 signaling pathways. *Cell* 85:489–500.

Huang, J.X., Potts, J.D., Vincent, E.B., Weeks, D.L., and Runyan, R.B. (1995). Mechanisms of cell transformation in the embryonic heart [review]. *Ann NY Acad Sci* 752:317–330.

Icardo, J.M., and Manasek, F.J. (1992). Cardiogenesis: development mechanisms and embryology. In: Fozzard, H.A., Harber, E., Jennings, R.B., Katz, A.M., and Morgan, H.E., eds. *The Heart and Cardiovascular System*, 2nd ed. Raven Press, New York, pp. 1563–1586.

Imamura, T., Takase, M., Nishihara, A., et al. (1997). Smad6 inhibits signalling by the TGF-beta superfamily. *Nature* 389:622–626.

Ishisaki, A., Yamato, K., Nakao, A., et al. (1998). Smad7 is an activin-inducible inhibitor of activin-induced growth arrest and apoptosis in mouse B cells. *J Biol Chem* 273:24293–24296.

Kitten, G.T., Markwald, R.R., and Bolender, D.L. (1987). Distribution of basement membrane antigens in cryopreserved early embryonic hearts. *Anat Rec* 217(4):379–390.

Kretzschmar, M., Liu, F., Hata, A., Doody, J., and Massague, J. (1997). The TGF-beta family mediator Smad1 is phosphorylated directly and activated functionally by the BMP receptor kinase. *Genes Dev* 11:984–995.

Krug, E.L., Mjaatvedt, C.H., and Markwald, R.R. (1987). Extracellular matrix from embryonic myocardium elicits an early morphogenetic event in cardiac endothelial differentiation. *Dev Biol* 120(2):348–355.

Krug, E.L., Runyan, R.B., and Markwald, R.R. (1985). Protein extracts from early embryonic hearts initiate cardiac endothelial cytodifferentiation. *Dev Biol* 112(2):414–426.

Labbe, E., Silvestri, C., Hoodless, P.A., Wrana, J.L., and Attisano, L. (1998). Smad2 and Smad3 positively and negatively regulate TGF beta-dependent transcription through the forkhead DNA-binding protein FAST2. *Mol Cell* 2:109–120.

Lagna, G., Hata, A., Hemmati-Brivanlou, A., and Massague, J. (1996). Partnership between DPC4 and SMAD proteins in TGF-beta signalling pathways. *Nature* 383:832–836.

Lakkis, M.M., and Epstein, J.A. (1998). Neurofibromin modulation of ras activity is required for normal endocardial-mesenchymal transformation in the developing heart. *Development* 125:4359–4367.

Lebrun, J.J., Takabe, K., Chen, Y., and Vale, W. (1999). Roles of pathway-specific and inhibitory Smads in activin receptor signaling [in process citation]. *Mol Endocrinol* 13:15–23.

Lechleider, R.J., de Caestecker, M.P., Dehejia, A., Polymeropoulos, M.H., and Roberts, A.B. (1996). Serine phosphorylation, chromosomal localization, and transforming growth factor-beta signal transduction by human bsp-1. *J Biol Chem* 271:17617–17620.

Lehnert, S., and Akhurst, R.J. (1988). Embryonic pattern of TGF beta type-1 RNA suggests both paracrine and autocrine mechanisms of action. *Development* 104(2):263–273.

Liu, F., Hata, A., Baker, J.C., et al. (1996). A human Mad protein acting as a BMP-regulated transcriptional activator [see comments]. *Nature* 381:620–623.

Loeber, C.P., and Runyan, R.B. (1990). A comparison of fibronectin, laminin, and galactosyltransferase adhesion mechanisms during embryonic cardiac mesenchymal cell migration in vitro. *Dev Biol* 140(2):401–412.

Lopez-Casillas, F., Payne, H.M., Andres, J.L., and Massagué, J. (1994). Betaglycan can act as a dual modulator of TGF-beta access to signaling receptors: mapping of ligand binding and GAG attachment sites. *J Cell Biol* 124:557–568.

Lopez-Casillas, F., Wrana, J.L., and Massagué, J. (1993). Betaglycan presents ligand to the TGF beta signaling receptor. *Cell* 73:1435–1444.

Lyons, R.M., Keski-Oja, J., and Moses, H.L. (1988). Proteolytic activation of latent transforming growth factor-β from fibroblast-conditioned medium. *J Cell Biol* 106:1597–1605.

Macias-Silva, M., Hoodless, P.A., Tang, S.J., Buchwald, M., and Wrana, J.L. (1998). Specific activation of Smad1 signaling pathways by the BMP7 type I receptor, ALK2. *J Biol Chem* 273:25628–25636.

Markwald, R.R., Runyan, R.B., Kitten, G.T., Funderburg, F.M., Bernanke, D.H., and Brauer, P.R. (1984). Use of collagen gel cultures to study heart development: proteoglycan and glycoprotein interactions during formation of endocardial cushion tissue. In: Trelstad, R.L., ed. *The Role of Extracelluar Matrix in Development*. New York, Alan R. Liss, pp. 323–350.

Massague, J. (1998). TGF-beta signal transduction. *Annu Rev Biochem* 67:753–791.

Massagué, J. (1990). The transforming growth factor-β family. *Annu Rev Cell Biol* 6:597–641.

Massagué, J., Andres, J.L., Attisano, L., et al. (1992). TGF-beta receptors [review]. *Mol Reprod Dev* 32:99–104.

McGuire, P.G., and Alexander, S.M. (1992). Urokinase expression during the epithelial-mesenchymal transformation of the avian somite. *Dev Dyn* 194:193–197.

McGuire, P.G., and Alexander, S.M. (1993a). Inhibition of urokinase synthesis and cell surface binding alters the motile behavior of embryonic endocardial-derived mesenchymal cells in vitro. *Development* 118:931–939.

McGuire, P.G., and Alexander, S.M. (1993b). Urokinase production by embryonic endocardial-derived cells: regulation by substrate composition. *Dev Biol* 155:442–451.

Millan, F.A., Denhez, F., Kondaiah, P., and Akhurst, R.J. (1991). Embryonic gene expression patterns of TGF β1, β2, and β3 suggest different developmental functions in vivo. *Development* 111:131–144.

Miyazono, K., Hellman, U., Wernstedt, C., and Heldin, C.H. (1988). Latent high molecular weight complex of transforming growth factor beta 1. Purification from human platelets and structural characterization. *J Biol Chem* 263:6407–6415.

Mjaatvedt, C.H., Lepera, R.C., and Markwald, R.R. (1987). Myocardial specificity for initiating endothelial-mesenchymal cell transition in embryonic chick heart correlates with a particulate distribution of fibronectin. *Dev Biol* 119:59–67.

Mjaatvedt, C.H., and Markwald, R.R. (1989). Induction of an epithelial-mesenchymal transition by an in vivo adheron-like complex. *Dev Biol* 136:118–128.

Moore, C.S., Mjaatvedt, C.H., and Gearhart, J.D. (1998). Expression and function of activin beta A during mouse cardiac cushion tissue formation. *Dev Dyn* 212:548–562.

Moses, H.L. (1992). TGF-beta regulation of epithelial cell proliferation. *Mol Reprod Dev* 32:179–184.

Moses, H.L., Yang, E.Y., and Pietenpol, J.A. (1990). TGF-beta stimulation and inhibition of cell proliferation: new mechanistic insights. *Cell* 63:245–247.

Nakajima, Y., Krug, E.L., and Markwald, R.R. (1994). Myocardial regulation of transforming growth factor-β expression by outflow tract endothelium in the early embryonic chick heart. *Dev Biol* 165:615–626.

Nakajima, Y., Miyazono, K., Kato, M., Takase, M., Yamagishi, T., and Nakamura, H. (1997). Extracellular fibrillar structure of latent TGF beta binding protein-1: role in TGF beta-dependent endothelial-mesenchymal transformation during endocardial cushion tissue formation in mouse embryonic heart. *J Cell Biol* 136:193–204.

Nakao, A., Afrakhte, M., Moren, A., et al. (1997a). Identification of Smad7, a TGFbeta-inducible antagonist of TGF-beta signalling [see comments]. *Nature* 389:631–635.

Nakao, A., Imamura, T., Souchelnytskyi, S., et al. (1997b). TGF-beta receptor-mediated signalling through Smad2, Smad3 and Smad4. *EMBO J* 16:5353–5362.

Nakayama, T., Gardner, H., Berg, L.K., and Christian, J.L. (1998). Smad6 functions as an intracellular antagonist of some TGF-beta family members during *Xenopus* embryogenesis. *Genes Cells* 3:387–394.

Panganiban, G.E., Reuter, R., Scott, M.P., and Hoffmann, F.M. (1990). A *Drosophila* growth factor homolog, decapentaplegic, regulates homeotic gene expression within and across germ layers during midgut morphogenesis. *Development* 110:1041–1050.

Potts, J.D., Dagle, J.M., Walder, J.A., Weeks, D.L., and Runyan, R.B. (1991). Epithelial-mesenchymal transformation of embryonic cardiac endothelial cells is inhibited by a modified antisense oligodeoxynucleotide to TGFβ3. *Proc Natl Acad Sci USA* 88(4):1516–1520.

Potts, J.D., and Runyan, R.B. (1989). Epithelial-mesenchymal cell transformation in the embryonic heart can be mediated, in part, by transforming growth factor β. *Dev Biol* 134(2):392–401.

Potts, J.D., Vincent, E.B., Runyan, R.B., and Weeks, D.L. (1992). Sense and antisense TGF beta 3 mRNA levels correlate with cardiac valve induction. *Dev Dyn* 193(4):340–345.

Proetzel, G., Pawlowski, S.A., Wiles, M.V., et al. (1995). Transforming growth factor-beta 3 is required for secondary palate fusion. *Nat Genet* 11:409–414.

Qu, R., Silver, M.M., and Letarte, M. (1998). Distribution of endoglin in early human development reveals high levels on endocardial cushion tissue mesenchyme during valve formation. *Cell Tissue Res* 292:333–343.

Ramsdell, A.F., and Markwald, R.R. (1997). Induction of endocardial cushion tissue in the avian heart is regulated, in part, by TGFβ3-mediated autocrine signaling. *Dev Biol* 188:64–74.

Romano, L.A., and Runyan, R.B. (1999). Slug is a mediator of epithelial-mesenchymal cell transformation in the developing chicken heart. *Dev Biol* 212:243–254.

Rosa, F., Roberts, A., Danielpour, D., Dart, L., Sporn, M.B., and Dawid, I. (1988). Mesoderm induction in amphibians: the role of TGFβ2-like factors. *Science* 239:783–786.

Runyan, R.B., and Markwald, R.R. (1983). Invasion of mesenchyme into three-dimensional collagen gels: a regional and temporal analysis of interaction in embryonic heart tissue. *Dev Biol* 95:108–114.

Runyan, R.B., Potts, J.D., and Weeks, D.L. (1992). TGF-β3-mediated tissue interaction during embryonic heart development. *Mol Reprod Dev* 32:152–159.

Runyan, R.B., Potts, J.D., Weeks, D.L., et al. (1990). Tissue interaction and signal transduction in the atrioventricular canal of the embryonic heart. *Ann NY Acad Sci* 588:442–443.

Sanford, L.P., Ormsby, I., Gittenberger-de Groot, A.C., et al. (1997). TGFβ2 knockout mice have multiple development defects that are non overlapping with other TGFβ knockout phenotypes. *Development* 124:2659–2670.

Savage, C., Das, P., Finelli, A.L., et al. (1996). *Caenorhabditis elegans* genes sma-2, sma-3, and sma-4 define a conserved family of transforming growth factor beta pathway components. *Proc Natl Acad Sci USA* 93:790–794.

Schneider, M.D., and Parker, T.G. (1990). Cardiac myocytes as targets for the action of peptide growth factors. *Circulation* 81:1443–1456.

Segarini, P.R., and Seyedin, S.M. (1988). The high molecular weight receptor to transforming growth factor-beta contains glycosaminoglycan chains. *J Biol Chem* 263:8366–8370.

Sekelsky, J.J., Newfeld, S.J., Raftery, L.A., Chartoff, E.H., and Gelbart, W.M. (1995). Genetic characterization and cloning of mothers against dpp, a gene required for decapentaplegic function in *Drosophila melanogaster*. *Genetics* 139:1347–1358.

Shull, M.M., Ormsby, I., Kier, A.B., et al. (1992). Targeted disruption of the mouse transforming growth factor-beta 1 gene results in multifocal inflammatory disease. *Nature* 359:693–699.

Sun, D., Vanderburg, C.R., Odierna, G.S., and Hay, E.D. (1998). TGFbeta3 promotes transformation of chicken palate medial edge epithelium to mesenchyme in vitro. *Development* 125:95–105.

Taipale, J., Miyazono, K., Heldin, C.H., and Keski-Oja, J. (1994). Latent transforming growth factor-beta 1 associates to fibroblast extracellular matrix via latent TGF-beta binding protein. *J Cell Biol* 124:171–181.

Taniguchi, A., Matsuzaki, K., Nakano, K., Kan, M., and McKeehan, W.L. (1998). Ligand-dependent and -independent interactions with the transforming growth factor type II and I receptor subunits reside in the aminoterminal portion of the ectodomain of the type III subunit. *In Vitro Cell Dev Biol Anim* 34:232–238.

Theiler, K. (1989). *The House Mouse: Atlas of Embryonic Development*. Springer-Verlag, New York.

Trelstad, R.L., Hayashi, A., Hayashi, K., and Donahoe, P.K. (1982). The epithelial-mesenchyme interface of the male rat mullerian duct: loss of basement membrane integrity and ductal regression. *Dev Biol* 92(1):27–40.

Tsuji, M., Shima, H., Yonemura, C.Y., Brody, J., Donahoe, P.K., and Cunha, G.R. (1992). Effect of human recombinant mullerian inhibiting substance on isolated epithelial and mesenchymal cells during mullerian duct regression in the rat. *Endocrinology* 131:1481–1488.

Vincent, E.B., Runyan, R.B., and Weeks, D.L. (1998). Production of the transforming growth factor-beta binding protein endoglin is regulated during chick heart development. *Dev Dyn* 213:237–247.

Wang, X.F., Lin, H.Y., Ng-Eaton, E., Downward, J., Lodish, H.F., and Weinberg, R.A. (1991). Expression cloning and characterization of the TGF-beta type III receptor. *Cell* 67:797–805.

Watanabe, T.K., Suzuki, M., Omori, Y., et al. (1997). Cloning and characterization of a novel member of the human Mad gene family (MADH6). *Genomics* 42:446–451.

Weeks, D.L., and Melton, D.A. (1987). A maternal mRNA localized to the vegetal hemisphere in *Xenopus* eggs codes for a growth factor related to TGF-β. *Cell* 51:861–867.

Wiersdorff, V., Lecuit, T., Cohen, S.M., and Mlodzik, M. (1996). Mad acts downstream of Dpp receptors, revealing a differential requirement for dpp signaling in initiation and propagation of morphogenesis in the *Drosophila* eye. *Development* 122:2153–2162.

Wrana, J.L., Attisano, L., Carcamo, J., et al. (1992). TGF beta signals through a heteromeric protein kinase receptor complex. *Cell* 71:1003–1014.

Wrana, J.L., Attisano, L., Wieser, R., Ventura, F., and Massagué, J. (1994). Mechanism of activation of the TGF-β receptor. *Nature* 370:341–347.

Wunsch, A.M., Little, C.D., and Markwald, R.R. (1994). Cardiac endothelial heterogeneity defines valvular development as demonstrated by the diverse expression of JB3, an antigen of the endocardial cushion tissue. *Dev Biol* 165:585–601.

Yamamoto, N., Akiyama, S., Katagiri, T., Namiki, M., Kurokawa, T., and Suda, T. (1997). Smad1 and smad5 act downstream of intracellular signalings of BMP-2 that inhibits myogenic differentiation and induces osteoblast differentiation in C2C12 myoblasts. *Biochem Biophys Res Commun* 18(2):238,574–580.

Yingling, J.M., Das, P., Savage, C., Zhang, M., Padgett, R.W., and Wang, X.F. (1996). Mammalian dwarfins are phosphorylated in response to transforming growth factor beta and are implicated in control of cell growth. *Proc Natl Acad Sci USA* 93:8940–8944.

Zhang, Y., Feng, X.H., and Derynck, R. (1998). Smad3 and Smad4 cooperate with c-Jun/c-Fos to mediate TGF-beta-induced transcription [published erratum appears in *Nature* 1998;396(6710):491]. *Nature* 394:909–913.

Zhang, Y., Feng, X., We, R., and Derynck, R. (1996). Receptor-associated Mad homologues synergize as effectors of the TGF-beta response. *Nature* 383:168–172.

CHAPTER **12**

Homocysteine and the *N*-Methyl-D-Aspartate (NMDA) Receptor: Are They Keys to Conotruncal Abnormalities?

Thomas H. Rosenquist and Daniel T. Monaghan

A great deal of what is understood about normal development of neural tube and neural crest derivatives, including the conotruncal region of the heart, has been inferred from the regulation (induction or prevention) of abnormal development. A series of epidemiologic studies showed that periconceptional folate supplementation was uniquely successful at preventing abnormal development of these areas, indicating that folic acid was of extraordinary and specific significance to the normal development of the neural tube and neural crest (Czeizel and Rode, 1984; Seller and Nevin, 1984; Eskes and Steegers-Theunissen, 1994; Shaw et al, 1995; Botto et al, 1996; Czeizel et al, 1996). Since long-term folate insufficiency may result in anemia or other signs of methyl group unavailability, it was postulated that methyl group unavailability also was the basis of the teratologic effect. However, there was no explanation for the special susceptibility of a small subset of cells, when virtually all cells in the embryo were dividing rapidly. Another hypothesis held that the amino acid homocysteine was the teratologic agent, since homocysteine always rises in folate insufficiency, and hyperhomocysteinemia was associated with the same neural tube and neural crest defects, and even the presence of normal folate concentrations. Our experiments with the chicken embryo model (Rosenquist et al, 1996) confirmed that homocysteine per se was highly specific for the neural crest and neural tube, inducing defects that simulated those that had been reported widely in association with folate insufficiency.

The cellular mechanism that was disrupted by homocysteine obviously was of profound importance for normal development of the neural crest and neural tube, but this mechanism was unknown. Homocysteine at very high concentrations was cytotoxic to various cell types, but there was no evidence that homocysteine at teratogenic concentrations was killing any embryonic cells, much less any evidence for a specific toxic effect on neural crest and neural tube cells. Homocysteic acid and some other derivatives of homocysteine were ligands for the *N*-methyl-D-aspartate (NMDA) type of glutamate receptor in the central nervous system (CNS). Furthermore, the NMDA receptor in the fully developed CNS is a unique

Based on these experimental results, the biologic basis for the various epidemiologic correlations among conotruncal, orofacial, and neural tube defects is now generally assumed to be found in the common origin of their primordial cells. In spite of the data offered in support of this generally held assumption, there is no consensus about the mechanism per se. That is, it is assumed that a given teratogen or mutation may affect both neural crest and neural tube cells because of their shared heritage, but the nature of the interaction between the cells and the teratogen or mutation generally is not known. Most reviews of the subject stress the hypothesis that the causes of conotruncal and other developmental defects must be multifactorial (e.g., Pexieder, 1995). Indeed, the apparently complex etiology of most congenital defects argues against the adduction of any unifying hypothesis.

Nevertheless, recent developments imply a surprising degree of homogeneous etiology underlying large portions of what appears to be a bewildering array of defects, teratogens, and mutations. For example, the 22q11 microdeletion is present in up to 30% of cases of isolated conotruncal heart defects and 88% of cases of DiGeorge syndrome (Leatherbury and Kirby, 1996). This deletion also is associated with orofacial and neural tube closure defects. Recently, the 22q11 microdeletion was shown to involve the deletion of a gene coding for a protein involved in degradation of ubiquinated proteins (Yamagishi et al, 1999).

In another example where a single phenomenon appears to apply to a large number of defects, recent evidence indicates folic acid concentration may have singular significance in conotruncal, orofacial, and neural tube development. In the early 1980s, intervention studies showed that periconceptional vitamin supplementation was protective for women with a history of pregnancy involving neural tube and conotruncal defects (e.g., Czeizel and Rode, 1984; Seller and Nevin, 1984). More recent reports confirmed that the apparently simple step of providing women with supplemental periconceptional folic acid was associated with as much as a 70% decrease in the rate of occurrence of neural tube defects, reduction by nearly two of conotruncal defects, and a significant decrease in facial closure defects as well (Eskes and Steegers-Theunissen, 1994; Shaw et al, 1995; Botto et al, 1996; Czeizel et al, 1996; Wald et al, 1996). Assuming that the missing protein has nothing to do with folic acid metabolism, one may calculate that the 22q11 deletion and folic acid insufficiency added together account for 70% or more of conotruncal defects, and an equally impressive number of other neural crest or neural tube–related defects. (But a potential relationship between ubiquination and the NMDA receptor is discussed below.)

The surprisingly high level of protection offered by folic acid supplementation was quite exciting for those who study the basic processes of neural tube closure and neural crest migration, since it implied that an unnamed process of extraordinary importance to these events was sensitively dependent on an adequate supply of folic acid. The effect of folate deficiency that has been cited as a potential cause of congenital defects is based on its central role in maintaining one-carbon groups and the levels of the primary methyl donor, S-adenosylmethionine, in transmethylation reactions. Folic acid is a key factor in transmethylation, and as such has the potential to affect embryonic cells by maintaining the availability of methyl groups for synthesis of nucleic acids. Because of this, folate insufficiency would be expected to limit the ability of embryonic cells to divide; however, it is not immediately obvious why such an effect would have special consequences to neural crest and neural tube cells, since embryonic cells in all areas of the embryo divide rapidly in early development. It appeared, therefore, that some other mechanism with a higher degree of tissue specificity must be responsible for the relationship between folate insufficiency and developmental defects originating in the neural crest and neural tube. A group of Dutch investigators led by T.K. Eskes was the first to hypothesize that the principal effector of these defects could be the amino acid homocysteine, since hyperhomocysteinemia results inevitably from folate insufficiency (Steegers-Theunissen et al, 1994, 1995; Thomas et al, 1994; van-der-Put et al, 1995). The relationship between folic acid and homocysteine is complex and has been the object of numerous studies over several decades. This relationship may be summarized briefly as follows:

- Dietary folic acid is reduced to tetrahydrofolate, which then is methylated to 5-methyl-tetrahydrofolate.
- 5-Methyl-tetrahydrofolate contributes a methyl group to homocysteine, converting it to methionine. Thus, with folate insufficiency, homocysteine becomes more concentrated in the serum and consequently in the follicular fluid (Steegers-Theunissen et al, 1994, 1995; Thomas et al, 1994).

Epidemiologic studies showed that elevated fasting serum homocysteine per se was a risk factor for neural tube and neural crest defects, including conotruncal defects (Eskes, 1998), even in the presence of normal levels of folate (Mills et al, 1995). Furthermore, as many as 25% of those with postprandial hyperhomocysteinemia are not identified by their fasting concentration of homocysteine (Bostom et al, 1995; Selhub and D'Angelo, 1998), indicating again that the link between homocysteine and the occurrence of conotruncal and other defects is likely to be even more numerically significant than generally supposed.

With these data, the homocysteine-as-a-teratogen hypothesis gained credibility. Since periconceptional folic acid supplementation was remarkably successful in preventing the occurrence of conotruncal, orofacial, and neural tube defects, it appeared that an exploration of the proposed role of homocysteine in disruption of development clearly was in order. If homocysteine per se in fact was the culprit, then it could be postulated that the cellular mechanism disrupted by homocysteine was of major significance to the normal development of those disrupted areas. We designed a test of the hypothesis that homocysteine could induce conotruncal, orofacial, and neural tube defects in the chicken embryo model. This model previously had been used to provide a great deal of the useful, general information about development of these areas, especially (in the present context) the conotruncus and other regions of the heart (see Chapters 1, 6, and 13).

EXPERIMENTAL EVIDENCE THAT HOMOCYSTEINE INDUCES CONOTRUNCAL AND OTHER NEURAL CREST-RELATED DEFECTS, AS WELL AS NEURAL TUBE DEFECTS

To provide a test of the hypothesis that homocysteine per se induces defects in organs that are derived from the neural tube or neural crest, we treated avian embryos with various concentrations of exogenous homocysteine at each of three time points during the processes of gastrulation and neurulation (6, 24,

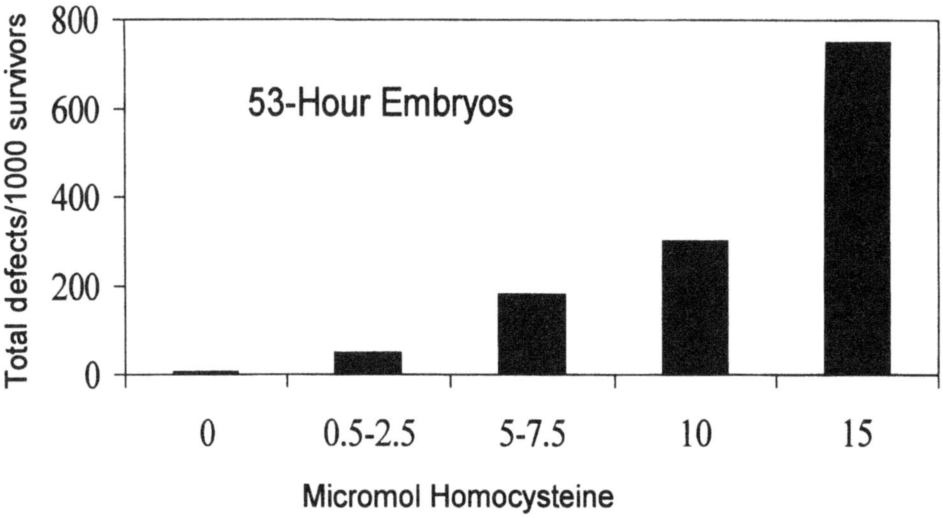

FIGURE 12.1. Chicken embryos were treated at 6, 24, and 48 hours with homocysteine doses that ranged from 0.5 to 15 µmol/dose. They were harvested after 53 hours (approximately stage 14) and examined for developmental abnormalities of the neural tube (see Figure 12.2) and of the orofacial region. The results show an obvious dose-dependency. (From Rosenquist et al., 1996, with permission.)

and 48 hours). The embryos were harvested upon the completion of neurulation at 53 hours (Hamilton-Hamburger stage 14). No congenital defects were found with any control amino acids including leucine, methionine, and cysteine; D-homocysteine was not effective in the presence of L-homocysteine. The results of this study showed that homocysteine was able to induce craniofacial and CNS defects in a dose-dependent fashion (Rosenquist et al, 1996) (Figures 12.1 and 12.2).

A second experiment was designed in which homocysteine treatment was limited to the period of neural crest migration, and the initial formation of the conotruncal septum as shown by our earlier studies. Embryos were treated at 24, 48, and 72 hours of incubation; migrating neural crest cells make initial contact with the myocardium, and initiate septum formation, between 72 and 84 hours of incubation. This window of time also includes the latter part of the process of neurulation. The embryos were harvested at day 9, upon the completion of conotruncal septation. Ventricular septal defects were found in about 23% of the embryos that survived 9 days (Rosenquist et al, 1996) (Figure 12.3).

Closure defects of both the dorsum (exencephaly and spina bifida) and the ventral wall were observed (Figure 12.4). The induction of ventral closure defects (Figure 12.4) was not predicted, but was consistent with the hypothesis, since ventral and dorsal midline closure defects are associated in the so-called schisis association described by Czeizel (1981). In this association, ventral and dorsal midline closure defects, including cleft palate, are associated statistically; the dorsal closure defects in turn are associated with conotruncal and craniofacial defects.

The above experiments supported the hypothesis that homocysteine would be a teratogen whether hyperhomocysteinemia was related to folic acid deficiency, faulty methionine metabolism, or exogenous homocysteine. However, these

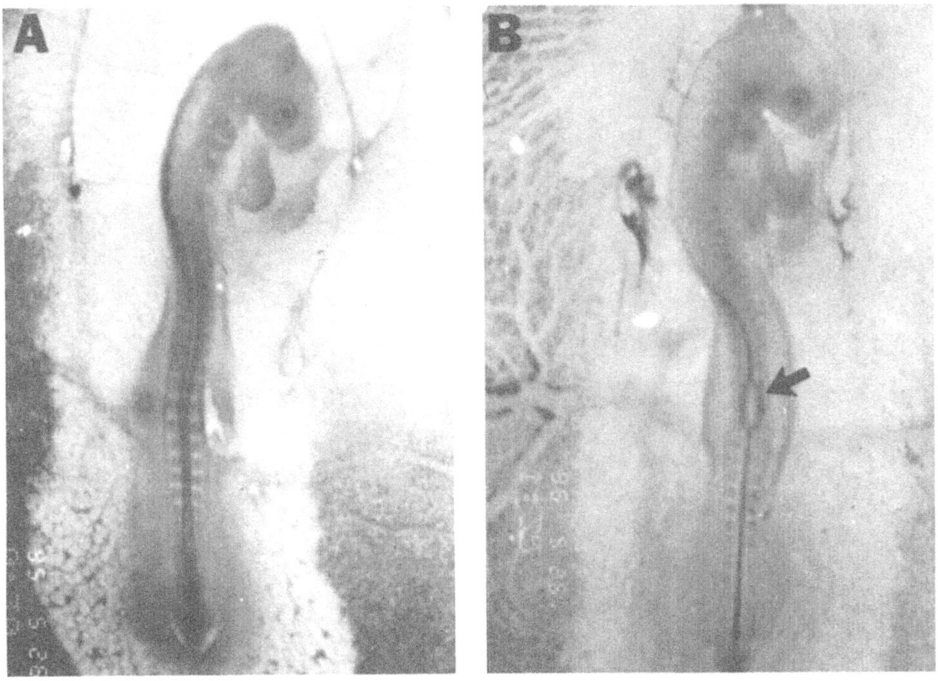

FIGURE 12.2. The control 53-hour embryo (approximately Hamilton-Hamburger stage 14) (A) Shows normal neural tube closure in the lumbar region, while the homocysteine-treated embryo (B) shows a failure of clusure in the same region (arrow) (nascent spina bifida). (From Rosenquist et al., 1996, with permission.)

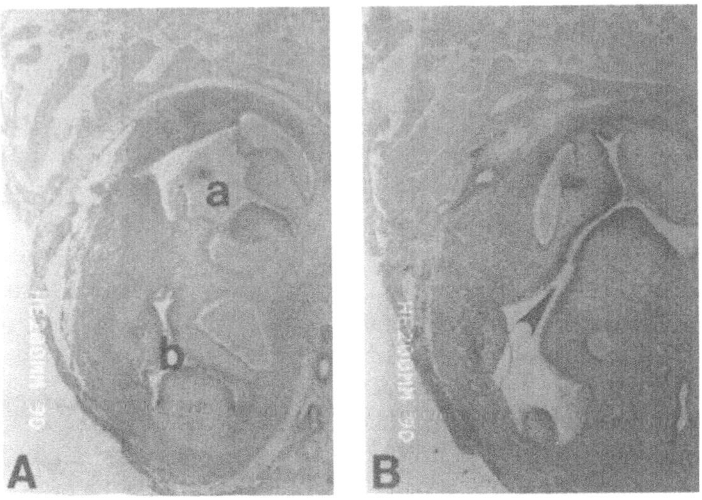

FIGURE 12.3. Nonconsecutive serial sections throught the cardiac outflow tract from an embryo day 9 (E9) embryo treated three times with 5 μmol/day homocysteine during incubation days 2 to 4. The aorticopulmonary septum separates the aorta (a) from the pulmonary artery (b) at the level in A, but a few sections downstream in B there is an obvious subaortic ventricular septal defect (VSD) (arrowhead). (From Rosenquist et al., 1996, with permission.)

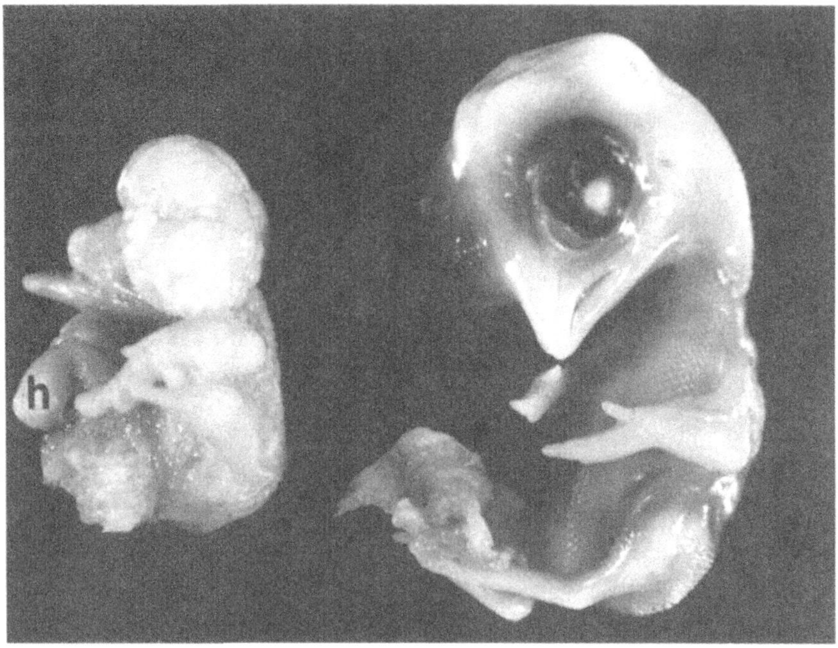

FIGURE 12.4. The homocysteine-treated E9 embryo on the left may be compared with the normal E9 embryo on the right. There is a lower beak but no upper beak, consistent with the closure of the face that frequently is a sequel of homocysteine treatment. The cranial vault is completely open and brain tissue has grown to cover the left side of the head and neck. The ventral wall of the thorax is open and the heart (h) protrudes (ectopia cordis). (From Rosenquist et al., 1996, with permission.)

experiments did not show the cellular or molecular mechanism(s) responsible for this effect. Homocysteine has been shown to be cytotoxic in some studies in vitro; but cytotoxicity resulted principally when the concentration of homocysteine was 50 to 100 times normal, or 5 to 10 times the highest concentration that was found to be teratogenic in our experiment. In our experiments we noted that there was an abundance rather than a paucity of tissue in the developmentally abnormal regions of homocysteine-treated embryos. For example, there was duplication of notochords rather than their absence, and each of the two was of normal size (Rosenquist et al, 1996). The brains of exencephalic embryos consisted of large folds of irregular neural tissue, resembling closely the brains of the exencephalic mice resulting from p53 knockout (Sah et al, 1995). The ectomesenchymal cells in the cardiac outflow tract were present and abundant, but ectopic. Taken together, these results did not indicate that the abnormal embryos had been subjected to some event that killed cells in their neural crest or neural tube.

A Hypothetical Mechanism for Homocysteine Interacting with the Neural Crest and Neural Tube: The *N*-Methyl-D-Aspartate Receptor

The presence of abundant neural crest and neural tube cells, and the duplicated notochords, implied an increase in mitosis; a decrease in embryonic apoptosis

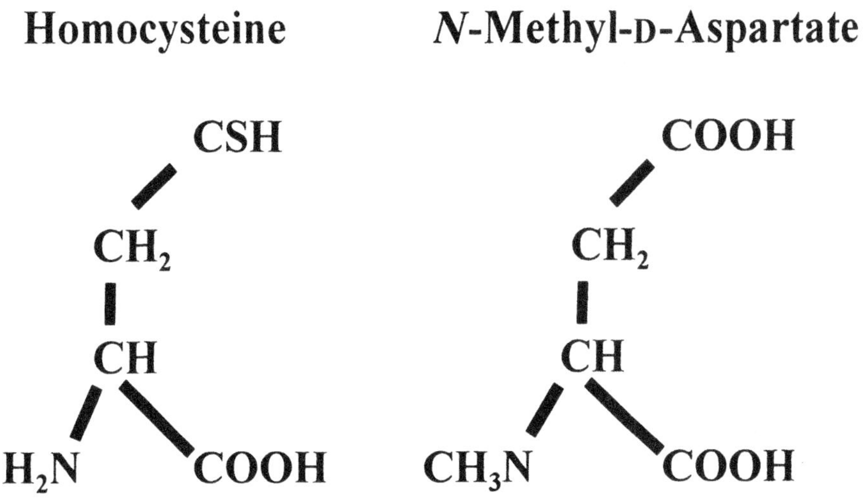

FIGURE 12.5. Two-dimensional representations of the molecular structures of homocysteine and *N*-methyl-D-aspartate (NMDA) show a high degree of similarity.

[a critical feature of both neural crest and neural tube development (Graham and Lumsden, 1996; Graham et al, 1996; Weil et al, 1997)]; or both a decrease in apoptosis and an increase in mitosis. With each of these hypothetical scenarios, the most likely mode of action of homocysteine would be via some receptor-mediated event. Although no receptor for homocysteine had been described, an oxidative product of homocysteine, homocysteic acid, was known to be an agonist of the NMDA receptor (e.g., Parsons et al, 1998). We noted also the structural similarity between the homocysteine molecule and the NMDA molecule (Figure 12.5).

Although little information had been published about the NMDA receptor in early development, it presented an attractive possibility in the context of regulation of neural crest migration and neural tube closure. NMDA receptor activation is a key component of the migration of neurons, neuronal plasticity, the outgrowth of neurites, and long-term potentiation of neurons. These in turn depend on the same functions that are central to neural tube closure and neural crest migration, that is, the cytoskeleton and key cell adhesion molecules (Balazs et al, 1989; Cornell-Bell et al, 1990; Rashid and Cambray-Deakin, 1992; Komuro and Rakic, 1993; Wang et al, 1996; Uberti et al, 1998). Based on the above, we hypothesized that our putative receptor for homocysteine would be an NMDA or NMDA-like receptor, and that homocysteine would disrupt normal development by overstimulation of this putative receptor.

To test this hypothesis, we applied a strategy based on the common anticonvulsant therapy; homocysteine was given to embryos in teratologic doses, along with a selected series of NMDA receptor antagonists (anticonvulsants) to reverse the predicted agonistic effect of homocysteine. The prediction was that the antagonists would restore the appropriate level of activity to the NMDA receptors, promote normal development, and protect neural crest and neural tube derivatives from abnormal development.

In fact this prediction was 180° wrong. When NMDA receptor antagonists were given to the embryos with homocysteine, there were more defects than with homocysteine alone, not fewer, and some of the well-known and widely used antagonists (such as Mk801) induced neural crest and neural tube defects, including conotruncal defects, even in the absence of homocysteine. Although they were puzzling initially, these results ultimately made a great deal of sense. A paper by Lipton et al (1997) published fortuitously at the same time as these experiments were being carried out showed that homocysteine was capable of acting as either an agonist or an antagonist of the NMDA receptor, depending on its concentration and other features of the solution. It was obvious, therefore, that the exogenous homocysteine we had given to the avian embryos in our model system was behaving as an NMDA receptor *antagonist*, following the rules for its behavior as determined by Lipton et al.

Although no other investigation had specifically identified teratogenicity as a general property of the NMDA receptor antagonists, the literature in fact was full of examples; the ability to act as an inhibitor of the NMDA receptor per se, or to induce an effect that worked in opposition to the NMDA receptor, was found to be a property of several of the major chemical risk factors cited in the Baltimore-Washington Infant Study of developmental abnormalities of the heart (Ferencz et al, 1997). Examples of compounds whose effects, include inactivation of the NMDA receptor, and also are associated with neural crest and neural tube defects, include ethanol (Bhave et al, 1996; Johnson et al, 1996; Munger et al, 1996); valproic acid (Nau, 1985; Gofflot et al, 1996; Finnell et al, 1997); benzodiazepines interacting with valproic acid (Laegreid et al, 1993); and antitussives (Ferencz et al, 1997; Andaloro et al, 1998). Ibuprofen, shown to be a risk factor for outflow tract defects in the Baltimore-Washington Infant Study (Ferencz et al, 1997), exerts its antinociceptive effect by interfering with NMDA receptor activation (Bjorkman, 1996). There are other associations that tempt speculation, for example diabetes is a major risk factor for congenital heart defects, but the mechanism is not known (Ferencz et al, 1997). However, hypoinsulinism is significantly related to NMDA receptor hypofunction (Liu et al, 1995). Thus, according to our hypothesis, hypoinsulinism would present a risk factor for conotruncal and related defects because it induces NMDA receptor hypofunction.

The hypothesis that the induction of neural tube–neural crest defects may occur as the result of NMDA receptor hypofunction is in a sense an elaboration of the 1995 hypothesis of Ferencz and Correa-Willasenor (1995), who said that there must be some neurotoxic or neuroteratogenic origin for conotruncal and related defects.

It is interesting to note that the 22q11 microdeletion [*UFD1L* haploinsufficency (Nickel and Magnis, 1996; Yamagishi et al, 1999)] contributes to conotruncal and craniofacial defects that are identical to those induced by NMDA receptor antagonists, and also is associated with learning and behavioral disorders, or psychoses that are closely associated with the NMDA receptor, for example schizophrenia (Vataja and Elomaa, 1998). A pathway that is common to both the mutation and the receptor antagonists is an effective sparing of ubiquinated proteins (Hirsch et al, 1997; Lachman et al, 1997). NMDA receptor activation leads to degradation of ubiquinated proteins in neurons (Okada et al, 1995), whereas NMDA receptor antagonists would cause the same ubiquinated proteins to be spared, and the presumed effect of insufficient *UFD1L* also is to spare ubiquinated proteins.

Two apparent ambiguities need to be addressed regarding the proposed relationship between NMDA receptor antagonists and conotruncal defects. First,

NMDA receptor antagonism is not the only, and may not be the principal, action of each of the compounds cited above. Indeed, a complexly diverse menu of receptors may be affected, for example by ethanol, including GABAA (the ionotropic GABA receptor), serotonergic, nicotinergic, muscarinic and glycine receptors (Crews et al, 1996). For ethanol as well for the other compounds cited above, the principal or best-known pharmacologic effect may involve some effect distinct from the NMDA receptor, and the range of these other effects may be quite disparate for the compounds cited above. But this disparity could be taken as support for the NMDA receptor antagonist teratogenicity hypothesis, since NMDA receptor antagonist activity is virtually the only common factor among a set of compounds with widely differing pharmacologic activities, all of which induce similar neural tube and neural crest defects. Furthermore, the NMDA receptor has the exceptional ability, via its calcium channel, to regulate genes for cell adhesion and cytoskeletal proteins that are critical for cell migration or morphogenetic movement (Rashid and Cambray-Deakin, 1992), as discussed above. Therefore, while other receptors also may be affected by these compounds, the NMDA receptor appears to be highly suited to regulate certain key functions, and also to be the receptor that is most relevant to the observed defects.

The second apparent ambiguity in the NMDA receptor antagonist teratogenicity hypothesis is that the same drugs also may be associated with heart or other defects that have no apparent relationship with neural crest migration. For example, ethanol ingestion is a risk factor for muscular interventricular septal defect (Ferencz et al, 1997), a defect that has not been linked with the cardiac neural crest. Thus, there may be a mechanism whereby ethanol may induce heart defects in addition to its effect on the cardiac neural crest, or the neural crest may have a role that has not yet been identified. There are three obvious possibilities. First, the well-described toxic and depressant effects of ethanol may have specific importance to the myocardial cells in the muscular septum (Combs and Acosta, 1990; Martinez and Penna, 1992; Richards, 1997). Second, embryonic myocardiocytes may express an NMDA or NMDA-like receptor that has not been described previously (see further discussion of NMDA receptor heterogeneity, below). Third, the observed effect may be indirect, whereby neural crest or neural tube cells that regulate normal development of the septum by a previously unrecognized mechanism are altered via the NMDA receptor mechanism, and this alteration impacts negatively on the development of the muscular septum. Recent results indicate that cardiac neural crest cells can affect the development of myocardial cells (Conway et al, 1997). These questions may be addressed by appropriate tests of the hypothesis.

Tests of the Hypothesis That Inhibition of the *N*-Methyl-D-Aspartate Receptor by Homocysteine Disrupts Development of the Neural Crest and Neural Tube

Significant corollaries of the above hypothesis that lend themselves to testing have been outlined above:

- If the ability of homocysteine to induce neural crest and neural tube defects is related to its ability to act as an antagonist of the NMDA receptor, then the ability to induce such defects must be a general property of NMDA receptor antagonists.

- If neural crest and neural tube defects arise when the activity of the NMDA receptor is inhibited, then the same defects will be reduced or prevented by activation of these receptors in the presence of teratogenic concentrations of the inhibitor. According to this corollary, neural crest and neural tube defects induced by homocysteine should be reduced or prevented by activation of the NMDA receptor during episodes of hyperhomocysteinemia. This corollary may apply as well to those compounds described above that are disruptive of neural tube and neural crest development, and also have some ability to act as NMDA receptor antagonists, such as ethanol or valproic acid. If the teratogenic effect of these compounds is related to their inhibition of the NMDA receptor, then activation of the receptor by agonists should reduce the rate of occurrence of the defects.
- If the teratogenic action of homocysteine is a function of its ability to inhibit the NMDA receptor, then homosyteine may interact additively with other NMDA receptor antagonists to induce defects of the neural crest and neural tube, or synergistically under certain conditions.
- The present hypothesis holds that the NMDA receptor is uniquely important to normal development of the neural crest and neural tube. Since the NMDA receptor can function only when the plasma membrane is depolarized by the activity of other neurotransmitter receptors, inactivation (e.g., by a GABA receptor agonist) would effectively inhibit the NMDA receptor. Thus, applying the present hypothesis, if a GABA receptor agonist is teratogenic (e.g., ethanol or valproic acid), it is most likely to be because of its secondary effect on the activity of the NMDA receptor.

Testing the Hypothesis

The results described above imply that the ability to induce abnormal neural tube and neural crest development may be a general characteristic of antagonists of the NMDA receptor, but no evidence to support this implication had been derived from direct tests of well-characterized and highly specific NMDA receptor antagonists. Therefore, we evaluated a set of such antagonists that were selected to represent three different classes, each acting at a different site on the hetero-oligomeric NMDA receptor: ion channel blockers, glycine site antagonists, and glutamate site antagonists (Andaloro et al, 1998). The results showed that all three classes of NMDA receptor antagonist were able to induce defects in the development of the neural tube and neural crest. Ion channel blockers were the most effective, while glutamate site antagonists were the least effective (Figure 12.6). These differences do not correlate with the relative efficacies of the various compounds, or with their respective affinities for the NMDA receptor, but they may correlate with their relative bioavailabity. The channel blocker antagonists were more highly concentrated in treated embryos than antagonists in the other classes, consistent with the fact that their structures generally are more lipophilic (Andaloro et al, 1998). Likewise, homocysteine is significantly more teratogenic than any of the well-characterized antagonists except the ion channel blocker memantine (Andaloro et al, 1998), and homocysteine has shown a remarkable ability to become concentrated in the embryos in our model system (Rosenquist et al, 1996; Andaloro et al, 1998).

Another possibility for the differing teratogenic potencies of the antagonists is suggested by NMDA receptor physiology during the early stages development,

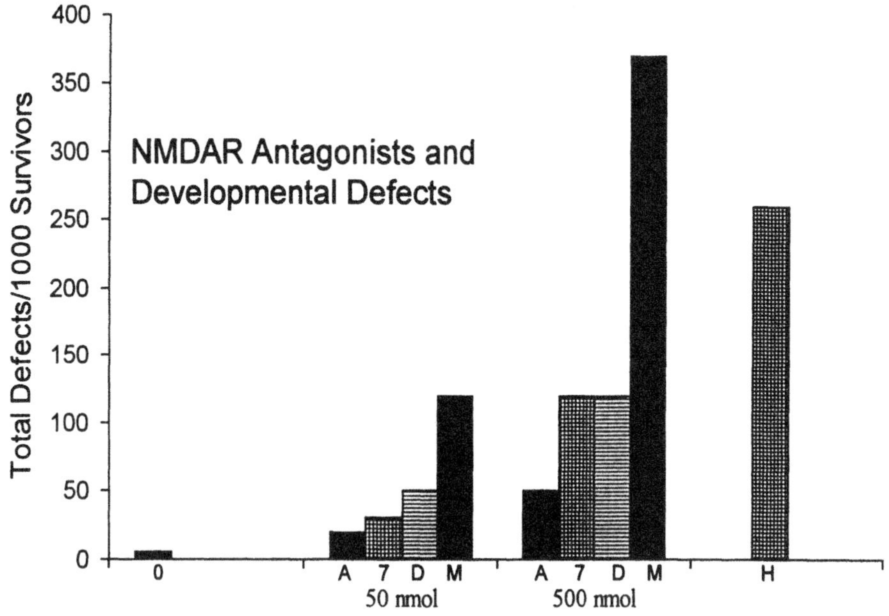

FIGURE 12.6. NMDA receptor (NMDAR) antagonists were given to chicken embryos at various dose levels, on 3 days during neurulation, and the embryos were evaluated after 76 hours of incubation for neural tube and neural crest defects. The NMDA receptor calcium channel blockers dextromethorphan (D) and memantine (M) were the most teratogenic. The glycine site antagonist 7-chlorokynurenic acid (7) also induced a significant number of defects at a dose of 500 nmol/day. The glutamate site antagonist aminocyclopropane-carboxylic acid (A) was not significantly teratogenic. H = the number of defects given by a daily 5 nmol dose of homocysteine. (From Andaloro et al., 1988, with permission.)

when the neural tube is closing and the neural crest cells are preparing to emigrate. If, for example, there are essential NMDA receptor–regulated functions that respond in the presence of high local concentrations of glutamate or glycine, then the noncompetitive ion channel blockers would have a greater impact than the same concentrations of glutamate site or glycine site antagonists (Andaloro et al, 1998). Unfortunately, the nature of the NMDA receptor in the early embryo is not known. Mark Thomas (personal communication) has shown by immunoblotting that one form of the generic NR1 portion of the NMDA receptor, containing the ion channel, is present in the chicken embryo at stage 8 to 10. This information is rudimentary however, since the NMDA receptor is a hetero-oligomer composed of $NR1_{(A\text{-}L)}$, $NR2_{(A\text{-}D)}$, and NR3 subunits (Monyer et al, 1992; Sugihara et al, 1992; Ishii et al, 1993; Buller et al, 1994), and various combinations of these subunits appear to manifest functional differences.

Many questions remain to be explored; however, the data obtained to date support the hypothesis that the ability to induce neural tube and neural crest defects is a general property of NMDA receptor antagonists.

According to the second corollary of this hypothesis, neural crest and neural tube defects that arise when the activity of the NMDA receptor is inhibited will be reduced or prevented by activation of these receptors in the presence of teratogenic concentrations of the inhibitor. Neural crest and neural tube defects

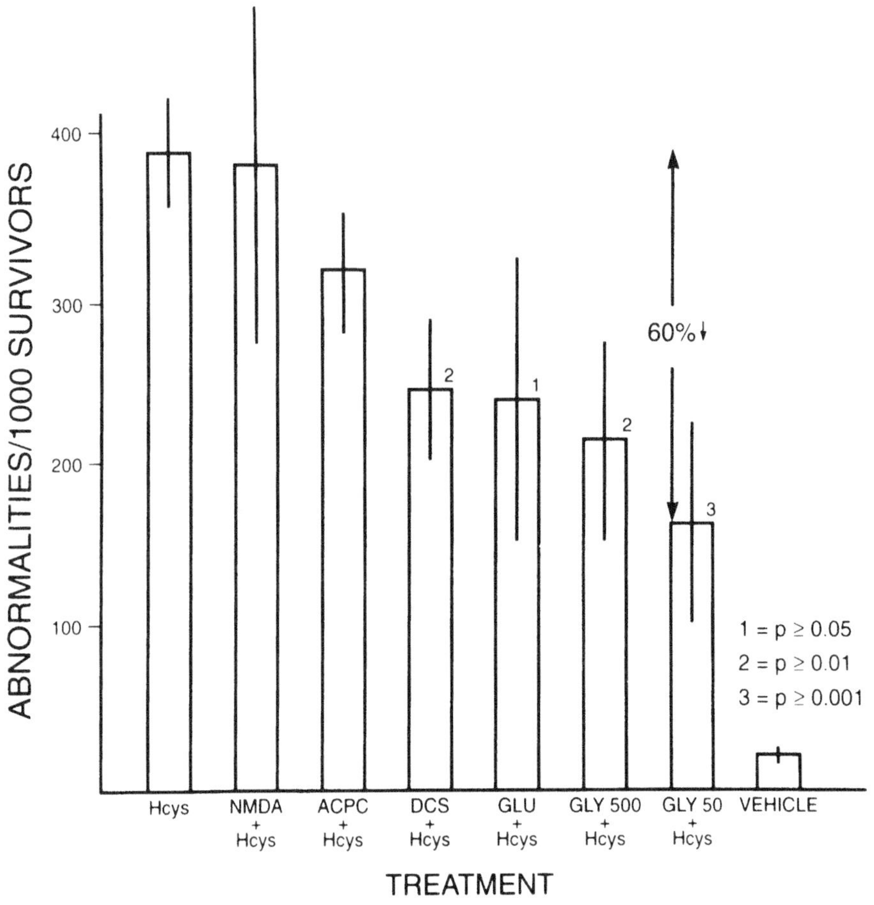

FIGURE 12.7. The total number of developmental defects that followed treatment of embryos with homocysteine is compared with the number that occurred following treatment with vehicle only, or with homocysteine and an NMDA receptor agonist. For each agonist, the dose represented here is the dose that gave the most protection: NMDA, 50 nmol/day; and glutamic acid (GLU), 50 nmol/day. Glycine at 500 nmol/day (GLY500) had a significant protective effect, but 50 nmol glycine/day (GLY 50) was the most effective protective agonist, reducing the occurrence of neural crest and neural tube defects by about 60%. Bars = standard error of the mean.

induced by homocysteine should be reduced or prevented by activation of the NMDA receptor during episodes of hyperhomocysteinemia.

To test this hypothesis, we treated embryos during the process of neural tube closure with homocysteine thiolactone sufficient to induce neural crest and neural tube defects in about 40% of survivors, or homocysteine thiolactone in combination with each of a selected set of NMDA receptor agonists in 0.05 to 5,000-nmol doses. The glutamate-site agonists selected were *L*-glutamate and *N*-methyl-D-aspartate. Glycine-site agonists were L-glycine, D-cycloserine, and aminocyclopropane-carboxylic acid. At their most effective dose, D-cycloserine and glutamate each provided significant protection. However, glycine was the most effective overall, reducing defects by over half ($p > .001$) (Figure 12.7). These results support the hypothesis that homocysteine may affect neural crest and neural tube development by its ability to inhibit the NMDA receptor. One potentially important consequence of this putative mechanism is that homocysteine may interact synergistically with other NMDA receptor antagonists to enhance its effect on development (see below).

These results predict that any agent or genetic event associated with NMDA receptor hypofunction should be looked at as a probable teratogen for conotruncal and related defects. Conversely, a teratogen or genetic defect that is associated with conotruncal defects, but whose mechanism is unknown, should be investigated for a potential effect on the NMDA receptor. For example, many of the miscellaneous solvents that were found in the Baltimore-Washington Infant Study to be risk factors for conotruncal defects, also have CNS effects (Ferencz et al, 1997). The molecular structures of many industrial solvents suggest the possibility of their ability to inhibit the NMDA receptor (Daniell, 1991; Martin et al, 1991, 1995; Aronstam et al, 1994; Martin and Aronstam, 1995). These might include volatile anesthetics or structurally similar compounds that are associated, both experimentally and epidemiologically, with an increased occurrence of neural crest and neural tube defects (e.g., chloroform or trichloroacetic acid).

It may be argued that valproic acid is the most widely studied of the above set of compounds, yet there is still no consensus on its mechanism of teratogenesis. Among its other pharmacologic effects, valproic acid activates the GABA receptor and therefore effectively inhibits the inhibits the activity of the NMDA receptor in fully differentiated neurons of the CNS (Ko et al, 1997). Therefore, according to our hypothesis, the NMDA receptor or its functional homologue in the early embryo would be inhibited by valproic acid. This inhibition would be the basis of the teratogenic effects of valproate, and these effects would be attenuated by providing agonists of the NMDA receptor when valproate is given to embryos. As a test of this hypothesis, we first showed that valproic acid induced neural tube and neural crest defects in the chicken embryo model in a dose-dependent manner (Figure 12.8). Most of the previous data for valproic acid referred only to mammalian models. From these data, we chose a concentration of 100 mM because it had induced a significant increase in neural crest and neural tube defects (approximately 75/1,000 survivors), but killed significantly fewer treated embryos than higher concentrations. When we gave 100 mM valproic acid with the natural NMDA agonist glycine, neural crest and neural tube defects were eliminated, strongly supporting the hypothesis (Figure 12.9).

According to the next corollary of the general hypothesis, homocysteine and other NMDA receptor antagonists may interact additively or synergistically to increase the rate of occurrence or severity of congenital defects of the neural crest and neural tube.

For an initial test of this corollary, we treated embryos with homocysteine and ethanol, or the widely used antitussive/anticonvulsant dextromethorphan. Ethanol is an ideal candidate since it is a potent GABA receptor agonist as well as an NMDA receptor antagonist, and one of its NMDA binding sites [the PCB binding site (Hundt et al, 1998)] is different from that of homocysteine [the glycine binding site (Lipton et al, 1997)]. Ethanol also is a well-described effector of neural tube and neural crest defects, and fetal alcohol syndrome is widely known about by both scientists and laymen. Human use of both ethanol and dextromethorphan is related to conotruncal defects, according to epidemiologic studies in humans as discussed above (e.g., Ferencz et al, 1997). Dextromethorphan blocks the NMDA receptor at the calcium channel, and it was a potent inducer of neural crest and

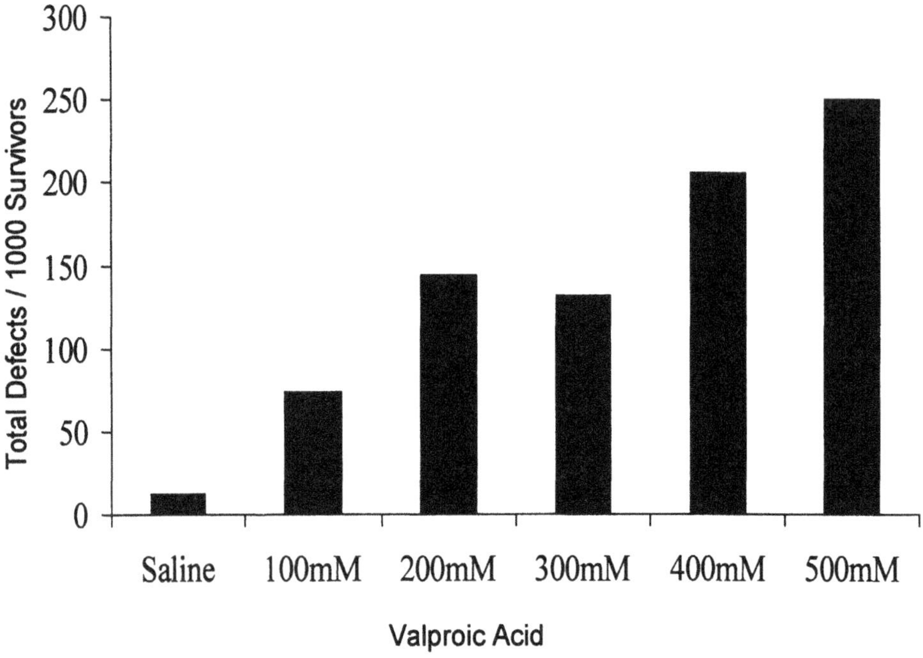

FIGURE 12.8. Chicken embryos were treated three times with 50 μL of valproic acid in concentrations that ranged from 100 to 500 mM. They were harvested after 53 hours (approximately stage 14) and examined for developmental abnormalities as described above. Valproic acid was teratogenic to the chicken embryos at all concentrations, and the effect was dose-dependent.

neural tube defects in our chicken embryo model (Andaloro et al, 1998). Another point that added interest to the use of ethanol and dextromethorphan is that they may be the NMDA receptor antagonists that are most readily available to most individuals.

We treated embryos during the process of neural tube closure with homocysteine thiolactone to induce neural crest and neural tube defects in about 40% of survivors; homocysteine thiolactone in combination with 50 μL 0.1% ethanol, 0.005 nmol dextromethorphan, or 0.5 nmol dextromethorphan; or the ethanol and dextromethorphan solutions alone. We had shown previously that these doses of dextromethorphan did not increase neural crest or neural tube defects when they were given alone (Andaloro et al, 1998). However, when they were given in combination with homocysteine, dextromethorphan as well as ethanol acted synergistically to increase the number of neural crest and neural tube defects to a level that was significantly higher than could be predicted from the behavior of the compounds when any was given independently of the others (Figure 12.10). The interaction of homocysteine with 0.5 nmol dextromethorphan was especially dramatic. These results predict that the combined effects of two or more NMDA antagonists may be teratogenic, even when the concentration of each is in a safe range.

Although the above results show that inhibition of an NMDA receptor leads to neural tube and neural crest defects, they do not show the role of the putative NMDA receptor in the regulation of normal neural crest cell migration or neural tube closure.

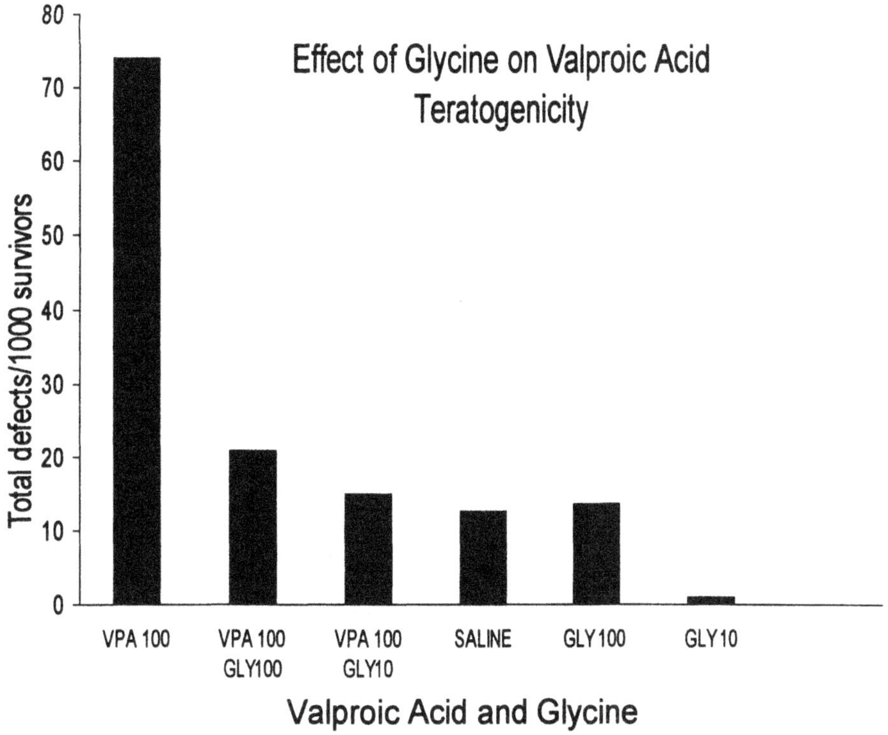

FIGURE 12.9. Chicken embryos were treated three times with 50 µL of a 100-mM solution of valproic acid to which was added glycine in concentrations of 10 or 100-mM. Both concentrations of glycine were highly effective in preventing valproic acid–induced defects, reducing the rate of defects to levels that were not different from saline treatment. It is interesting as well that 10 mM glycine actually improved the rate of defects over that obtained with saline.

What Cellular Process or Pathway That Is Important to Normal Neural Crest Migration and Neural Tube Closure Is Disrupted by Homocysteine?

Beneficial Cell Death

From the above, it is clear that homocysteine and other NMDA receptor antagonists pose a major threat to normal development of neural tube and neural crest derivatives. The results described above show that regulation of an NMDA receptor or an NMDA-like receptor must be important for normal development of these areas, but they do not show what aspect of the physiology of neural crest and neural tube cells is subject to this regulation. Although few published studies deal with the NMDA receptor or its ligands in early development, it is informative to review some of the roles that have been described for the NMDA receptor in later prenatal and in postnatal development of the CNS, and then to compare these functions with the requirements of neural crest during migration and neural tube cells during closure of the neural tube.

One of the most widely studied roles of NMDA antagonists in the postnatal CNS is their protective function. Antagonists limit the ability of agonists to overactivate the NMDA receptor and consequently to initiate the cell death program

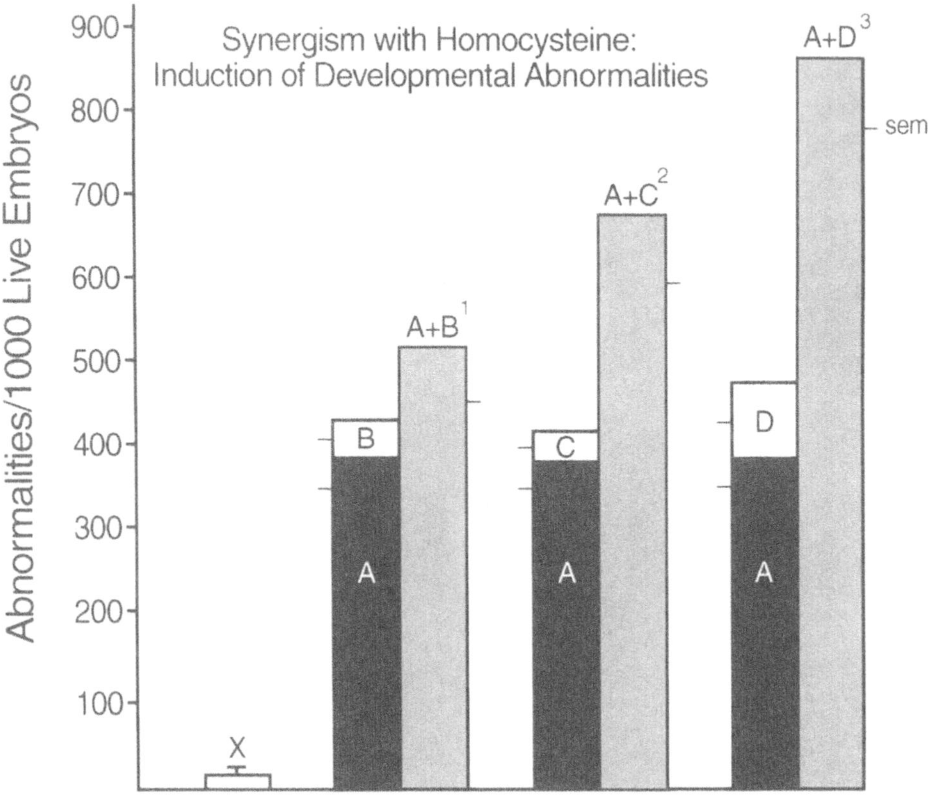

X = Vehicle Control
A = 5 μmol Homocysteine Thiolactone alone
B = 0.1% Ethanol alone
C = 0.005 nmol Dextromethorphan Hydrobromide alone
D = 0.5 nmol Dextromethorphan Hydrobromide alone
A+B = Homocysteine given with Ethanol
A+C = Homocysteine given with 0.005 nmol Dextromethorphan
A+D = Homocysteine given with 0.5 nmol Dextromethorphan

[1] $p > 0.01$ [2] $p \leq 0.01$ [3] $p \leq 0.01$

FIGURE 12.10. Homocysteine given at a rate of 5 μmol/day for the first 3 days of incubation resulted in approximately 381 defects/1,000 surviving embryos (A). Addition of 1% ethanol (B), 0.005 nmol dextromethorphan (C) or 0.5 nmol dextromethorphan (D) resulted in a synergistic interaction with homocysteine (A+B, A+C, A+D). This synergism was highly significant for both concentrations of dextromethorphan.

of neurons (Bonfoco et al, 1996; Tenneti et al, 1998). Programmed cell death is an essential and beneficial feature of the neural tube during closure, and of the neural crest during segmentation (Graham and Lumsden, 1996; Graham et al, 1996; Weil et al, 1997). When this beneficial cell death is inhibited, as in p53 loss-of-function mutations in mice (Sah et al, 1995), neural tube defects result. Based on the above, we tested the hypothesis that antagonists of the NMDA receptor including

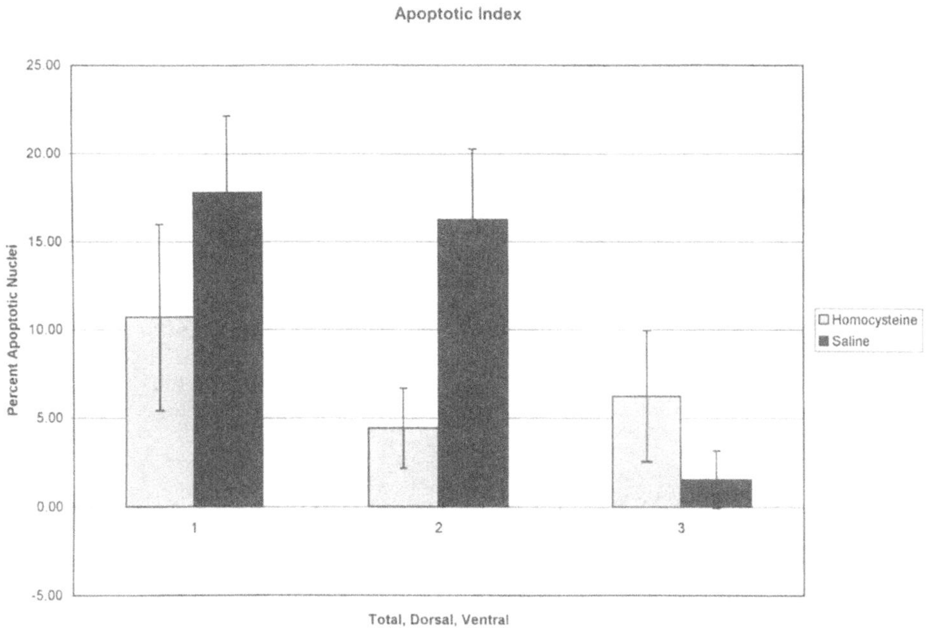

FIGURE 12.11. Chicken embryos were given a single dose of 5 μmol homocysteine after 4 hours of incubation, or a saline control solution. Then they were harvested at stage 10. Sectioned embryos were assayed for programmed cell death by terminal deoxynucleotidyl transferage dUTP nick end-labelling (TUNEL). For the neural tube, total nuclei shown by propidium iodide were compared with TUNEL-positive nuclei in the whole neural tube (1), in the dorsal half of the neural tube (2), or in the ventral half of the neural tube (3). Equivalent levels of the neural tube were determined by counting somites. In the saline-treated embryos, nearly 90% of the TUNEL-positive cells were in the dorsal half of the neural tube at all levels, whereas in the homocysteine-treated embryos there was no difference in distribution between the two halves. The total number of TUNEL-positive cells was reduced following homocysteine treatment, but the reduction did not achieve any statistical significance. (From the doctoral thesis of G.L. Witt, University of Nebraska Medical Center, July 1999.)

homocysteine may be teratogenic because they interfere with an essential program for beneficial cell death.

The results of experiments with homocysteine treatment during early development in the chicken embryo model indicated that hyperhomocysteinemia was related to reduction of beneficial programmed cell death in the neural tube, and a disruption of the signal that marks specifically the dorsal-most cells for programmed death (Figure 12.11). During normal development, most of the apoptotic nuclei in the neural tube are located near the midline in the dorsal half of the neural tube, but treatment with homocysteine was associated with a more random distribution of apoptotic nuclei within the neural tube (Figure 12.12). Quantitative and geographic changes in beneficial cell death during neurulation would be detrimental to the highly regulated process of neural tube closure and neural crest cell migration.

The results of the above experiment implied that homocysteine treatment was associated with a reduction in the dorsalizing signal, or in the reception of such a signal, during neural tube closure. The experiment described in the next section

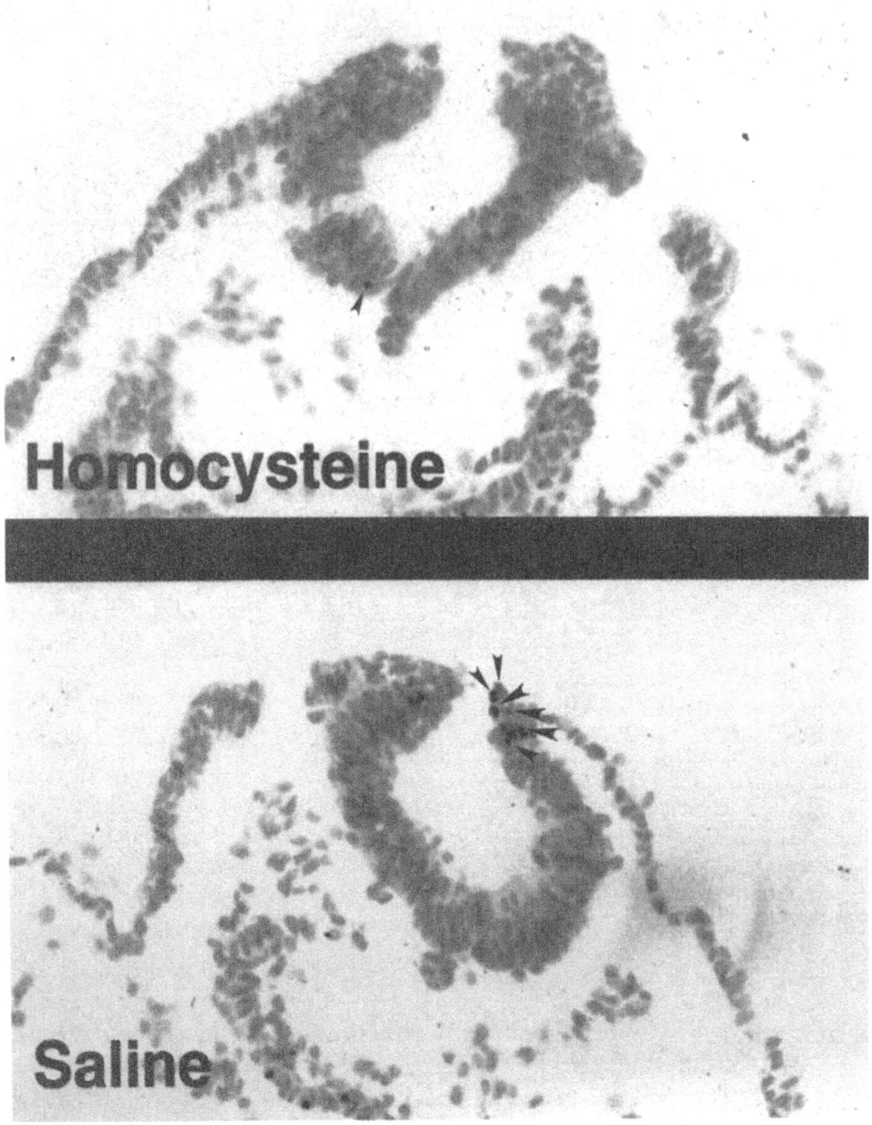

FIGURE 12.12. Embryos were treated with saline or homocysteine after 4 hours of incubation, harvested after 34 hours, fixed, embedded, and sectioned. The presence of apoptotic nuclei was analyzed with the TUNEL assay. In these cross sections taken near the spinal cord–brain junction, TUNEL-positive nuclei in the saline treated embryo are concentrated in the neural crest (arrows), but the only TUNEL-positive nucleus in the section from the homocysteine-treated embryo is ventral (arrow). The neural crests from the treated embryo have a somewhat enlarged appearance compared with those of the saline treated embryo. (From the doctoral thesis of G.L. Witt, University of Nebraska Medical Center, July 1999.)

was designed to examine genes associated with establishment of the dorsal-ventral axis in the neural tube.

Expression of PAX-3

Data from the various studies cited above for homocysteine and other NMDA receptor antagonists indicate a pattern of abnormal development that can be compared with patterns that arise with perturbation of the expression of the principal regulatory gene for dorsalization of the neural tube, *pax-3* (Wehr and Gruss, 1996). When the expression of *pax-3* has been disrupted experimentally, or with spontaneous mutations of *pax-3*, dramatic changes have occurred in the development of the neural crest including the conotruncus and the neural tube, for example in Waardenburg syndrome in humans, or in the Splotch mouse. In chicken embryos, the *pax-3* gene is expressed in the primitive streak at stage 7, and then in the neural folds and early somites at stage 8 (Goulding and Paquette, 1994). *Pax-3* expression becomes restricted to the dorsal neural folds and migrating neural crest by stage 10. Alterations in the expression of the *pax-3* gene have been induced chemically with valproic acid (Wlodarczyk et al, 1996), which also disrupts neural tube closure resulting in exencephaly and spina bifida (Nau, 1985; Finnell et al, 1997). Because of the apparent alteration in the dorsalization signal as well as the array of other homocysteine-induced defects that simulated *pax-3* perturbation, Georgianna Witt tested the hypothesis that homocysteine may interfere with the proper expression of *pax-3*. Her results indicated that homocysteine treatment during early embryonic development in the chicken embryo resulted in a repression of *pax-3* gene expression (Witt et al, 1997). Although we postulate that this apparent repression, like the more downstream teratogenic effects of homocysteine, is a mediated by an NMDA or NMDA-like receptor, the whole signaling pathway for activating transcription of the *pax-3* gene has yet to be uncovered. It is known that the dorsal restriction of *pax-3* expression in the neural tube appears to be related to signals mediated by Sonic Hedgehog (SHH) (Fan and Tessier-Lavigne, 1994).

In summary, it appears that homocysteine given to embryos very early in development may interfere in some way with the process of dorsalization that is central to neural tube closure. Evidence in support of this interference is a decrease in *pax-3* expression and an increased entropy in the distribution of apoptosis.

Other Potential Roles for Homocysteine and an Nmda or NMDA-Like Receptor in Heart Development

A teratogen may affect the development of an embryonic structure if it is administered at a critically sensitive period, for example, when the tissue is susceptible to its action. Cells of the cardiac neural crest migrate into the branchial arches and differentiate into the vascular smooth muscle cells of the great arteries (see discussion in Chapter 6). Their sister cells in the outflow septum also express features of the smooth muscle phenotype (Beall and Rosenquist, 1990; Rosenquist et al, 1990b). We have found that homocysteine is a potent mitogen for neural crest–derived ectomesenchymal cells that express the smooth muscle phenotype in the early chicken embryo, and we have described several components of the signal transduction cascade for the growth promoting effects of homocysteine (Gadson et al, 1997). These effects were found to be inhibited by NMDA recep-

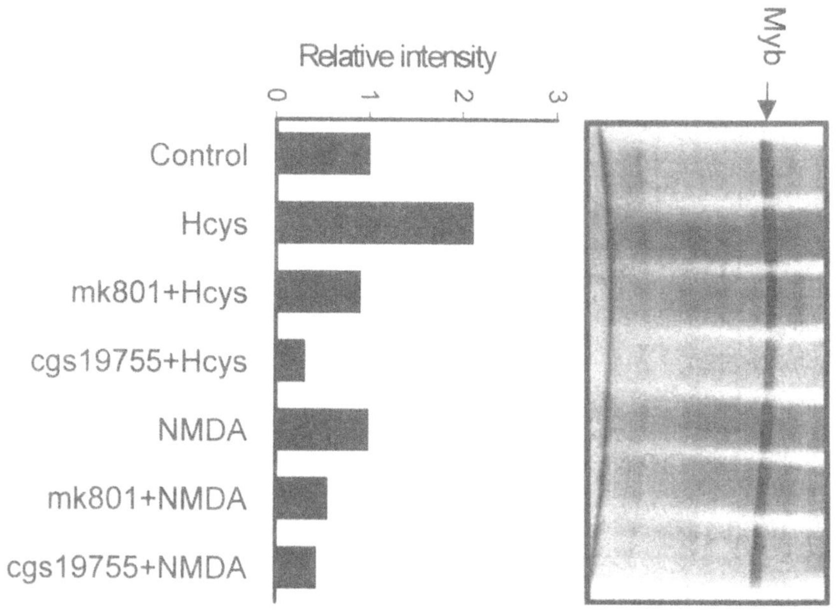

FIGURE 12.14. Embryonic neural crest–derived smooth muscle cells were treated with homocysteine (Hcys); with NMDA; and with Hcys and NMDA in the presence of a noncompetitive NMDA receptor blocker (MK801) or a competitive NMDA receptor blocker (CGS19755). Then the expression of Myb was measured from immunoblots. Homocysteine induced about a twofold increase in Myb. This increase was completely abolished by both of the NMDA receptor blockers. (From Dalton et al., 1997, with permission.)

neural tube closure. Indeed, they seldom occur with any other malformations, and rarely are part of a syndrome or chromosomal abnormality.

The biologic mechanism for TGA is not well understood (Ferencz and Correa-Willasenor, 1995); however, in experimental model systems, TGA has been associated with changes in the extracellular matrix in the outflow tract (Nakajima et al, 1997; Yasui et al, 1997). Since hyperhomocysteinemia has a profound effect on the expression of the matrix regulatory gene c-*myb* as discussed above, it is conceivable that homocysteine and related compounds could induce or exacerbate TGA by disrupting the coordinated expression of elastin and collagen in the highly ordered outflow tract matrix (Rosenquist et al, 1988, 1990a,b). Matrix changes that might not induce TGA by themselves may exacerbate the effects of other phenomena associated with TGA, for example dysgenesis of the cushions (Nakajima et al, 1997; Yasui et al, 1997).

We hypothesize that the relationship between TGA defects and folate insufficiency is based on a growth factor–like effect of homocysteine, acting on ectomesenchymal cells in the outflow septum, and that this effect is normalized by folate supplementation.

SUMMARY

The data that have been obtained to date support the general hypothesis that NMDA receptor antagonists, including homocysteine, induce conotruncal, orofacial, and neural tube defects. It appears that this effect could be the mechanism of action for many of the most common risk factors for these defects, and that these risk factors may be able to interact synergistically to increase their pernicious effect on development.

REFERENCES

Andaloro, V.J., Monaghan, D.T., and Rosenquist, T.H. (1998). Dextromethorophan and other N-methyl-D-asparate receptor antagonists are teratogenic in the avian embryo model. *Pediatr Res* 43:1–7.

Aronstam, R.S., Martin, D.C., and Dennison, R.L. (1994). Volatile anesthetics inhibit NMDA-stimulated 45Ca uptake by rat brain microvesicles. *Neurochem Res* 19:1515–1520.

Balazs, R., Hack, N., Jorgensen, O.S., and Cotman, C.W. (1989). N-methyl-D-aspartate promotes the survival of cerebellar granule cells: pharmacological characterization. *Neurosci Lett* 101:241–246.

Beall, A.C., and Rosenquist, T.H. (1990). Smooth muscle cells of neural crest origin form the aorticopulmonary septum in the avian embryo. *Anat Rec* 226:360–366.

Been, W., and Lieuw-Kie-Song, S.H. (1978). Harelip and cleft palate conditions in chick embryos following local destruction of the cephalic neural crest. A preliminary note. *Acta Morphol Neerl Scand* 16:245–255.

Begum, N., Song, Y., Rienzie, J., and Ragolia, L. (1998). Vascular smooth muscle cell growth and insulin regulation of mitogen-activated protein kinase in hypertension. *Am J Physiol* 275:C42–49.

Bhave, S.V., Snell, L.D., Tabakoff, B., and Hoffman, P.L. (1996). Mechanism of ethanol inhibition of NMDA receptor function in primary cultures of cerebral cortical cells. *Alcohol Clin Exp Res* 20:934–941.

Bjorkman, R. (1995). Central antinociceptive effects of non-steroidal anti-inflammatory drugs and paracetamol. Experimental studies in the rat. *Acta Anaesthesiol Scand Suppl* 103:1–44.

Bjorkman, R., Hallman, K.M., Hedner, J., and Henning, M. (1996). Nonsteroidal anti-inflammatory drug modulation of behavioral responses to intrathecal N-methyl-D-aspartate, but not to substance P and amino-methyl-isoxazole-propionic acid in the rat. *J Clin Pharmacol* 36:20S–26S.

Bonfoco, E., Leist, M., Zhivotovsky, B., Orrenius, S., Lipton, S.A., and Nicotera, P. (1996). Cytoskeletal breakdown and apoptosis elicited by NO donors in cerebellar granule cells require NMDA receptor activation. *J Neurochem* 67:2484–2493.

Bostom, A.G., Jacques, P.F., Nadeau, M.R., Williams, R.P., Ellison, R.C., and Selhub, J. (1995). Post-methionine load hyperhomocysteinemia in persons with normal fasting total plasma homocysteine: initial results from the NHLBI Family Heart Study. *Atherosclerosis* 116:147–151.

Botto, L.D., Khoury, M.J., Mulinare, J., and Erickson, J.D. (1996). Periconceptional multivitamin use and the occurrence of conotruncal heart defects: results from a population-based, case-control study. *Pediatrics* 98:911–917.

Bronner-Fraser, M., and Fraser, S. (1989). Developmental potential of avian trunk neural crest cells in situ. *Neuron* 3:755–766.

Brown, J.C., Rosenquist, T.H., and Monaghan, D.T. (1998). ERK2 activation by homocysteine in vascular smooth muscle cells. *Biochem Biophys Res Commun* 251:669–676.

Buller, A.L., Larson, H.C., Schneider, B.E., Beaton, J.A., Morrisett, R.A., and Monaghan, D.T. (1994). The molecular basis of NMDA receptor subtypes: native receptor diversity is predicted by subunit composition. *J Neurosci* 14:5471–5484.

Combs, A.B., and Acosta, D. (1990). Toxic mechanisms of the heart: a review. *Toxicol Pathol* 18:583–596.

Conway, S.J., Godt, R.E., Hatcher, C.J., et al. (1997). Neural crest is involved in development of abnormal myocardial function. *J Mol Cell Cardiol* 29:2675–2685.

Cornell-Bell, A.H., Thomas, P.G., and Smith, S.J. (1990). The excitatory neurotransmitter glutamate causes filopodia formation in cultured hippocampal astrocytes. *Glia* 3:322–324.

Crews, F.T., Morrow, A.L., Criswell, H., and Breese, G. (1996). Effects of ethanol on ion channels. *Int Rev Neurobiol* 39:283–367.

Czeizel, A. (1981). Schisis-association. *Am J Med Genet* 10:25–35.

Czeizel, A., and Rode, K. (1984). Trial to prevent first occurrence of neural tube defects by periconceptional multivitamin supplementation [letter]. *Lancet* 2:40.

Czeizel, A.E., Toth, M., and Rockenbauer, M. (1996). Population-based case control study of folic acid supplementation during pregnancy. *Teratology* 53:345–351.

Dalton, M., Gadson, P., Wrenn, R., and Rosenquist, T.H. (1997). Homocysteine signal cascade: production of phospholipids, activation of protein kinase C and the induction of c-*fos* and c-*myb* in smooth muscle cells. *FASEB J* 11:703–711.

Daniell, L.C. (1991). Effect of CGS 19755, a competitive N-methyl-D-aspartate antagonist, on general anesthetic potency. *Pharmacol Biochem Behav* 40:767–769.

Eskes, T.K. (1998). Open or closed? A world of difference: a history of homocysteine research. *Nutr Rev* 56:236–244.

Eskes, T.K., and Steegers-Theunissen, R.P. (1994). Primary prevention of neural tube defects with folic acid. *Eur J Obstet Gynecol Reprod Biol* 53:147–152.

Facchinetti, F., Ciani, E., Dall'Olio, R., Virgili, M., Contestabile, A., and Fonnum, F. (1993). Structural, neurochemical and behavioral consequences of neonatal blockade of NMDA receptor through chronic treatment with CGP 39551 or MK-801. *Dev Brain Res* 74:219–224.

Fan, C.M., and Tessier-Lavigne, M. (1994). Patterning of mammalian somites by surface ectoderm and notochord: evidence for sclerotome induction by a hedgehog homolog. *Cell* 79:1175–1186.

Ferencz, C., and Correa-Willasenor, A. (1995). Overview: the epidemiologic approach to the study of congenital cardiovascular malformations. In: Clark, E.B., Markwald, R.R., and Takao, A., eds. *Developmental Mechanisms of Heart Disease*. Futura, Armonk, NY, pp. 629–638.

Ferencz, C., Loffredo, C.A., Correa-Villasenor, A., and Wilson, P.D. (1997). *Perspectives in pediatric cardiology*. Genetic and environmental risk factors of major cardiovascular malformations. In: *The Baltimore-Washington Infant Study 1981–89*, vol 5. Futura, Armonk, NY. pp. 362–363.

Finnell, R.H., Wlodarczyk, B.C., Craig, J.C., Piedrahita, J.A., and Bennett, G.D. (1997). Strain-dependent alterations in the expression of folate pathway genes following teratogenic exposure to valproic acid in a mouse model. *Am J Med Genet* 70:303–311.

Gadson, P.F., Dalton, M., Patterson, E., et al. (1997). Differential response of mesoderm and neural crest-derived smooth muscle cells to TGF-β1: c-*myb* and procollagen genes. *Exp Cell Res* 230:160–180.

Gofflot, F., van-Maele-Fabry, G., and Picard, J.J. (1996). Cranial nerves and ganglia are altered after in vitro treatment of mouse embryos with valproic acid (VPA) and 4-en-VPA. *Brain Res Dev Brain Res* 93:62–69.

Goulding, M., and Paquette, A. (1994). Pax genes and neural tube defects in the mouse. *Ciba Found Symp* 181:103–113; discussion 113–117.

Graham, A., Koentges, G., and Lumsden, A. (1996). Neural crest apoptosis and the establishment of craniofacial pattern: an honorable death. *Mol Cell Neurosci* 8:76–83.

Graham, A., and Lumsden, A. (1996). Patterning the neural crest. *Biochem Soc Symp* 62:77–83.

Hirsch, S.R., Das, I., Garey, L.J., and de-Belleroche, J. (1997). A pivotal role for glutamate in the pathogenesis of schizophrenia, and its cognitive dysfunction. *Pharmacol Biochem Behav* 56:797–802.

Hundt, W., Danysz, W., Holter, S.M., and Spanagel, R. (1998). Ethanol and n-methyl-D-asparate receptor complex interactions: a detailed drug discrimination study in the rat. *Psychopharmacol Berl* 135:44–51.

Ishii, T., Moriyoshi, K., Sugihara, H., et al. (1993). Molecular characterization of the family of the n-methyl-D-aspartate receptor subunits. *J Biol Chem* 268:2836–2843.

Johnson, V.P., Swayze, V.W. II, Sato, Y., and Andreasen, N.C. (1996). Fetal alcohol syndrome: craniofacial and central nervous system manifestations. *Am J Med Genet* 61:329–339.

Keenan, C., and Kelleher, D. (1998). Protein kinase C and the cytoskeleton. *Cell Signal* 10:225–232.

Kirby, M.L., Gale, T.F., and Stewart, D.E. (1983). Neural crest cells contribute to normal aorticopulmonary septation. *Science* 220:1059–1061.

Kirby, M.L., Turnage, K.L., and Hays, B.M. (1985). Characterization of conotruncal malformations following ablation of "cardiac neural crest." *Anat Rec* 213:87–93.

Kirby, M.L., and Waldo, K.L. (1990). Role of neural crest in congenital heart disease. *Circulation* 82:332–340.

Kirby, M.L., and Waldo, K.L. (1995). Neural crest and cardiovascular patterning. *Circ Res* 77:211–215.

Ko, G.Y., Brown Croyts, L.M., and Teyler, T.J. (1997). The effects of anticonvulsant drugs on NMDA-EPSP, AMPA-EPSP, and GABA-IPSP in the rat hippocampus. *Brain Res Bull* 42:297–302.

Komuro, H., and Rakic, P. (1993). Modulation of neuronal migration by NMDA receptors. *Science* 260:95–97.

Lachman, H.M., Kelsoe, J.R., Remick, R.A., et al. (1997). Linkage studies suggest a possible locus for bipolar disorder near the velo-cardio-facial syndrome region on chromosome 22. *Am J Med Genet* 74:121–128.

Laegreid, L., Kyllerman, M., Hedner, T., Hagberg, B., and Viggedahl, G. (1993). Benzodiazepine amplification of valproate teratogenic effects in children of mothers with absence epilepsy. *Neuropediatrics* 24:88–92.

Leatherbury, L., and Kirby, M.L. (1996). Cardiac development and perinatal care of infants with neural crest-associated conotruncal defects. *Semin Perinatol* 20:473–481.

Lipton, S.A. Kim, W.K., Choi, Y.B., et al. (1997). Neurotoxicity associated with dual actions of homocysteine at the N-methyl-D-aspartate receptor. *Proc Natl Acad Sci USA* 94:5923–5928.

Liu, L., Brown, J.C., Webster, W.W., Morrisett, R.A., and Monaghan, D.T. (1995). Insulin potentiates N-methyl-D-asparate receptor activity in *Xenopus* oocytes and rat hippocampus. *Neurosci Lett* 192:5–8.

Martin, D.C., Abraham, J.E., Plagenhoef, M., and Aronstam, R.S. (1991). Volatile anesthetics and NMDA receptors. Enflurane inhibition of glutamate-stimulated [3H]MK-801 binding and reversal by glycine. *Neurosci Lett* 132:73–76.

Martin, D.C., and Aronstam, R.S. (1995). Spermidine attenuation of volatile anesthetic inhibition of glutamate-stimulated [3H] 5D, 10S)-(+)-methyl-10,11-dihydro-5H-dibenzo [a,d]cyclohepten-5,10-imine ([3H]MK-801) binding to N-methyl-D-Asparate (NMDA) receptors in rat brain. *Biochem Pharmacol* 50:1373–1377.

Martin, D.C., Plagenhoef, M., Abraham, J., Dennison, R.L., and Aronstam, R.S. (1995). Volatile anesthetics and glutamate activation of N-methyl-D-aspartate receptors. *Biochem Pharmacol* 49:809–817.

Martinez, J.L., and Penna, M. (1992). Influences of changes in calcium concentration and verapamil on the cardiac depressant effect of ethanol in cat papillary muscle. *Gen Pharmacol* 23:1051–1056.

Mills, J.L., McPartlin, J.M., Kirke, P.N., et al. (1995). Homocysteine metabolism in pregnancies complicated by neural-tube defects. *Lancet* 345:149–151.

Monyer, H., Sprengel, R., Schoepfer, R., et al. (1992). Heteromeric NMDA receptors: molecular and functional distinction of subtypes. *Science* 256:1217–1221.

Munger, R.G., Romitti, P.A., Daack-Hirsch, S., Burns, T.L., Murray, J.C., and Hanson, J. (1996). Maternal alcohol use and risk of orificial cleft birth defects. *Teratology* 54:27–33.
Nakajima, Y., Morishima, M., Nakazawa, M., Momma, K., and Nakamura, H. (1997). Distribution of fibronectin, type I collagen, type IV collagen, and laminin in the cardiac jelly of the mouse embryonic heart with retinoic acid-induced complete transposition of the great arteries. *Anat Rec* 249:478–485.
Nau, H. (1985). Teratogenic valproic acid concentrations: infusion by implanted minipumps vs conventional injection regimen in the mouse. *Toxicol Appl Pharmacol* 80:243–250.
Nickel, R.E., and Magnis, R.E. (1996). Neural tube defects and deletions of 22q11. *Am J Med Genet* 66:25–27.
Noden, D.M. (1975). An analysis of migratory behavior of avian cephalic neural crest cells. *Dev Biol* 42:106–130.
Noden, D.M. (1978a). The control of avian cephalic neural crest cytodifferentiation. I. Skeletal and connective tissues. *Dev Biol* 67:296–312.
Noden, D.M. (1978b). The control of avian cephalic neural crest cytodifferentiation. II. Neural *Tissues Dev Biol* 67:313–329.
Noden, D.M. (1983). The embryonic origins of avian cephalic and cervical muscles and associated connective tissues. *Am J Anat* 68:257–276.
Okada, M., Sakaguchi, T., and Kawasaki, K. (1995). Correlation between anti-ubiquitin immunoreactivity and region-specific neuronal death in N-methyl-D-aspartate-treated rat hippocampal organotypic cultures. *Neurosci Res* 22:359–366.
Parsons, R.B., Waring, R.H., Ramsden, D.B., and Williams, A.C. (1998). In vitro effect of the cysteine metabolites homocysteic acid, homocysteine and cysteic acid upon human neuronal cell lines. *Neurotoxicology* 19:599–603.
Rashid, N.A., and Cambray-Deakin, M.A. (1992). N-methyl-D-aspartate effects on the growth, morphology and cytoskeleton of individual neurons in vitro. *Brain Res Dev Brain Res* 67:301–308.
Richards, I.S. (1997). Ethanol potentiates the depressant effects of cocaine in human fetal myocardium in vitro. *J Toxicol Clin Toxicol* 35:365–369.
Rosenberg, R.D. (1993). Vascular smooth muscle cell proliferation: basic investigations and new therapeutic approaches. *Thromb Haemost* 70:10–16.
Rosenquist, T.H., and Beall, A.C. (1990). Elastogenic cells in the developing cardiovascular system: smooth muscle, non-muscle and cardiac neural crest. *Ann NY Acad Sci* 588:106–119.
Rosenquist, T.H., Beall, A.C., Modis, L., and Fishman, R. (1990a). Impaired elastic matrix development in the great arteries after ablation of the cardiac neural crest. *Anat Rec* 226:347–359.
Rosenquist, T.H., Fray-Gavalas, C.A., Waldo, K., and Beall, A.C. (1990b). Development of the musculoelastic septation complex in the avian truncus arteriosus. *Am J Anat* 180:339–356.
Rosenquist, T.H., McCoy, J.R., Waldo, K., and Kirby, M.L. (1988). Origin and propagation of elastogenesis in the cardiovascular system. *Anat Rec* 221:860–871.
Rosenquist, T.H., Ratashak, S.A., and Selhub, J. (1996). Homocysteine induces congenital heart and neural tube defects. Effect of folic acid. *Proc Natl Acad Sci USA* 93:15227–15232.
Sah, V.P., Attardi, L.D., Mulligan, G.C., Williams, B.O., Bronson, R.T., and Jacks, T. (1995). A subset of p53-deficient embryos exhibit exencephaly. *Nat Genet* 10:175–180.
Scherson, T., Serbedzija, G., Fraser, S., and Bronner-Fraser, M. (1993). Regulative capacity of the cranial neural tube to form neural crest. *Development* 118:1049–1062.
Selhub, J., and D'Angelo, A. (1998). Relationship between homocysteine and thrombotic disease. *Am J Med Sci* 316:129–141.
Seller, M.J., and Nevin, N.C. (1984). Periconceptional vitamin supplementation and the prevention of neural tube defects in south-east England and Northern Ireland. *J Med Genet* 21:325–330.

Shaw, G.M., O'Malley, C.D., Wasserman, C.R., Tolarova, M.M., and Lammer, E.J. (1995). Maternal Periconceptional use of multivitamins and reduced risk for conotruncal heart defects and limb deficiencies among offspring. *Am J Med Genet* 59:536–545.

Steegers-Theunissen, R.P., Boers, G.H., Blom, H.J., et al. (1995). Neural tube defects and elevated homocysteine levels in anmiotic fluid. *Am J Obstet Gynecol* 172:1436–1441.

Steegers-Theunissen, R.P., Boers, G.H., Trijbels, F.J., et al. (1994). Maternal hyperhomocysteinemia: a risk factor for neural-tube defects? *Metabolism* 43:1475–1480.

Sugihara, H., Moriyoshi, K., Ishii, T., Masu, M., and Nakanishi, S. (1992). Structures and properties of seven isoforms of the NMDA receptor generated by alternative splicing. *Biochem Biophys Res Commun* 185:826–832.

Tenneti, L., D'Emilia, D.M., Troy, C.M., and Lipton, S.A. (1998). Role of caspases in N-methyl-D-aspartate-induced apoptosis in cerebrocortical neurons. *J Neurochem* 71:946–959.

Thomas, C.M., Borm, G.F., Wouters, M.G., and Eskes, T.K. (1994). Maternal hyperhomocysteinemia: a risk factor for neural-tube defects? *Metabolism* 43:1475–1480.

Touyz, R.M., He, G., Deng, L.Y., and Schiffrin, E.L. (1999). Role of extracellular signal-regulated kinases in angiotensin II–stimulated contraction of smooth muscle cells from human resistance arteries. *Circulation* 99:392–399.

Uberti, D., Belloni, M., Grilli, M., Spano, P., and Memo, M. (1998). Induction of tumour-suppressor phosphoprotein p53 in the apoptosis of cultured rat cerebellar neurones triggered by excitatory amino acids. *Eur J Neurosci* 10:246–254.

van der Put, N.M., Steegers-Theunissen, R.P., Frosst, P., et al. (1995). Mutated methylenetetrahydrofolate reductase as a risk factor for spina bifida. *Lancet* 346:1070–1071.

Vataja, R., and Elomaa, E. (1998). Midline brain anomalies and schizophrena in people with CATCH 22 syndrome. *Br J Psychiatry* 172:518–520.

Wald, N.J., Hackshaw, A.D., Stone, R., and Sourial, N.A. (1996). Blood folic acid and vitamin B_{12} in relation to neural tube defects. *Br J Obstet Gynaecol* 103:319–324.

Waldo, K., Zdanowicz, M., Burch, J., Kumiski, D.H., Stadt, H.A., Godt, R.E., Creazzo, T.L., and Kirby, M.L. (1999). A novel role for cardiac neural crest in heart development. *J Clin Invest* 103:1499–1507.

Wang, C., Pralong, W.F., Schulz, M.F., et al. (1996). Functional N-methyl-D-asparate receptors in O-2A glial precursor cells: a critical role in regulating polysialic acid-neural cell adhesion molecule expression and cell migration. *J Cell Biol* 135:1565–1581.

Wehr, R., and Gruss, P. (1996). Pax and vertebrate development. *Int J Dev Biol* 40:369–377.

Weil, M., Jacobson, M.D., and Raff, M.C. (1997). Is programmed cell death required for neural tube closure? *Curr Biol* 7:281–284.

Witt, G., Rosenquist, T.H., and Gadson, P.F. (1997). Homocysteine inhibition of PAX-3 expression. *FASEB J* 11:A427.

Wlodarczyk, B.C., Craig, J.C., Bennett, G.D., Calvin, J.A., and Finnell, R.H. (1996). Valproic acid–induced changes in gene expression during neurulation in a mouse model. *Teratology* 54:284–297.

Yamagishi, H., Garg, V., Matsuoka, R., Thomas, T., and Srivastava, D. (1999). A molecular pathway revealing a genetic basis for human cardiac and craniofacial defects. *Science* 283:1158–1161.

Yasui, H., Nakazawa, M., Morishima, M., Ando, M., Takao, A., and Aidawa, E. (1997). Cardiac outflow tract septation process in the mouse model of transposition of the great arteries. *Teratology* 55:353–363.

CHAPTER 13

Function and Biomechanics of Developing Cardiovascular Systems

Bradley B. Keller

MODEL SELECTION

Clinicians and scientists have searched for the mechanisms underlying normal and altered cardiovascular (CV) development for millennia. Scientific discovery regarding the CV system has occurred coincident with major advances in our understanding of CV anatomy, embryology, physiology, biomechanics, and, most recently, cellular and molecular biology. These investigations have occurred in a wide range of invertebrate and vertebrate species, representing the diverse adaptation of the CV system to environmental and life-cycle demands (Keller, 1997). The primary experimental model for the investigation of cardiac morphogenesis has been the chick embryo. Despite the small size of the embryonic chick heart, numerous investigators have developed experimental methods to accurately measure blood pressure, blood flow, chamber size, and altered cardiac function or form (Clark and Hu, 1982; Nakazawa et al, 1988; Keller, 1995). These investigations have described individual parameters and integrated measures of embryonic CV function comparable to the analysis of the mature heart (Table 13.1).

The investigation of mammalian models of CV development has accelerated due to technical advances in defining embryonic CV morphology and physiology (Nakazawa et al, 1985; Keller et al, 1996) and the rapid development of targeted genetic models in mice that are associated with altered CV structure and function (Dyson et al, 1995; Keller, 1998). Confounding the investigation of mammalian embryonic CV function are the technical challenges associated with supporting or replacing maternal metabolic support of the embryo (Furukawa et al, 1998). Most investigations of the murine embryo involve maternal anesthesia followed by placement of the embryo into either a suffusion or culture system (Dyson et al, 1995; Yasui et al, 1995; Hofman et al, 1997; Tanaka et al, 1997). However, we know very little regarding the impact of maternal anesthesia or altered metabolic support on the growth and function of the mammalian CV system (Figure 13.1).

Alternate vertebrate CV models are also responsible for major contributions to our understanding of both normal and aberrant structural and functional development (Burggren and Fritsche, 1997; Rombough, 1997). By applying large-scale mutagenesis protocols to the translucent developing zebrafish, investigators have

TABLE 13.1. Examples of embryonic cardiovascular functional parameters

	Individual parameters	Integrated parameters
Time interval	Cycle length	Rate of ventricular pressure decay (tau)
Pressure	End-diastolic pressure	End-diastolic pressure-volume relation
Flow	Stroke volume	Cardiac output
Vascular tone	Vessel diameter	Vascular impedance
Surface deformation	Epicardial strain	Stress-strain relations
Isolated muscle	Time to peak tension	Force-length relations

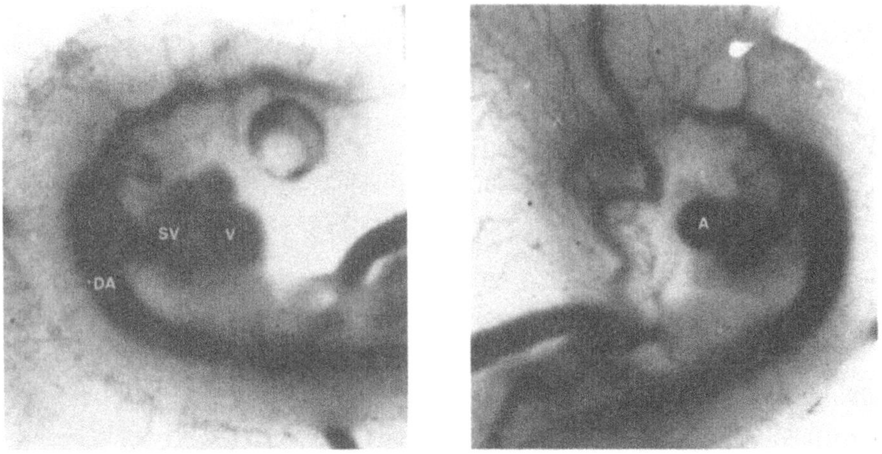

FIGURE 13.1. Right-lateral and left-lateral views of a stage 18 chick embryo. From the right-lateral view the embryonic ventricle is visible and from the left-lateral view the embryonic atrium is visible. DA, dorsal aorta; SV, sinus venosus; V, ventricle; A, atrium.

identified numerous single gene targets associated with altered CV structure and function (Stainier et al, 1996). Many of these genes will have mammalian homologs that will provide insight into the molecular mechanisms responsible for human CV malformations. Technical advances have also allowed the investigation of CV physiology in *Xenopus* embryos (Fritsche and Burggren, 1996). As will be shown in this chapter, these embryonic CV models share similar functional properties early in development despite a wide spectrum of final CV phenotypes (Keller, 1997).

ANALYTICAL METHODS

One challenge in the investigation of embryonic CV function and biomechanics is the application of quantitative analysis to a system that is undergoing coincident geometric growth and fundamental changes in structure. The first step in quantifying CV function is to define at which time points during the developmental process for each organism data will be acquired for analysis. The two basic approaches are to divide the developmental process by hours (or days) of gestation, or to divide the developmental process by distinct morphologic stages (Pexieder, 1984a,b). It is important to summarize functional data on individual embryos at comparable stages of development. The use of external and internal morphologic features such as somite number, limb size, and cardiac morphology to define developmental stage is preferable to the use of duration of gestation because significant variations in the rate of development can occur (Hamburger and Hamilton, 1951; Pexieder, 1984a,b; Sissman, 1970; Theiler, 1989; Kaufman, 1992). We frequently note significant stage variation in chick embryos incubated for comparable periods. Likewise, regional variations in uterine blood flow influence the rate of murine embryonic development (Vuillemin and Pexieder, 1989a,b). Thus, external morphologic staging is a critical first step in quantifying embryonic CV development.

The next issue for defining embryonic CV function is the selection of specific developmental stages for analysis. One standard method of quantifying changes in CV performance is the correlation of incremental changes in body mass to changes in functional performance. Allometric analysis has been applied to most metabolic processes across a wide range of developing and mature species (Burggren and Pinder, 1991). The correlation of embryonic growth (doublings of embryonic wet weight) with measures of CV function has simplified the analysis of functional maturation despite complex changes in CV morphology (Clark and Hu, 1982; Nakazawa et al, 1988; Keller et al, 1990; Burggren and Pinder, 1991).

Most major advances in the definition of embryonic CV function and biomechanics have resulted from the application of analysis paradigms developed for mature CV systems. The analytic methods to define CV performance include single parameters of function such as heart rate, blood pressure, and blood flow as well as integrated measures that depend on the simultaneous measurement (Table 13.1). These functional parameters are described in detail in this chapter. Important insights have also been gained by the application of biomechanical modeling to the developing CV system (Taber et al, 1992; Miller et al, 1997a,b; Taber, 1998). However, it is important to note that while developing embryonic CV systems share some functional characteristics of mature systems, there are also important developmental stage and species specific distinctions.

STRUCTURAL ISSUES

Chamber Morphogenesis

Embryonic CV systems undergo simultaneous structural and functional maturation of atrial, ventricular, and vascular segments. The morphometry of the developing heart has been extensively studied by scanning electron microscopy and confocal laser microscopy in several species (Terracio and Borg, 1988; Pexieder, 1986). Magnetic resonance microscopy (Smith et al, 1996), optical coherence tomography (Boppart et al, 1997), and high-resolution ultrasound imaging (Srinivasan et al, 1998) have been applied to the embryonic CV system. Distinct venous, atrial, ventricular, and arterial segments are apparent shortly after the onset of cardiac contraction (Pexieder, 1984a). In the chick embryo, the embryonic right and left atria increase in size and have a large interatrial communication until stage 29 (Hay and Low, 1970). Video microscopy and planimetry of the beating embryonic chick left atrium reveal a smooth-walled endocardium, a spherical shape, and a distinct atrial contraction phase (Campbell et al, 1992).

The embryonic ventricle undergoes species-specific structural morphogenesis at the global and local levels. The external morphology of the embryonic ventricle is smooth in the chick embryo (Keller et al, 1990), while the murine embryonic ventricle contains a prominent interventricular sulcus (Vuillemin and Pexieder, 1989a,b). Initially the embryonic ventricle has a smooth endocardial surface; then the ventricular wall becomes trabeculated (Sedmera et al, 1997). The time course and pattern of ventricular trabeculation and compaction differ between the developing left and right ventricles and are species specific (Pexieder, 1984b; Sedmera et al, 1997). Changes in chamber morphology may both result from and influence alterations in chamber function (Sedmera et al, 1999).

Cardiomyocyte Maturation

At the onset of cardiac contraction, the embryonic myocyte bears little resemblance to the mature cell. The molecular commitment of cells in the lateral plate mesoderm to the cardiomyocyte lineage occurs prior to gastrulation. At the onset of cardiac contraction, the spherical, immature cardiomyocyte contains basic intracellular organelles and a small volume-fraction of immature myofibers (Hirakow, 1980). The embryonic heart initially depolarizes from the site of earliest activation in the sinus venosus (Kamino et al, 1981) toward the arterial pole, with variations in conduction velocity at the atrioventricular (AV) and conotruncal regions (Moorman et al, 1998). Following maturation of specialized conducting cells within ventricular trabeculae (Mikawa and Fischman, 1996), the embryonic ventricular contraction pattern shifts to the apex to base pattern found in the mature heart (Chuck et al, 1997; Moorman et al, 1998).

Maturation of the embryonic cardiomyocyte excitation-contraction apparatus results in increased force production and relaxation during shorter cycle lengths (Vetter and Will, 1986). The immature cardiomyocyte depends initially on sarcolemmal ion channels for the calcium entry (Chin et al, 1990; Josephson and Sperelakis, 1990; Davies et al, 1996). Subsequent maturation of the sarcoplasmic reticulum results in greater and more rapid calcium release and reuptake with commensurate changes in contractile performance (Godt et al, 1991; Protasi et al, 1996).

Maturation of the extracellular environment surrounding each cardiomyocyte has a dramatic impact on functional performance (Terracio and Borg, 1988; Keller, 1995). Interactions between developing myocytes and between myocytes and the extracellular matrix affect the passive mechanical characteristics of the ventricle (Miller et al, 1997a,b). Prior to the trabeculation of the heart, the biomechanical force from the myocardial mantle is transmitted through cardiac jelly to the blood contained in the ventricular lumen (Patten et al, 1948). As the embryo grows, myocytes are woven together by intricate cell-matrix and cell-cell adhesions, limiting wall motion and thus altering the local myocardial properties of the heart (Taber et al, 1992).

Vasculogenesis

The embryonic vascular bed undergoes dramatic expansion and morphogenesis concurrent to cardiac morphogenesis (Rosenquist and Modis, 1991). At the onset of cardiac contraction the embryonic ventricle ejects into the common arterial trunk. Alterations in outflow tract septation induced by neural crest ablation are associated with altered cardiac function, though the mechanism remains unclear (Leatherbury et al, 1993; Conway et al, 1997). Finally, altered vascular dimensions and function dramatically influence cardiac performance (Yoshigi, 1996b).

FUNCTIONAL MATURATION

Atrium

Atrial conduit, reservoir, pump, and endocrine functions directly affect ventricular performance (Keller, 1995). Although the posterosuperior embryonic chick atrium is obscured by the anteroinferior and rightward ventricle early in development, the embryo can be repositioned left side up in order to image the beating atrium (Campbell et al, 1992). Heart rate and activation sequence are crucial determinants of ventricular function, and each of the cardiac time intervals shortens during cardiac development. The embryonic sinus venosus functions as the dominant pacemaker during cardiac development (Kamino et al, 1981). At the onset of cardiac contraction, the embryonic chick heart rate is approximately 60 beats per minute (bpm). From stage 12 to 29 the heart rate increases from 103 to 208 bpm (Hu and Clark, 1989).

Heart rate and activation sequence can be acutely altered in the chick embryo by environmental hyperthermia and hypothermia (Nakazawa et al, 1985), sinus venosus pacing (Benson et al, 1989), thermal probe application to the sinus venosus (Cuneo et al, 1991), or by digoxin-induced heart block (Paff et al, 1948). Environmental hypothermia decreases heart rate (Nakazawa et al, 1985), and hyperthermia increases heart rate (Nakazawa et al, 1986) in the chick embryo, both resulting in reduced cardiac output. Sinus venosus pacing above the intrinsic rate also decreases cardiac output by reducing passive ventricular filling, and the altered activation sequence produced by ventricular pacing drastically reduces output in both chick and mouse embryos (Dunnigan et al, 1987; MacLennan and Keller, 1998). Of interest, systolic and diastolic time intervals change proportionately with cycle length (Cuneo et al, 1991; Casillas et al, 1994).

Maximum cardiac output occurs at the intrinsic heart rate of each embryonic stage due to a dynamic optimization of atrial/ventricular and ventricular/vascular coupling. In contrast to the mature circulation, the preinnervated embryonic heart does not rely on heart rate regulation to acutely adjust cardiac output (Benson et al, 1989). The lack of compensatory cycle length change during acute preload alteration suggests that embryonic pacemaker function may be developmentally regulated and may not be acutely load-sensitive (Keller et al, 1994). The specific mechanisms responsible for matching embryonic heart rate to the hemodynamic demand and mechanical characteristics of the rapidly developing cardiovascular system are largely unknown.

The AV cushions function as a valve to prevent retrograde flow during ventricular systole (Patten et al, 1948; Hu et al, 1991). Atrial developed pressure increases with stage, and atrial pressure exceeds ventricular pressure throughout ventricular filling (Hu, 1995b). Multiple separate flow streams course from the venous circulation through the atria and AV canal into the developing ventricle (Hogers et al, 1995) and alterations in this venous return pattern result in structural malformations of the heart and great vessels (Hogers et al, 1997).

As in the mature heart, ventricular filling can be partially characterized by measuring the velocity profile of flow across the AV cushions. Ventricular filling has distinct passive and active phases, and in the chick embryo the ratio of passive to active filling increases sharply from stage 12 to 18, then gradually to stage 27 (Hu et al, 1991). In rat embryos at comparable developmental stages, pulsed-Doppler measures of blood flow reveal similar results (Nakazawa et al, 1988); however, the mouse embryo has significantly less passive ventricular filling (Keller et al, 1996). Pulsed-Doppler studies of human embryos at comparable stages also reveal distinct passive and active ventricular filling phases and a geometric increase in blood flow with development (Wladimiroff et al, 1992).

Ventricular diastolic filling is influenced by diastolic filling time and the AV pressure gradient (Hu, 1995). In the chick embryo, peak atrial pressure increases geometrically from 0.38 to 1.21 mmHg, while ventricular end-diastolic pressure increases linearly from 0.18 to 0.55 mmHg from stage 16 to 27, respectively (Hu, 1995b). Thus, a pressure gradient across the AV orifice may be a resistance point that regulates blood flow during ventricular development.

Embryonic atrial pump function can be defined using video microscopy (Campbell et al, 1992) and servonull pressure measurement (Hu, 1995b). When left atrial pressure of stage 24 chick embryo is plotted against simultaneous left atrial volume, the result is a dual pressure-volume (PV) loop. This loop consists of a clockwise passive filling curve corresponding to atrial filling followed by atrial contraction, similar to the mature atrium. These PV loops can be analyzed to determine the filling and ejection characteristics of the developing atrium, similar to the PV loops used to define embryonic ventricular function (Keller et al, 1994).

Finally, the developing myocardium produces and releases vasoactive substances. In the mature circulation, atrial natriuretic polypeptide regulates renal and cardiovascular function (Pegram et al, 1986). The developing cardiovascular system is also sensitive to exogenous atrial natriuretic peptide exposure. The endocrine effect of this protein is predominantly via peripheral venodilation (Nakazawa et al, 1990). In response to atrial natriuretic peptide, passive ventricular filling decreases, active filling remains relatively constant, and dorsal aortic stroke volume decreases (Hu, 1995a). It is interesting to note that this vasoactive substance is present in the embryonic circulation prior to formation of a functional kidney or nervous system. Thus, embryonic CV regulatory systems may include elements that persist in the mature circulation, but with unique effects on the developing heart (Keller, 1997).

Ventricle

Ventricular performance during embryogenesis depends on the same variables of pacemaker rate, filling volume, contractility, and vascular load that influence the mature circulation (Keller, 1997). At the onset of cardiac contraction, metabolic supply to the rapidly developing tissues occurs primarily via simple diffusion. However, survival of the developing embryo depends on adequate cardiac performance when metabolic demand exceeds the local supply of nutrients (Pelster and Burggren, 1996). Thus, the structural and functional maturation of the embryonic CV system begins slightly ahead of embryonic metabolic demand as one example of the functional reserve inherent in the system.

Ventricular pump function is directly dependent on adequate ventricular filling, which in turn is related to ventricular relaxation, diastolic filling time, passive myocardial properties, and dynamic AV coupling (Gaasch et al, 1987). Diastolic ventricular relaxation is determined by the mechanisms responsible for sarcomere recovery to slack length and myocardial viscoelastic properties. Sarcomere relaxation following shortening requires calcium removal from the thick and thin filaments followed by reduction of unbound intracellular calcium via the sarcoplasmic reticulum and sarcolemma. There have been few direct investigations of diastolic calcium kinetics in the embryonic myocardium (Protasi et al, 1996). Evidence for a reduced rate of diastolic removal of intracellular calcium from the developing myocyte includes increased diastolic pressure during a shortened cycle length (Zimmerman et al, 1991; Casillas et al, 1994) and increased tonic tension during increased pacing frequency or increased extracellular calcium in isolated embryonic myocardial strips (Keller, 1998). This reduced rate of calcium removal is also predicted by the immaturity of the sarcoplasmic reticulum in the developing cardiomyocyte and the profile of ion channels in the developing sarcolemma (Vetter and Will, 1986; Chin et al, 1990; Artman, 1992; Davies et al, 1996).

The rate of ventricular pressure decay following end-systole, termed "tau," is an integrated index of ventricular relaxation in the mature ventricle (Weiss et al, 1976). At each developmental stage the maximum rate of ventricular pressure decay is related to cycle length, consistent with maturation of calcium reuptake or changes in the time course of viscoelastic relaxation (Cheanvechai et al, 1992; Braunstein et al, 1994). As with the mature ventricle, the rate of embryonic ventricular pressure decay can be defined using a monoexponential decay function with or without a pressure asymptote (Keller, 1997). The decay rate is influenced by acute changes in ventricular afterload (Keller, 1997). As in the mature heart, the time course of embryonic ventricular pressure decay is most likely dependent on the maturation of intracellular calcium kinetics.

Following ventricular "relaxation," ventricular filling is influenced by venous return, atrial pump function, and the duration of diastole. As noted above, acute alterations in venous return affect passive ventricular filling (Hu et al, 1991; Keller et al, 1994; Hu, 1991; Bowers et al, 1996). Chronic alterations in venous return also influence endocardial cushion (Icardo, 1989; Hogers et al, 1997) and ventricular morphogenesis (Harh et al, 1973; Sweeney, 1981; Hogers et al, 1997; Sedmera et al, 1998). Dynamic changes in ventricular filling volume are accurately measured by cinephotography and videography (Faber et al, 1974; Keller et al, 1990, 1996), and end-diastolic volume is linearly related to diastolic filling time (Casillas et al, 1994). In contrast to the mature, innervated heart, the embryonic heart does not increase heart rate to maintain cardiac output in response to reduced ventricular filling (Keller et al, 1994), providing evidence for the lack of autonomic regulation of early cardiac function.

Ventricular systolic function depends on active force generation, and the similarity between embryonic and mature ventricular pressure waveforms has been shown in a wide range of vertebrate and invertebrate species (Keller, 1997). Intraventricular pressure is negative at the onset of filling in the chick embryo. The ventricular pressure waveform displays an atrial "kick" followed by a rapid rise and then decay in pressure (Clark and Hu, 1982). Peak-systolic and end-systolic ventricular pressure increase during cardiac development (Hu and Clark, 1989; Keller et al, 1994) coincident with ventricular morphogenesis. Simultaneous measures of ventricular pressure and chamber dimensions display the same pressure-volume loop characteristics noted for the mature heart (Figure 13.2) (Keller et al, 1994). As in the mature heart, active force generation displays preload dependence in the intact embryo (Keller et al, 1994) and in isolated myocardial specimens (Keller, unpublished data). Embryonic ventricular force production is also sensitive to changes in myofilament calcium availability due to changes in extracellular calcium

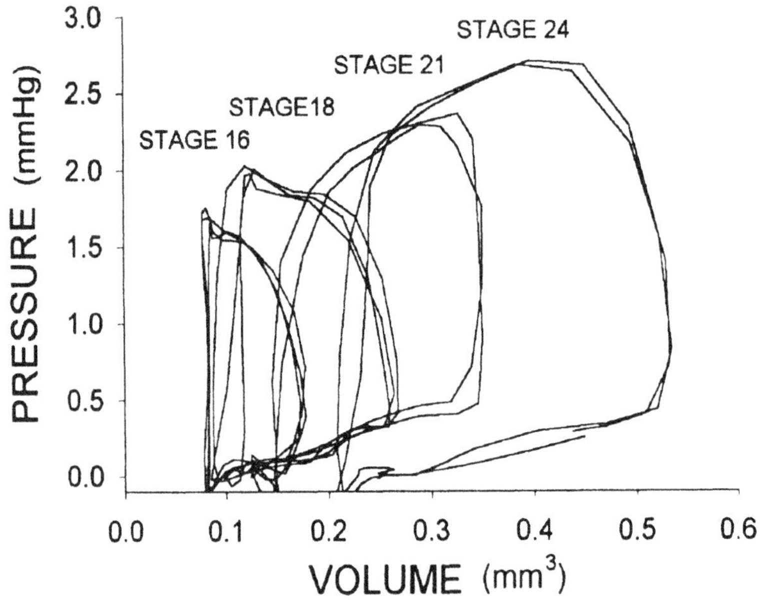

FIGURE 13.2. Relationship of simultaneous ventricular pressure and volume in stage 16 to 24 chick embryos. Note the increase in end-diastolic filling volume, developed pressure, and stroke volume with increasing stage.

concentration or calcium transport through the sarcolemmal (Clark et al, 1991) or sarcoplasmic reticulum (Keller, 1997).

Ventricular systolic function can be defined using load-dependent and load-independent indices. Peak ventricular systolic pressure, the rate of ventricular pressure generation, and stroke volume are both preload- and afterload-dependent indices in the embryo, and each measure of ventricular performance increases during cardiac development (Keller, 1995). Despite the structural immaturity of the developing ventricle, ventricular stroke volume varies linearly with diastolic filling volume and is sensitive to changes in contractility and afterload, as also occurs in the mature heart (Keller et al, 1994; Bowers et al, 1996; Keller, 1997). The relatively load-insensitive end-systolic pressure-volume relationship used to define ventricular contractility in the mature ventricle (Sagawa et al, 1988; Kass et al, 1989) is extremely curvilinear in the embryonic heart (Keller, 1997). However, embryonic ventricular contractility can be quantified using a load-independent measure of end-systolic maximum stiffness (Mirsky et al, 1987; Tobita and Keller, 1999, 2000a). As discussed below, the curvilinear end-systolic pressure-volume relation in the ejecting embryonic heart is due to the dynamic coupling between the ventricle and the arterial circulation (Figure 13.3) (Keller, 1997).

Embryonic heart failure has become a relatively common term applied to a number of transgenic mouse models associated with an embryo-lethal phenotype (Dyson et al, 1995; Ya et al, 1998). Obviously, the rapidly developing embryo depends on adequate cardiac output to maintain critical metabolic systems. However, as with "heart failure" in the mature circulation, the underlying mechanisms responsible for altered CV performance relate to specific abnormalities in components of the heart and/or vascular system. In mammalian embryos, altered

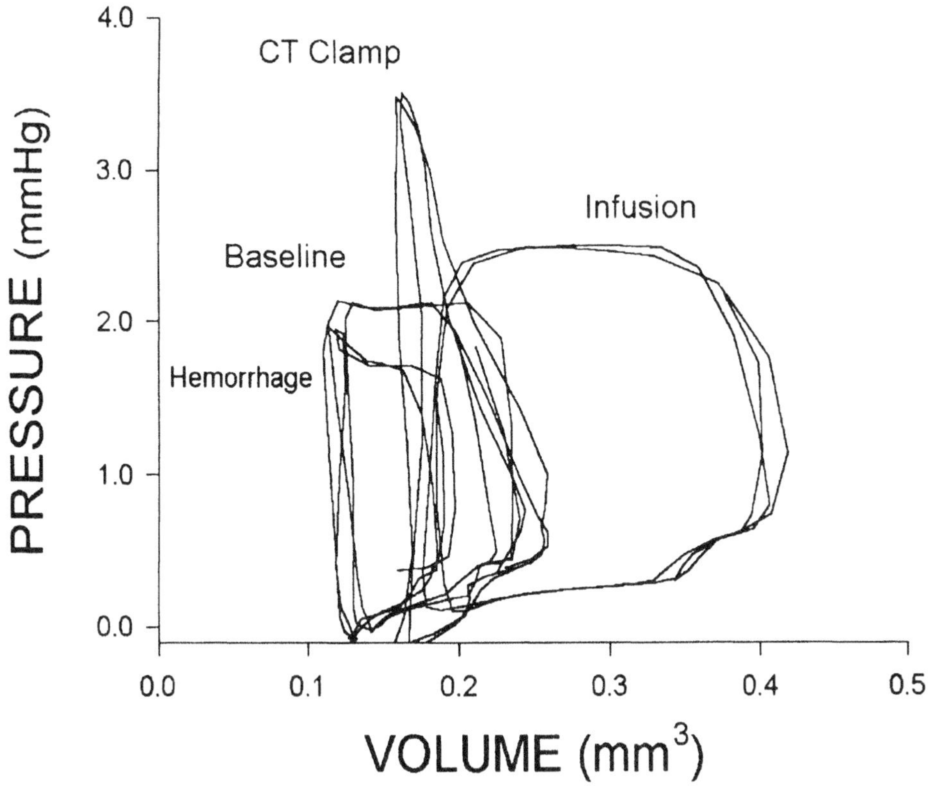

FIGURE 13.3. Relationship of simultaneous ventricular pressure and volume in a stage 21 chick embryo during volume infusion to increase preload, venous hemorrhage to decrease preload, and acute conotruncal (CT) occlusion to uncouple the ventricle from the arterial circulation. Note that end-diastolic volume changes significantly during volume infusion and venous hemorrhage with subsequent changes in stroke volume. During CT occlusion, developed pressure increases significantly despite reduced stroke volume consistent with a contractile reserve that is not evident during ventricular ejection.

maternal-embryo coupling via the placenta may influence embryonic CV function. Thus, the presence of a dilated and poorly contractile embryonic ventricle is important but insufficient to define the specific mechanism(s) responsible for CV failure.

Conotruncus

It is important to note that in contrast to the mature circulation, the developing ventricle ejects into a pulsatile chamber, the conotruncus, which then actively contributes to forward blood flow. As with the developing ventricle, the conotruncus contracts and contributes to cardiac output while undergoing simultaneous structural maturation (Figure 13.4). Migrating cardiac neural crest cells are one critical component of normal conotruncal septation (Kirby and Waldo, 1990; Poelmann et al, 1998) and altered neural crest cell migration results in depressed ventricular function (Conway et al, 1997). Increased resistance to ventricular ejection caused by experimental conotruncal banding stimulates the embryonic myocardium to

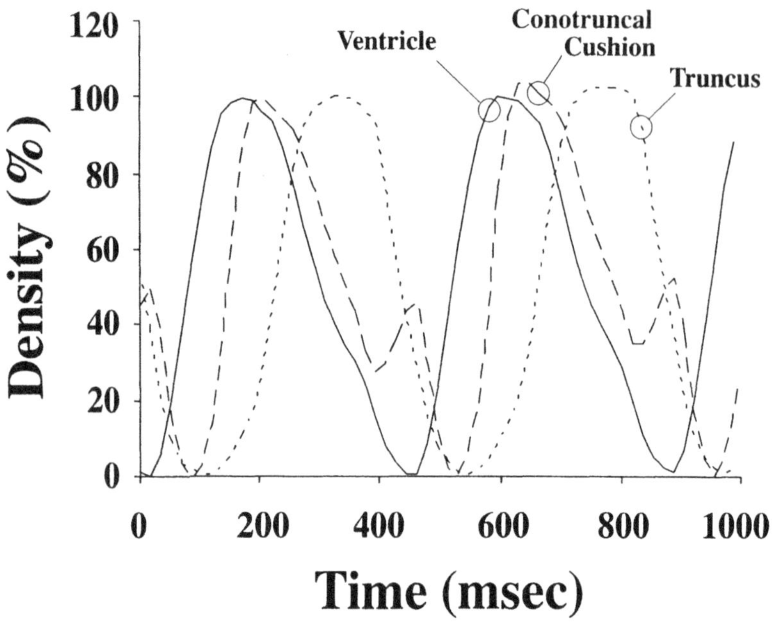

FIGURE 13.4. Time course of ventricular, conotruncal cushion, and truncal shortening determined from video densitometry of a stage 21 chick embryo. Chamber dimension was determined using automated densitometry of a region of interest. Note the time delay from ventricular to truncal contraction, which facilitates efficient forward blood flow in the absence of formed valves.

increase myocardial mass (Clark et al, 1989). However, we really know very little about the basic mechanisms by which changes in conotruncal morphogenesis alter either conotruncal or proximal ventricular function.

Systemic Circulation

The embryonic vascular system develops coincident with the developing heart, though we usually think of CV performance primarily with respect to cardiac function. As in the mature circulation, arterial loading conditions are a critical determinant of embryonic systolic performance (Maughan et al, 1985; Sunagawa et al, 1987; Colon et al, 1992; Yoshigi et al, 1996a,b; Keller, 1997). Arterial systolic and diastolic blood pressure increase as ventricular systolic pressure rises (Clark and Hu, 1982). Despite increase arterial blood pressure, total arterial vascular resistance actually declines during development due to the rapid expansion of new capillary beds (Zahka et al, 1989). Mathematical representation of the embryonic arterial circulation as a lumped parameter system including proximal and distal resistance, capacitance, and inductance elements provides a very accurate description of dynamic arterial load (Yoshigi, 1997a).

Ventricular-vascular coupling allows the low-pressure, energetically limited embryonic cardiovascular system to rapidly adapt to acute and chronic alterations in its biochemical (nutrient) and biomechanical (loading) microenvironment (Yoshigi, 1996a; Keller, 1997). In response to alterations in cardiac function the embryonic vasculature acutely adjusts vascular impedance to optimize cardiac output (Yoshigi, 1996a). Diffusible mediators of altered embryonic vascular tone

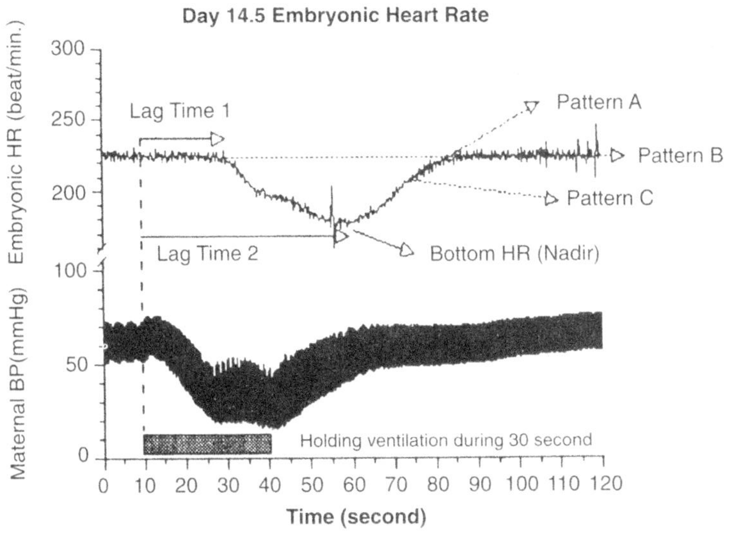

FIGURE 13.5. Representative maternal blood pressure (BP) and embryonic heart rate (HR) before and after maternal hypoxia. Lag time 1 is the time from the start of hypoxia to the onset of embryonic bradycardia, and lag time 2 is the time from the start of hypoxia to the embryonic heart rate minimum. Patterns of embryonic HR recovery are defined as pattern A, posthypoxia tachycardia; pattern B, normal embryonic HR posthypoxia; and pattern C, persistent embryonic bradycardia posthypoxia.

include atrial natriuretic peptide (Hu, 1995a) and nitric oxide (Bowers et al, 1996), though there are likely many additional regulators of global and regional vascular tone and permeability. As in the mature heart, stroke volume is more sensitive to changes in peripheral vascular resistance than to acute changes in intrinsic myocardial contractility.

Maternal Coupling

One obvious and important aspect of mammalian CV development is the coupling of the embryonic CV system to the placental circulation and thus to the maternal circulation (MacLennan and Keller, 1998; Ursem et al, 1998). The developing myocardium depends on a higher supply of blood per milligram of tissue than the mature myocardium due to the lower intrauterine partial pressure of oxygen (Thornburg et al, 1997). Alterations in maternal metabolic state associated with diabetes are associated with altered fetal CV function (Tsyvian et al, 1998) as well as structural CV malformations (Meyer-Wittkopf et al, 1996). Murine embryonic heart rate is rapidly influenced by acute maternal hypoxia produced by suspending mechanical ventilation for 30 seconds (Figure 13.5) (Furukawa, unpublished data). The anesthetized, pregnant mouse becomes hypotensive within 10 seconds of hypoxia and embryonic bradycardia rapidly follows. The embryonic response to hypoxia is gestation specific, representing increased oxygen demand of the developing embryonic myocardium and maturation of neurohumoral regulation

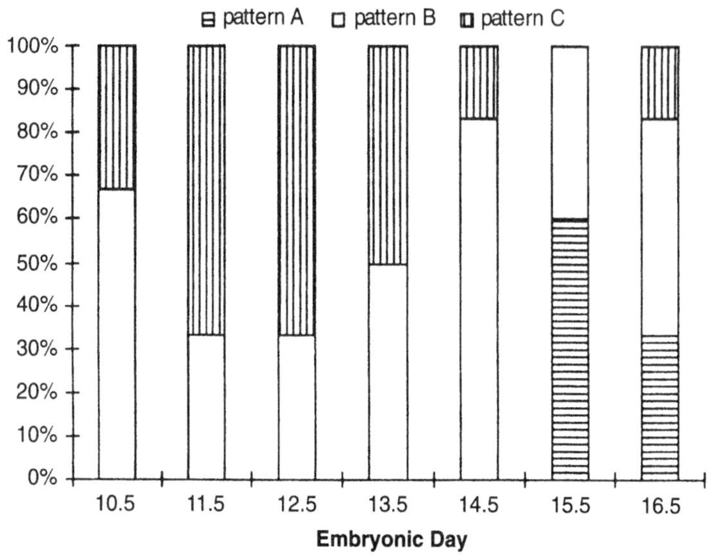

FIGURE 13.6. Distribution of posthypoxic embryonic heart rate recovery patterns following a 30-second suspension of maternal mechanical ventilation: pattern A, posthypoxic embryonic tachycardia; pattern B, normal embryonic heart rate posthypoxia; pattern C, persistent embryonic bradycardia posthypoxia.

of embryonic heart rate. However, the impact of transient alterations in maternal metabolic support of the developing embryo on final CV function or phenotype is unknown (Figure 13.6).

BIOMECHANICS

The integration of biologic experimental data with bioengineering theory has greatly increased our understanding of the biomechanical processes that are fundamental to CV structural organization and function (Fung, 1997). Bioengineering methods developed for the mature circulation have been applied to the embryonic heart (Taber et al, 1992; Miller et al, 1997a,b; Taber, 1998). These engineering approaches will become an increasingly critical component of the biotechnology of tissue replacement and organogenesis (Ranu, 1998).

Passive Material Properties

The embryonic heart must fill prior to generating forward blood flow and ventricular filling is directly influenced by myocardial passive material properties (Miller et al, 1997a,b). Surprisingly, the immature embryonic myocardium is actually "stiffer" than the mature heart, which may explain the dependence of diastolic filling volume on diastolic filling time (Keller et al, 1994). The viscoelastic properties of the developing myocardium are also time dependent, and myocardial relaxation is incomplete at experimentally increased heart rates (Casillas et al, 1994; Miller et al, 1997a,b). In contrast to the passive adult myocardium, the embryonic ventricular diastolic pressure-volume curve reaches a pressure asymp-

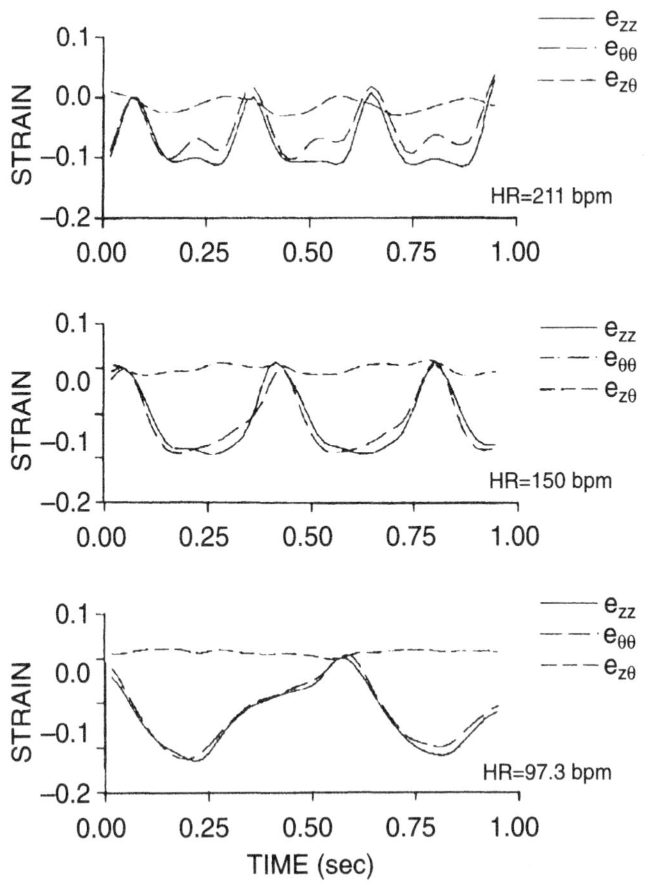

FIGURE 13.7. Epicardial ventricular strains for a stage 21 chick embryo at baseline heat rate (HR) of 150 beats per minute (bpm) and following thermal probe application to the sinus venosus to increase HR to 211 bpm (upper panel) or decrease HR to 97.3 bpm (lower panel). Note that longitudinal strain (e_{zz}) and circumferential strain ($e_{\theta\theta}$) change inversely with heart rate. Epicardial torsion ($e_{z\theta}$) increases with HR.

tote resembling the inflation of an elastic balloon, consistent with immature cardiomyocyte architecture and limited extracellular support (Terracio and Borg, 1988).

Epicardial Strains

The dynamic deformation of the embryonic myocardium during systole and diastole can be quantified in chick and mouse embryos using epicardial markers (Taber et al, 1994; Keller, unpublished data). At the onset of cardiac contraction the immature myocardium contracts isotropically (equally in all directions), and then gradually the contraction pattern becomes anisotropic, reflecting maturation of cardiomyocyte contractile elements and ventricular architectural heterogeneity (Taber et al, 1994; Sedmera et al, 1997). Epicardial deformation also reflects acute changes in heart rate (Figure 13.7) most likely due to changes in ventricular preload

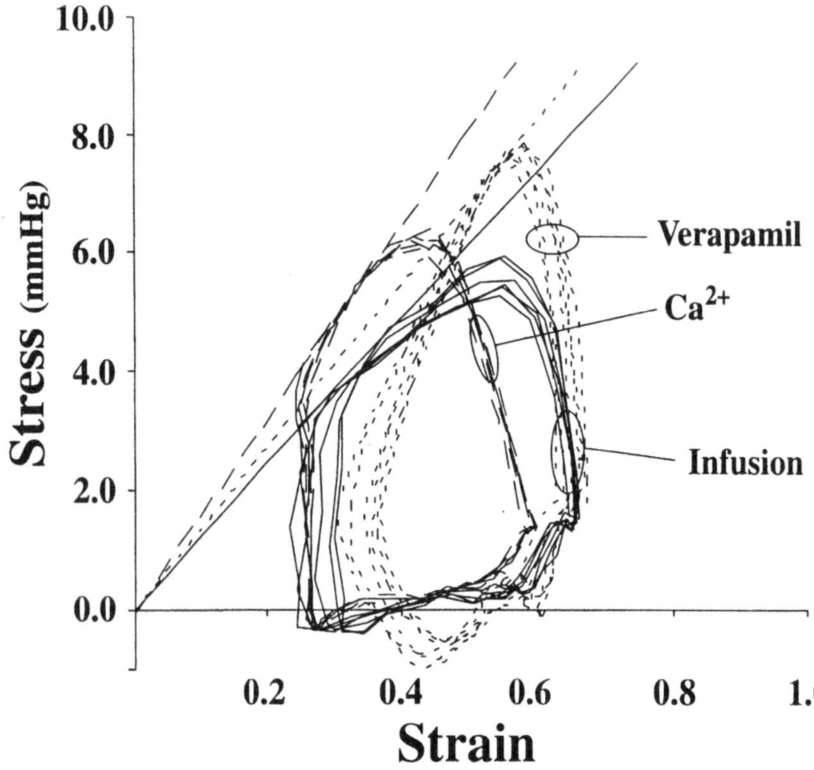

FIGURE 13.8. Relationship between end-systolic stress and end-systolic strain for stage 24 chick embryos during increased contractility produced by increased extracellular calcium (Ca^{2+}), decreased contractility produced by sarcolemmal calcium channel blockade with verapamil, or increased ventricular preload produced by volume infusion. Note that the slope of the end-systolic stress-strain relationship is relatively linear.

(Keller, unpublished data). Regional alterations in myocardial deformation can also be defined following experimentally induced altered cardiac morphogenesis (Tobita, 2000b). Data on regional changes in myocardial deformation are critical for the mathematical modeling of changes in myocardial stress, strain, and shear that my regulate cardiac morphogenesis (Taber, 1998).

End-Systolic Wall Stress

The combination of measured ventricular pressure with mathematical models of ventricular geometry and estimates of myocardial properties can be used to estimate the "stress" within the developing myocardium throughout the cardiac cycle (Tobita and Keller, 1999). As in the mature heart, embryonic end-systolic stress-strain relations are linear, load-independent, and correlate with acute changes in embryonic myocardial contractility (Figure 13.8) (Tobita are Keller, 1999). We know that the embryonic myocardium alters growth to compensate for chronically increased wall stress (Clark et al, 1989); however, the mechanisms responsible for transducing altered mechanical forces within the developing myocardium remain unclear.

Force-Interval Relations

As with the analysis of passive embryonic myocardial properties using in vitro muscle preparations, active myocardial force-length and force-interval relations can be defined during the period of CV morphogenesis (Schroder and Keller, unpublished data). Isolated embryonic myocardial specimens display linear force-length relations similar to the load dependence of the in vivo embryonic heart (Keller et al, 1994). Force production normalized to myocardial mass increases with development, consistent with maturation of excitation-contraction coupling. Due to immature calcium transport in the developing myocardium, sudden increases in heart rate are associated with impaired relaxation and reduce force generation. Of note, there are important species differences that may reflect differences in contractile protein isoforms related to final CV phenotype (Keller, 1998). The measurement of force-interval and velocity-force relations during selective pharmacologic stimulation or blockade of components in the excitation-contraction cascade will provide insight into the mechanisms responsible for force generation and functional reserve in the developing myocardium.

Numerical Models

Numerical approaches to bioengineering modeling of developing CV systems have depended on the application of methods developed in the mature circulation to the embryo (Taber et al, 1992; Yang et al, 1994; Yoshigi, 1996a,b; Miller et al, 1997a,b; Tobita and Keller, 1999). These mathematical approaches have also benefited from the accessibility of the embryonic heart for the acquisition of experimental data for the refinement and validation of mathematical models (Keller et al, 1991; Taber et al, 1993, 1994; Miller et al, 1997a,b). Future directions in developmental biomechanics will include the refinement of biomechanical analysis to the cellular level (Emery and Omens, 1997; Taber, 1998).

SUMMARY

The past 25 years have resulted in dramatic progress in the investigation of embryonic CV function and links between changes in function and morphogenesis (Pexieder, 1984a; Clark et al, 1989; Nakazawa et al, 1988; Keller, 1997). The embryonic CV system generates forward blood flow and blood pressure to support the metabolic demands of the rapidly growing embryo (Clark and Hu, 1982; Hu and Clark, 1989; Keller, 1995). Embryonic CV function is defined using quantitative measures of venous return, atrial function (Hu, 1995b), ventricular performance (Casillas et al, 1994; Keller et al, 1994), and arterial load (Yoshigi, 1996a,b). These functional indices include individual measures (heart rate, blood pressure and flow, mean vascular resistance) and integrated measures (end-diastolic volume vs. stroke volume relations and pressure-volume relations (Keller et al, 1996), stress-strain relations, and arterial impedance (Yoshigi, 1997a). Embryonic CV function has been investigated in a broad range of species, including chick (Clark and Hu, 1982; Keller et al, 1990), rat (Nakazawa et al, 1988), mouse (Keller et al, 1996), *Xenopus* (Fritsche and Burggren, 1996), and zebrafish (Stainier et al, 1996). These developing CV systems can adjust CV function acutely to compensate for changes chamber filling or vascular tone (Keller, 1997). The developing CV system is now the subject of biomechanical analysis of passive and active myocardial function and remodeling. Quantitative measures of performance and biomechanics in the developing myocardium and vasculature can provide insights into the mechanisms that link morphogenesis and function in this complex system.

REFERENCES

Artman, M. (1992). Sarcolemmal Na^+-Ca^{2+} exchange activity and exchanger immunoreactivity in developing rabbit hearts. *Am J Physiol* 263:H1506–H1513.

Benson, D.W. Jr., Hughes, S.F., Hu, N., and Clark, E.B. (1989). Effect of heart rate increase on dorsal aortic flow before and after volume loading in the stage 24 chick embryo. *Pediatr Res* 26:245–249.

Boppart, S.A., Tearney, G.J., Bouma, B.E., Southern, J.F., Brezinski, M.E., and Fujimoto, J.G. (1997). Noninvasive assessment of the developing *Xenopus* cardiovascular system using optical coherence tomography. *Proc Natl Acad Sci USA* 94:4256–4261.

Bowers, P.N., Tinney, J.P., and Keller, B.B. (1996). Nitroprusside selectively reduces ventricular preload in the stage 21 chick embryo. *Cardiovasc Res* 31:E132–E138.

Braunstein, J.B., Donovan, M., Hughes, S., and Benson, D.W. Jr. (1994). Assessment of ventricular relaxation in the developing chick embryo using a monoexponential model. *Am J Physiol* 267:H631–H635.

Burggren, W.W., and Fritsche, R. (1997). Amphibian cardiovascular development. In: Burggren, W., and Keller, B.B., eds. *Development of Cardiovascular Systems: Molecules to Organisms.* Cambridge University Press, New York, pp. 166–183.

Burggren, W.W., and Pinder, A.W. (1991). Ontogeny of cardiovascular and respiratory physiology in lower vertebrates. *Annu Rev Physiol* 53:107–135.

Campbell, K.A., Hu, N., Clark, E.B., and Keller, B.B. (1992). Analysis of dynamic atrial dimension and function during early cardiac development in the chick embryo. *Pediatr Res* 32:333–337.

Casillas, C.B., Tinney, J.P., and Keller, B.B. (1994). Influence of acute alterations in cycle length on ventricular function in the chick embryo. *Am J Physiol* 267:H905–911.

Cheanvechai, V., Hughes, S.F., and Benson, D.W. Jr. (1992). Relationship between cardiac cycle length and ventricular relaxation rate in the chick embryo. *Pediatr Res* 31:480–482.

Chin, T.K., Friedman, W.F., and Klitzner, T.S. (1990). Developmental changes in cardiac myocyte calcium regulation. *Circ Res* 67:574–579.

Chuck, E.T., Freeman, D.M., Watanabe, M., and Rosenbaum, D.S. (1997). Changing activation sequence in the embryonic chick heart: implications for the development of the His-Purkinje system. *Circ Res* 81:470–476.

Clark, E.B., and Hu, N. (1982). Developmental hemodynamic changes in the chick embryo from stages 18 to 27. *Circ Res* 51:810–815.

Clark, E.B., Hu, N., Frommelt, P., Vandekieft, J.L., Dummett, J.L., and Tomanek, R.J. (1989). Effect of increased ventricular pressure on heart growth in chick embryo. *Am J Physiol* 257:H55–H61.

Clark, E.B. (1990). Hemodynamic control of the embryonic circulation. In: Clark, E.B., and Takao, A., eds. *Developmental Cardiology: Morphogenesis and Function.* Futura, Mount Kisco, NY, pp. 291–304.

Clark, E.B., Hu, N., Turner, D.R., Litter, J.E., and Hansen, J. (1991). Effect of chronic verapamil treatment on the ventricular function and growth in chick embryos. *Am J Physiol* 261:H166–171.

Colon, S.D., Parness, I.P., Spevak, P.J., and Saunders, S.P. (1992). Developmental modulation of myocardial mechanics: age- and growth-related alterations in afterload and contractility. *J Am Coll Cardiol* 19:619–629.

Conway, S.J., Godt, R.E., Hatcher, C.J., et al. (1997). Neural crest is involved in development of abnormal myocardial function. *J Mol Cell Cardiol* 29:2675–2685.
Cuneo, B., Hughes, S., and Benson, D.W. Jr. (1991). Heart rate perturbation in the chick embryo: a comparison of two methods. *Am J Physiol* 260:H1864–1869.
Davies, M.P., An, R.H., Doevendans, P., Kubalak, S., Chien, K.R., and Kass, R.S. (1996). Developmental changes in ionic channel activity in the embryonic murine heart. *Circ Res* 78:15–25.
Dunnigan, A., Hu, N., Benson, D.W. Jr., and Clark, E.B. (1987). Effect of heart rate increase on dorsal aortic flow in the stage 24 chick embryo. *Pediatr Res* 22:442–444.
Dyson, E., Sucov, H.M., Kubalak, S.W., et al. (1995). Atrial-like phenotype is associated with embryonic ventricular failure in retinoid X receptor I –/– mice. *Proc Natl Acad Sci USA* 92:7386–7390.
Emery, J.L., and Omens, J.H. (1997). Mechanical regulation of myocardial growth during volume-overload hypertrophy in the rat. *Am J Physiol* 273:H1198–1204.
Faber, J.J., Green, T.J., and Thornburg, K.L. (1974). Embryonic stroke volume and cardiac output in the chick. *Dev Biol* 41:14–21.
Fritsche, R., and Burggren, W.W. (1996). Development of cardiovascular responses to hypoxia in larvae of the frog *Xenopus laevis*. *Am J Physiol* 271:R912–917.
Fung, Y.C. (1997). *Biodynamics: Circulation.* Springer-Verlag, New York.
Furukawa, S., MacLennan, M.J., and Keller, B.B. (1998). Hemodynamic response to anesthesia in pregnant and nonpregnant ICR mice. *Lab Anim Sci* 48:357–363.
Gaasch, W.H., Zile, R.M., Blaustein, A.S., and Bing, O.H.L. (1987). Loading conditions and left ventricular relaxation. In: Grossman, W., and Lorell, B.H., eds. *Diastolic Relaxation of the Heart.* Martinus Nijhoff, New York, pp. 133–142.
Godt, R.E., Fogaca, R.T.H., and Nosek, T.M. (1991). Changes in force and calcium sensitivity in the developing avian heart. *Can J Physiol Pharmacol* 69:1692–1697.
Gui, Y.H., Linask, K.K., Khowsathit, P., and Huhta, J.C. (1996). Doppler echocardiography of normal and abnormal embryonic mouse heart. *Pediatr Res* 40:633–642.
Hamburger, V., and Hamilton, H.L. (1951). A series of normal stages in the development of the chick embryo. *J Morphol* 88:49–92.
Harh, J.Y., Paul, M.H., Gallen, W.J., Friedberg, D.Z., and Kaplan, S. (1973). Experimental production of hypoplastic left heart syndrome in the chick embryo. *Am J Cardiol* 31:51–56.
Hay, D.A., and Low, F.N. (1970). The fusion of dorsal and ventral endocardial cushions in the embryonic chick heart: a study in fine structure. *Am J Anat* 133:1–24.
Hirakow, R. (1980). Quantitative studies on the ultrastructural differentiation and growth of mammalian cardiac muscle cells. *Acta Anat (Basel)* 108:114–152.
Hofman, P.L., Hiatt, K., Yoder, M.C., and Rivkees, S.A. (1997). A1 adenosine receptors potently regulate heart rate in mammalian embryos. *Am J Physiol* 273:R1374–1380.
Hogers, B., DeRuiter, M.C., Baasten, A.M.J., Gittenberger-de Groot, A.C., and Poelmann, R.E. (1995). Intracardiac blood flow patterns related to the yolk sac circulation of the chick embryo. *Circ Res* 76:871–877.
Hogers, B., DeRuiter, M.C., Gittenberger-de Groot, A.C., and Poelmann, R.E. (1997). Unilateral vitelline vein ligation alters intracardiac blood flow patterns and morphogenesis in the chick embryo. *Circ Res* 80:473–481.
Hu, N., and Clark, E.B. (1989). Hemodynamics of the stage 12 to stage 29 chick embryo. *Circ Res* 65:1665–1670.
Hu, N., Connuck, D.M., Keller, B.B., and Clark, E.B. (1991). Diastolic filling characteristics in the stage 12 to 27 chick embryo ventricle. *Pediatr Res* 29(4):334–337.
Hu, N., Hansen, A.L., Clark, E.B., and Keller, B.B. (1995a). Atrial natriuretic peptide reduces diastolic filling in the stage 21 chick embryo. *Pediatr Res* 37:465–468.
Hu, N., and Keller, B.B. (1995b). Relationship of simultaneous atrial and ventricular pressures in the stage 16 to 27 chick embryo. *Am J Physiol* 269:H1359–H1362.

Icardo, J.M. (1989). Endocardial cell arrangement: role of hemodynamics. *Anat Rec* 225:150–155.

Josephson, I.R., and Sperelakis, N. (1990). Developmental increases in the inwardly-rectifying K^+ current of embryonic chick ventricular myocytes. *Biochem Biophys Acta* 1052:123–127.

Kamino, K., Hirota, A., and Fujii, S. (1981). Localization of pacemaking activity in early embryo heart monitored using voltage-sensitive dye. *Nature* 260:595–597.

Kass, D.A., Beyar, R., Lankford, E., Heard, M., Maughan, W.L., and Sagawa K. (1989). Influence of contractile state on curvilinearity of in situ end-systolic pressure-volume relations. *Circulation* 79:167–178.

Kaufman, M.H. (1992). *The Atlas of Mouse Development.* Academic Press, New York.

Keller, B.B. (1995). Functional maturation and coupling of the embryonic cardiovascular system. In: Clark, E.B., Markwald, R.R., and Takao, A., eds. *Developmental Mechanisms of Heart Disease*. Futura, Mount Kisco, NY, pp. 367–386.

Keller, B.B. (1997). Embryonic cardiovascular function, coupling, and maturation: a species view. In: Burggren, W., and Keller, B.B., eds. *Development of Cardiovascular Systems: Molecules to Organisms.* Cambridge University, Press, New York, pp. 65–87.

Keller, B.B. (1998). Methods to detect cardiovascular phenotypic changes during development. In: Hoit, B.D., and Walsh, R.A., eds. *Cardiovascular Physiology in the Genetically Engineered Mouse.* Kluwer Academic, New York, pp. 260–275.

Keller, B.B., Hu, N., and Clark, E.B. (1990). Correlation of ventricular area, perimeter, and conotruncal diameter with mass and function in the stage 12 to 24 chick embryo. *Circ Res* 66:109–114.

Keller, B.B., Hu, N., Serrino, P.J., and Clark, E.B. (1991). Ventricular pressure-area loop characteristics in the stage 16 to 24 chick embryo. *Circ Res* 68:226–231.

Keller, B.B., Hu, N., and Tinney, J.P. (1994). Embryonic ventricular diastolic and systolic pressure-volume relation. *Cardiology in the Young* 4:19–27.

Keller, B.B., MacLennan, M.J., Tinney, J.P., and Yoshigi, M. (1996). In vivo assessment of embryonic cardiovascular dimensions and function in day 10.5 to 14.5 mouse embryos. *Circ Res* 79:247–255.

Keller, B.B., Yoshigi, M., and Tinney, J.P. (1997). Ventricular-vascular uncoupling by acute conotruncal occlusion in the stage 21 chick embryo. *Am J Physiol* 273:H2861–H2866.

Kirby, M.L., and Waldo, K.L. (1990). Role of neural crest in congenital heart disease. *Circulation* 82:332–340.

Leatherbury, L., Connuck, D.M., and Kirby, M.L. (1993). Neural crest ablation versus sham surgical effects in a chick embryo model of defective cardiovascular development. *Pediatr Res* 33:628–631.

MacLennan, M.J., and Keller, B.B. (1999). Umbilical arterial blood flow in the mouse embryo during development and following acutely increased heart rate. *Ultrasound Med Biol* 25(3):361–370.

Maughan, W.L., Sunagawa, K., Burkhoff, D., and Sagawa, K. (1985). Effect of arterial impedance changes on the end-systolic pressure-volume relation. *Circ Res* 54:595–602.

Meyer-Wittkopf, M., Simpson, J.M., and Sharland, G.K. (1996). Incidence of congenital heart defects in fetuses of diabetic mothers: a retrospective study of 326 cases. *Ultrasound Obstet Gynecol* 8:8–10.

Mikawa, T., and Fischman, D.A. (1996). The polyclonal origin of myocyte lineages. *Annu Rev Physiol* 58:509–521.

Miller, C.E., Vanni, M.A., and Keller, B.B. (1997a). Characterization of passive embryonic myocardium by quasi-linear viscoelasticity theory. *J Biomech* 30:985–988.

Miller, C.E., Vanni, M.A., Taber, L.A., and Keller, B.B. (1997b). Passive stress-strain measurements in the stage 16 and stage 18 embryonic chick heart. *J Biomech Eng* 119:445–451.

Mirsky, I., Tajimi, T., and Peterson, K.L. (1987). The development of the entire end-systolic pressure-volume and ejection fraction-afterload relations: a new concept of systolic myocardial stiffness. *Circulation* 76:343–356.

Moorman, A.F., de Jong, F., Denyn, M.M., and Lamers, W.H. (1998). Development of the cardiac conduction system. *Circ Res* 82(6):629–644.

Nakazawa, M., Clark, E.B., Hu, N., and Wispe, J. (1985). Effect of environmental hypothermia on vitelline artery blood pressure and vascular resistance in the stage 18, 21, and 24 chick embryo. *Pediatr Res* 19:651–654.

Nakazawa, M., Kajio, F., Ikeda, K., and Takao, A. (1990). Effect of atrial natriuretic peptide on hemodynamics of the stage 21 chick embryo. *Pediatr Res* 27:557–560.

Nakazawa, M., Miyagawa, S., Ohno, T., Miura, S., and Takao, A. (1988). Developmental hemodynamic changes in rat embryos at 11 to 15 days of gestation: normal data of blood pressure and the effect of caffeine compared to data from chick embryo. *Pediatr Res* 23:200–205.

Nakazawa, M., Miyagawa, S., Takao, A., Hu, N., and Clark, E.B. (1986). Hemodynamic effects of environmental hyperthermia in the stage 18, 21, and 24 chick embryo. *Pediatr Res* 20:1213–1215.

Paff, G.H., Boucek, R.J., and Klopfenstein, H.S. (1948). Experimental heart-block in the chick embryo. *Anat Rec* 149:217–224.

Patten, B.M., Kramer, T.C., and Barry, A. (1948). Valvular action in the embryonic chick heart by localized apposition of endocardial masses. *Anat Rec* 102:299–311.

Pegram, B.L., Trippodo, N.C., Natsume, T., et al. (1986). Hemodynamic effects of atrial natriuretic hormone. *Fed Proc* 45:2382–2386.

Pelster, B., and Burggren, W.W. (1996). Disruption of hemoglobin oxygen transport does not impact oxygen-dependent physiological processes in developing embryos of zebra fish (Danio rerio). *Circ Res* 79:358–362.

Pexieder, T. (1986). Standardized method for study of normal and abnormal cardiac development in chick, rat, mouse, dog and human embryos. *Teratology* 33:91C–92C.

Pexieder, T., and Janecek, P. (1984a). Organogenesis of the human embryonic and early fetal heart as studied by microdissection and SEM. In: Nora, J.J., and Takao, A., eds. *Congenital Heart Disease: Causes and Processes.* Futura, Mt. Kisco, NY, pp. 401–421.

Pexieder, T., Christen, Y., Vuillemin, M., and Patterson, D.F. (1984b). Comparative morphometric analysis of cardiac organogenesis in chick, mouse, and dog embryos. In: Nora, J.J., and Takao, A., eds. *Congenital Heart Disease: Causes and Processes.* Futura, Mt. Kisco, NY, pp. 423–438.

Protasi, F., Sun, X.H., and Franzini-Armstrong, C. (1996). Formation and maturation of the calcium release apparatus in developing and adult avian myocardium. *Dev Biol* 173:265–278.

Ranu, H.S. (1998). Bioengineering in the millennium. *J Biomater Appl* 13:100–110.

Rombough, P.J. (1997). Piscine cardiovascular development. In: Burggren, W., and Keller, B.B., eds. *Development of Cardiovascular Systems: Molecules to Organisms.* Cambridge University Press, New York, pp. 145–165.

Rosenquist, T.H., and Modis, L. (1991). Spatial disorder of collagens in the great vessels, associated with congenital heart defects. *Anat Rec* 229:116–124.

Sagawa, K., Maughan, W.L., Suga, H., and Sunagawa, K. (1988). Physiologic determinants of pressure-volume relations. In: *Cardiac Contraction and the Pressure-Volume Relationship.* Oxford University Press, New York, pp. 110–172.

Sedmera, D., Pexieder, T., Hu, N., and Clark, E.B. (1997). Developmental changes in the myocardial architecture of the chick. *Anat Rec* 248:421–432.

Sedmera, D., Pexieder, T., Hu, N., and Clark, E.B. (1998). A quantitative study of the ventricular myoarchitecture in the stage 21–29 chick embryo following decreased loading. *Eur J Morphol* 36:105–119.

Sedmera, D., Pexieder, T., Rychterova, V., Hu, N., and Clark, E.B. (1999). Remodeling of chick embryonic ventricular myoarchitecture under experimentally changed loading conditions. *Anat Rec* 254(2):238–252.

Sissman, N.J. (1970). Developmental landmarks in cardiac morphogenesis: comparative chronology. *Am J Cardiol* 25:141–148.

Smith, B.R., Linney, E., Huff, D.S., and Johnson, G.A. (1996). Magnetic resonance microscopy of embryos. *Comput Med Imaging Graph* 20(6):486–490.

Srinivasan, S., Baldwin, H.S., Aristizabal, O., et al. (1998). Noninvasive, in utero imaging of mouse embryonic heart development with 40-MHz echocardiography. *Circulation* 98:912–918.

Stainier, D.Y., Fouquet, B., Chen, J.N., et al. (1996). Mutations affecting the formation and function of the cardiovascular system in the zebrafish embryo. *Development* 123:285–292.

Sunagawa, K., Sagawa, K., and Maughan, W.L. (1987). Ventricular interaction with the vascular system in terms of pressure-volume relationships. In: Yin, F.C.P., ed. *Ventricular Vascular Coupling: Clinical, Physiological, and Engineering Aspects.* Springer-Verlag, New York, pp. 210–239.

Sweeney, L.J. (1981). Morphometric analysis of an experimental model of left heart hypoplasia in the chick. Ph.D. thesis, Omaha, Nebraska, University of Nebraska Medical Center.

Taber, L.A. (1998). Mechanical aspects of cardiac development. *Prog Biophys Mol Biol* 69:237–255.

Taber, L.A., Hu, N., Pexieder, T., Clark, E.B., and Keller, B.B. (1993). Residual strain in the embryonic ventricle of the stage 16 to 24 chick embryo heart. *Circ Res* 72:455–462.

Taber, L.A., Keller, B.B., and Clark, E.B. (1992). Cardiac mechanics in the stage 16 chick embryo. *J Biomech Eng* 114:427–434.

Taber, L.A., Sun, H., Cartmell, J.S., Clark, E.B., and Keller, B.B. (1994). Epicardial strains in the stage 16–24 embryonic chick ventricle. *Circ Res* 75:896–903.

Tanaka, N., Mao, L., Delano, F.A., et al. (1997). Left ventricular volumes and function in the embryonic heart. *Am J Physiol* 273:H1368–H1376.

Terracio, L., and Borg, T.K. (1988). Factors affecting cardiac cell shape. *Heart Failure* 4:114–124.

Theiler, K. (1989). *The House Mouse: Development and Normal Stages from Fertilization to 4 Weeks of Age*, 2nd ed. Springer-Verlag, Berlin.

Thornburg, K.L., Giraud, G.D., Reller, M.D., and Morton M.J. (1997). Mammalian cardiovascular development: physiology and functional reserve of the fetal heart. In: Burggren, W.W., and Keller, B.B., eds. *Development of Cardiovascular Systems: Molecules to Organisms.* Cambridge University Press, Cambridge, UK, pp. 211–226.

Tobita, K., and Keller, B.B. (1999). End-systolic myocardial stiffness is a load independent index of contractility in the stage 24 chick embryonic heart. *Am J Physiol* 276: H2102–H2108.

Tobita, K., and Keller, B.B. (2000a). Maturation of end-systolic stress-strain relations in chick embryonic myocardi. *Am J Physiol Heart Circ Physiol* 279(1):H216–224.

Tobita, K., and Keller, B.B. (2000b). Right and left ventricular wall deformation patterns in normal and left heart hypoplasia chick embryos. *Am J Physiol Heart Circ Physiol.* 279(3):H959–H969.

Tsyvian, P., Malkin, K., Artemieva, O., and Wladimiroff, J.W. (1998). Assessment of left ventricular filling in normally grown fetuses, growth-restricted fetuses and fetuses of diabetic mothers. *Ultrasound Obstet Gynecol* 12:33–38.

Ursem, N.T.C., Brinkman, H.J.F., Struijk, P.C., et al. (1998). Umbilical artery waveform analysis based on maximum, mean, and mode velocity in early human pregnancy. *Ultrasound Med Biol* 24:1–7.

Vetter, R., and Will, H. (1986). Sarcolemmal Na-Ca exchange and sarcoplasmic reticulum calcium uptake in developing chick heart. *J Mol Cell Cardiol* 18:1267–1275.

Vuillemin, M., and Pexieder, T. (1989a). Normal stages of cardiac organogenesis in the mouse, I: development of the external shape of the heart. *Am J Anat* 184:101–113.

Vuillemin, M., and Pexieder, T. (1989b). Normal stages of cardiac organogenesis in the mouse, II: development of the internal relief of the heart. *Am J Anat* 184:114–128.

Weiss, J.L., Frederiksen, J.W., and Weisfeldt, M.L. (1976). Hemodynamic determinants of the time-course of fall in canine left ventricular pressure. *J Clin Invest* 58:751–760.

Wladimiroff, J.W., Huisman, T.W.A., Stewart, P.A., and Stijnen, T. (1992). Normal fetal Doppler inferior vena cava, trans tricuspid and umbilical artery flow velocity waveforms between 11 and 16 weeks of gestation. *Am J Obstet Gynecol* 3:921–924.

Ya, J., Schilham, M.W., de Boer, P.A., Moorman, A.F., Clevers, H., and Lamers, W.H. (1998). Sox4-deficiency syndrome in mice is an animal model for common trunk. *Circ Res* 83:986–994.

Yang, M., Taber, L.A., and Clark, E.B. (1994). A nonlinear poroelastic model for the trabecular embryonic heart. *J Biomech Eng* 116:213–223.

Yasui, H., Nakazawa, M., Morishima, M., Miyagawa-Tomita, S., and Momma, K. (1995). Morphological observations on the pathogenetic process of transposition of the great arteries induced by retinoic acid in mice. *Circulation* 91:2478–2486.

Yoshigi, M., and Keller, B.B. (1997a). Characterization of the embryonic aortic impedance with lumped parameter models. *Am J Physiol* 273:H19–H27.

Yoshigi, M., Ettel, J.M., and Keller, B.B. (1997b). Developmental changes in flow wave propagation velocity in the embryonic chick vascular system. *Am J Physiol* 273:H1523–H1529.

Yoshigi, M., Hu, N., and Keller, B.B. (1996). Dorsal aortic impedance in the stage 24 chick embryo following acute changes in circulating blood volume. *Am J Physiol* 270:H1597–1606.

Yoshigi, M., and Keller, B.B. (1996). Linearity of pulsatile pressure-flow relations in the embryonic chick vascular system. *Circ Res* 79:864–870.

Zahka, K.G., Hu, N., Brin, K.P., Yin, F.C.P., and Clark, E.B. (1989). Aortic impedance and hydraulic power in the chick embryo from stages 18 to 29. *Circ Res* 64:1091–1095.

Zimmerman, F.J., Hughes, S.F., Cuneo, B., and Benson, D.W. Jr. (1991). The effect of cardiac cycle length on ventricular end-diastolic pressure and maximum time derivative of pressure in the stage 24 chick embryo. *Pediatr Res* 29:338–341.

Index

GPSR Compliance

The European Union's (EU) General Product Safety Regulation (GPSR) is a set of rules that requires consumer products to be safe and our obligations to ensure this.

If you have any concerns about our products, you can contact us on ProductSafety@springernature.com

In case Publisher is established outside the EU, the EU authorized representative is:

Springer Nature Customer Service Center GmbH
Europaplatz 3
69115 Heidelberg, Germany

Batch number: 10732772

Printed by Printforce, the Netherlands